Nonlinear Second Order Parabolic Equations

Nonlinear Second Order Parabolic Equations

Mingxin Wang

CRC Press
Taylor & Francis Group
Boca Raton London New York

CRC Press is an imprint of the
Taylor & Francis Group, an **informa** business

First edition published 2021
by CRC Press
6000 Broken Sound Parkway NW, Suite 300, Boca Raton, FL 33487-2742

and by CRC Press
2 Park Square, Milton Park, Abingdon, Oxon, OX14 4RN

Library of Congress Cataloging-in-Publication Data

Names: Wang, Mingxin, 1957- author.
Title: Nonlinear second order parabolic equations / Mingxin Wang.
Description: First edition. | Boca Raton : CRC Press, 2021. | Includes
 bibliographical references and index. | Summary: "The parabolic partial
 differential equations model is one of the most important processes in
 the process of diffusion in nature. Whether it is the diffusion of
 energy in space-time, the diffusion of species in ecology, the diffusion
 of chemicals in biochemical processes, or the diffusion of information
 in social networks, diffusion processes are ubiquitous and crucial in
 the physical and natural world as well as our everyday lives. In this
 graduate textbook, the author starts with classical parabolic problems
 as well as basic methods and techniques to guide students into the field
 of second order nonlinear parabolic equations. The author then moves on
 to discuss key topics such as theory and Schauder theory, the maximum
 principle and the comparison principle. Topics additionally discussed
 include periodic boundary value problems, free boundary problems, and
 semigroup theory. This book is based on tried and tested teaching
 materials used at the Harbin Institute of Technology over the past 10
 years. Special care is taken to make the book suitable for classroom
 teaching as well as for self-study among graduate students"-- Provided
 by publisher.
Identifiers: LCCN 2020042644 (print) | LCCN 2020042645 (ebook) | ISBN
 9780367711986 (hardcover) | ISBN 9780367712846 (paperback) | ISBN
 9781003150169 (ebook)
Subjects: LCSH: Differential equations, Parabolic. | Differential
 equations, Partial.
Classification: LCC QA377 .W36 2021 (print) | LCC QA377 (ebook) | DDC
 515/.3534--dc23
LC record available at https://lccn.loc.gov/2020042644
LC ebook record available at https://lccn.loc.gov/2020042645

ISBN: 978-0-367-71198-6 (hbk)
ISBN: 978-0-367-71284-6 (pbk)
ISBN: 978-1-003-15016-9 (ebk)

Typeset in Computer Modern font
by KnowledgeWorks Global Ltd.

Contents

Preface

Parabolic partial differential equations model one of the most important processes in the nature world: diffusion. Whether it is the diffusion of energy in space-time, diffusion of species in ecology, diffusion of chemicals in biochemical processes, or the diffusion of information in social networks, diffusion processes are ubiquitous and crucial in the physical and natural world as well as our everyday lives.

The aim of this book is to provide graduate students with an introductory and self-contained textbook on second order nonlinear parabolic partial differential equations. We start with classical parabolic problems, basic methods and techniques to guide students into the exciting field of second order nonlinear parabolic equations as quickly and painlessly as possible. The book covers key topics such as L^p theory and Schauder theory, maximum principle, comparison principle, regularity and uniform estimate, initial-boundary value problems of semilinear parabolic scalar equations and weakly coupled parabolic systems, upper and lower solutions method, monotone properties and longtime behaviours of solutions, convergence of solutions and stability of equilibrium solutions, global solutions, and finite time blowup. It also touches on periodic boundary value problems, free boundary problems, and semigroup theory.

As a preliminary, Chapter 1 reviews the L^p theory, Schauder theory, Hopf boundary lemma and maximum principle, which are some of the main tools in the study of parabolic partial differential equations. As some of these materials may already be known to the reader from a basic course in linear partial differential equations, we only present the conditions and conclusions and give appropriate explanations and references, without giving the complete proofs for the better-known results.

The comparison principle based on the maximum principle will be frequently used in this textbook. The regularity and uniform estimates based on the L^p theory and Schauder theory are important ingredients for the study of convergence of solutions as time goes to infinity and the stability of equilibrium solutions. Chapter 2 is devoted to these topics.

Chapters 3 and 4 deal with initial-boundary value problems of semilinear parabolic scalar equations and weakly coupled parabolic systems, respectively. We first introduce the upper and lower solutions method, and then discuss the monotone property and longtime behaviour of solutions. In addition, the use of L^p theory, Schauder theory, and Schauder fixed-point theorem to prove the existence and uniqueness of local solutions is also included.

In Chapter 5, we first consider local stability of equilibrium solutions by using the linearized principal eigenvalue and the upper and lower solutions method. We establish that the linear stability implies the local stability. Then we introduce three additional methods: the Lyapunov functional method, the iteration method, and the average method for the investigation of stability of equilibrium solutions, and illustrate them with concrete examples.

Chapter 6 involves global solutions and finite time blowup. We first provide four basic methods for dealing with finite time blowup: they are the comparison method, Kaplan's first eigenvalue method, the energy method, and the concavity method. Then, through some concrete examples, we show how to determine conditions for the global existence and finite time blowup, and discuss the critical exponent, blowup estimates, and blowup on the boundary.

In Chapter 7, we focus on time-periodic parabolic boundary value problems. We first introduce the upper and lower solutions method, and then study a diffusive logistic equation (including the unbounded domain case) and a diffusive competition model as applications. For the diffusive logistic equation, the existence and uniqueness of positive solutions are established. Especially in the bounded domain case, we prove that the unique positive solution is stable. For the diffusive competition model, the existence of positive solutions is also established. Meanwhile, some important properties of time-periodic parabolic eigenvalue problems are provided.

In Chapter 8, we discuss some free boundary problems coming from ecology. The main content includes existence and uniqueness, regularity and uniform estimates, some technical lemmas, longtime behaviour of solutions, criteria and dichotomy for spreading and vanishing, and the spreading speed of the free boundary. Concrete examples are given as applications.

Finally, in Chapter 9, we first give a brief review of the basic results of semigroup theory, and then show how it can be used to prove the existence, uniqueness, extension, and regularity of the mild solution.

In the past ten years, materials in this book have been used for the course "Nonlinear Parabolic Equations" for graduate students at Harbin Institute of Technology. Its completion is benefitted by many students who have attended these classes.

While working on this book, I was partially supported by NSFC Grants 11071049, 11371113, and 11771110. I am grateful to my students and colleagues for their feedback and helpful comments on the manuscript. The materials presented and the references quoted here are mainly based on the author's taste and familiarity, which are inevitably biased with many important topics and references not included here. No offence is intended on account of such omissions or oversight.

Mingxin Wang, Harbin

CHAPTER 1

Preliminaries

There are many monographs and textbooks on parabolic equations, please refer to [36, 91, 60, 105, 88, 31, 16, 72, 93, 110, 124, 48]. The book [16] received the excellent textbook award from the Ministry of Education of China.

This chapter reviews some basic theories and results such as L^p theory, Schauder theory and maximum principle for linear parabolic equations, which are fundamental tools in the study of nonlinear parabolic equations. As some of these materials may be already known to the reader from a basic course in linear partial differential equations, we only present the conditions and conclusions, and give appropriate explanations and references, but not complete proofs for the better known results. The chapter starts with the L^p theory and Schauder theory for the first initial-boundary value problems, second initial-boundary value problems, and mixed initial-boundary value problems. It then moves on to topics such as the Hopf boundary lemma, maximum principle of classical solutions, and strong and weak solutions for scalar equations and systems including problems with non-classical boundary conditions and non-local terms.

1.1 NOTATIONS, AGREEMENTS AND BASIC ASSUMPTIONS

Notations

Given two sets $A, B \subset \mathbb{R}^n$, we use $\overline{A}$ to represent the closure of A, and use $d(A, B)$ or $\mathrm{dist}(A, B)$ to represent the distance between A and B. Notation $A \Subset B$ means that A is bounded and $\overline{A} \subset B$.

Given a set $A \subset \mathbb{R}^n$ and a function $f : A \to \mathbb{R}^m$, we use $f(A)$ to represent the image of A under f.

Let $k \geqslant 0$ be an integer and $0 \leqslant \beta < 1$ be a constant. We say that a domain $\Omega \subset \mathbb{R}^n$ is of class $C^{k+\beta}$, or Ω has a $C^{k+\beta}$ boundary $\partial\Omega$ if for each point $x_0 \in \partial\Omega$ there exist a neighbourhood U of x_0 and a function $\Phi \in C^{k+\beta}(\overline{U})$ such that

(i) the inverse function Φ^{-1} exists and $\Phi^{-1} \in C^{k+\beta}(\Phi(\overline{U}))$, and

(ii) if we set $y = (y_1, \ldots, y_n) = \Phi(x)$, then $\Phi(\partial\Omega \cap U) \subset \{y \in \mathbb{R}^n : y_n = 0\}$ and $\Phi(\Omega \cap U) \subset \{y \in \mathbb{R}^n : y_n > 0\}$.

We say that a domain Ω has *interior ball property* at point $x \in \partial\Omega$, if there exists a ball B with radius $r > 0$ for which $B \subset \Omega$ and $\overline{B} \cap \partial\Omega = \{x\}$. If Ω has the interior ball property at each point $x \in \partial\Omega$, then we say that Ω has the *interior ball property*. If Ω is of class C^2, then it has the interior ball property. However, if Ω is only of class $C^{1+\alpha}$ with $0 < \alpha < 1$, then it may not have the interior ball property. For example, curve $x_2 = |x_1|^{1+\alpha}$ does not have the interior ball property at point $(0,0)$.

Let $Q \subset \mathbb{R}^{n+1}$ and operator $\mathscr{L}$ be given by

$$\mathscr{L} := \partial_t - a_{ij}(x,t)D_{ij} + b_i(x,t)D_i + c(x,t),$$

where symbols $a_{ij}D_{ij} := \sum_{i,j=1}^n a_{ij}D_{ij}$, $b_iD_i := \sum_{i=1}^n b_iD_i$. The operator $\mathscr{L}$ is called *parabolic* in Q if matrix $(a_{ij})_{n \times n}$ is symmetric and positive definite in Q, i.e., $a_{ij} = a_{ji}$ in Q and

$$a_{ij}(x,t)y_iy_j > 0 \quad \text{for all } (x,t) \in Q,\ 0 \neq y \in \mathbb{R}^n.$$

A parabolic operator $\mathscr{L}$ in Q is called *strongly parabolic* if there are positive constants λ and Λ such that

$$\lambda|y|^2 \leqslant a_{ij}(x,t)y_iy_j \leqslant \Lambda|y|^2 \quad \text{for all } (x,t) \in Q,\ y \in \mathbb{R}^n.$$

For $x \in \mathbb{R}^n$ and $t \in \mathbb{R}$, we write $\mathbf{x} = (x,t)$. Given $\mathbf{x}_0 = (x_0, t_0)$ and $r > 0$, we define a lower hemisphere

$$Q(\mathbf{x}_0, r) := \{\mathbf{x} \in \mathbb{R}^{n+1} : |\mathbf{x} - \mathbf{x}_0| < r,\ t < t_0\}.$$

For a given domain $Q \subset \mathbb{R}^{n+1}$, we define the *parabolic boundary* $\partial_p Q$ of Q to be a set of all points $\mathbf{x}_0 \in \partial Q$ satisfying that for any $r > 0$, the lower hemisphere $Q(\mathbf{x}_0, r)$ contains points that do not belong to Q, i.e., $Q(\mathbf{x}_0, r) \not\subset Q$ for any $r > 0$. In Fig. 1.1, the solid line represents the parabolic boundary and the dashed line does not belong to the parabolic boundary.

Fig. 1.1

Let $Q \subset \mathbb{R}^{n+1}$ be a bounded domain and $\mathbf{x}_0 \in \overline{Q}$. We define $S(\mathbf{x}_0)$ to be a set of all $\mathbf{x} \in \overline{Q} \setminus \partial_p Q$ for which we can find a continuous function $g : [0,1] \to \overline{Q} \setminus \partial_p Q$ such that $g(0) = \mathbf{x}_0$, $g(1) = \mathbf{x}$ and t-component of g is monotonically decreasing. The shaded part in Fig. 1.2 is $S(\mathbf{x}_0)$.

Fig. 1.2

Let $\Omega \subset \mathbb{R}^n$ and $Q \subset \mathbb{R}^{n+1}$ be domains, $0 < \alpha < 1$ be a constant and $k \geqslant 0$ be an integer. Function spaces $C^{k+\alpha}(\overline{\Omega})$ and $C^{k+\alpha,\,(k+\alpha)/2}(\overline{Q})$ are the standard Hölder spaces. Notation $u \in C^{k+\alpha}(\Omega)$ means that $u \in C^{k+\alpha}(\overline{\Omega}_0)$ for any $\Omega_0 \Subset \Omega$, and notation $u \in C^{k+\alpha,\,(k+\alpha)/2}(Q)$ means that $u \in C^{k+\alpha,\,(k+\alpha)/2}(\overline{Q}_0)$ for each $Q_0 \Subset Q$.

Let Ω and Q be as above, k be a positive integer and $p \geqslant 1$. Function spaces $W_p^k(\Omega)$ and $W_p^{2k,k}(Q)$ are the standard Sobolev spaces. We denote the closure of $C_0^\infty(\Omega)$ in $W_p^k(\Omega)$ by $\overset{\circ}{W}{}_p^k(\Omega)$, and use $L^p(\Omega)\,(L^p(Q))$ instead of $L_p(\Omega)\,(L_p(Q))$. To simplify notations, sometimes we write

$$|\cdot|_{i+\alpha,\overline{\Omega}} = \|\cdot\|_{C^{i+\alpha}(\overline{\Omega})} \quad \text{and} \quad |\cdot|_{i+\alpha,\overline{Q}} = \|\cdot\|_{C^{i+\alpha,\,(i+\alpha)/2}(\overline{Q})}, \quad i = 0, 1, 2.$$

In case of no confusions, for simplicity, we write

$$\| \cdot \|_p = \| \cdot \|_{p, D} = \| \cdot \|_{L^p(D)} \quad \text{for} \quad D = \Omega \text{ or } D = Q$$

and

$$\| \cdot \|_{W_p^2} = \| \cdot \|_{W_p^2(\Omega)}, \quad \| \cdot \|_{W_p^{2,1}} = \| \cdot \|_{W_p^{2,1}(Q)}.$$

Sometimes, we use $\int_\Omega f(x)$ instead of $\int_\Omega f(x)\mathrm{d}x$ to save space.

Let X be a Banach space, and $f, g \in X$. We write

$$\|f, g\|_X = \|f\|_X + \|g\|_X.$$

Agreements Throughout this book, unless it is clearly stated, we shall make the following agreements.

- All functions are real-valued.

- We usually refer to a vector-function (and also a matrix of functions) briefly as a function.

- There is a function $b(x,t)$ in the boundary condition $\partial_{\boldsymbol{n}} u + bu = g$ (it is customary to use this notation). However, sometimes in the same problem, this symbol b will also appear in the differential equation. Careful readers are easy to distinguish and will not cause confusion.

- Constant $0 < \alpha < 1$ may be different in different places.

- We use C, C', C_i and c_i to represent the generic constants.

- $\Omega \subset \mathbb{R}^n$ and $Q \subset \mathbb{R}^{n+1}$ are bounded domains, $\boldsymbol{n}$ and $\boldsymbol{\nu}$ are the outward normal vectors of $\partial\Omega$ and ∂Q, respectively. For simplicity, we denote the outer normal derivatives $\frac{\partial}{\partial \boldsymbol{n}}$ and $\frac{\partial}{\partial \boldsymbol{\nu}}$ by $\partial_{\boldsymbol{n}}$ and $\partial_{\boldsymbol{\nu}}$, respectively.

- $Q_T = \Omega \times (0, T]$, $S_T = \partial\Omega \times (0, T]$ and $\partial_p Q_T = S_T \cup (\overline{\Omega} \times \{0\})$ with $0 < T < \infty$. Obviously, $\partial_p Q_T$ is the *parabolic boundary* of Q_T. We call S_T the *lateral boundary* of Q_T.

- $Q_\infty = \Omega \times (0, \infty)$, $S_\infty = \partial\Omega \times (0, \infty)$, $\mathbb{R}_+ = (0, \infty)$ and $\overline{\mathbb{R}}_+ = [0, \infty)$.

- Constant $1 < p < \infty$. For a given function g defined on S_T, we say $g \in W_p^{2,1}(Q_T)$ or $g \in C^{2+\alpha, 1+\alpha/2}(\overline{Q}_T)$ if g can be extended to a new function $\hat{g}$ so that $\hat{g} \in W_p^{2,1}(Q_T)$ or $\hat{g} \in C^{2+\alpha, 1+\alpha/2}(\overline{Q}_T)$.

Basic assumptions Throughout this book, for simplicity, in some places (theorems and lemmas), we will use the following assumptions and abbreviations.

(A) The operator $\mathscr{L}$ is strongly parabolic in Q_T, and $a_{ij} \in C(\overline{Q}_T)$. Let $\mathbf{x} = (x, t)$, $\mathbf{y} = (y, s)$ and $\delta(\mathbf{x}, \mathbf{y}) = \sqrt{|x - y|^2 + |t - s|}$. Then

$$\omega(R) = \max_{i,j} \sup_{\substack{\delta(\mathbf{x},\mathbf{y}) \leqslant R \\ \mathbf{x}, \mathbf{y} \in Q_T}} |a_{ij}(\mathbf{x}) - a_{ij}(\mathbf{y})| \to 0 \quad \text{as} \quad R \to 0.$$

Function $\omega(R)$ is called *modulus of continuity* of $a_{ij}(x, t)$.

(B) Condition **(A)** holds, and $b_i, c \in L^\infty(Q_T)$, $f \in L^p(Q_T)$.

(C) Condition **(A)** holds, and $a_{ij}, b_i, c, f \in C^{\alpha, \alpha/2}(\overline{Q}_T)$.

(D) $\varphi \in W_p^2(\Omega)$ and $g \in W_p^{2,1}(Q_T)$.

(E) $\varphi \in C^{2+\alpha}(\overline{\Omega})$ and $g \in C^{2+\alpha, 1+\alpha/2}(\overline{Q}_T)$.

1.2 THE L^p AND SCHAUDER THEORIES

The so-called L^p (Schauder) theory includes existence, uniqueness and estimates of $W_p^{2,1}$ ($C^{2+\alpha, 1+\alpha/2}$) *solutions*. They are very important in the field of second order parabolic equations.

This section is concerned with the following initial-boundary value problems:

$$\begin{cases} \mathscr{L}u = f(x,t) & \text{in } Q_T, \\ u = g & \text{on } S_T, \\ u(x,0) = \varphi(x) & \text{in } \Omega \end{cases} \tag{1.1}$$

and

$$\begin{cases} \mathscr{L}u = f(x,t) & \text{in } Q_T, \\ \partial_{\boldsymbol{n}} u + bu = g, \quad b \geqslant 0 & \text{on } S_T, \\ u(x,0) = \varphi(x) & \text{in } \Omega, \end{cases} \tag{1.2}$$

as well as

$$\begin{cases} \mathscr{L}u = f(x,t) & \text{in } Q_T, \\ \partial_{\boldsymbol{n}} u + bu = g_1, \quad b \geqslant 0 & \text{on } \Gamma_1 \times (0,T], \\ u = g_2 & \text{on } \Gamma_2 \times (0,T], \\ u(x,0) = \varphi(x) & \text{in } \Omega, \end{cases} \tag{1.3}$$

where Γ_1 and Γ_2 are two disjoint closed subsets of $\partial\Omega$ with $\Gamma_1 \cup \Gamma_2 = \partial\Omega$. For problem (1.3), we define $\mathscr{B}_1 u = \partial_{\boldsymbol{n}} u + bu$ on $\Gamma_1 \times (0,T]$ and $\mathscr{B}_2 u = u$ on $\Gamma_2 \times (0,T]$.

We call (1.1), (1.2), and (1.3) the *first initial-boundary value problem*, *second initial-boundary value problem*, and *mixed initial-boundary value problem*, respectively.

Recall that $b_i, c \in L^\infty(Q_T)$, and $\Omega \subset \mathbb{R}^n$ is a bounded domain.

1.2.1 L^p theory

For convenience, we denote

$$\mathcal{A} := \{n, \lambda, \Lambda, T, \|b_i, c\|_\infty, \omega(R)\}.$$

Theorem 1.1 (L^p theory for the first initial-boundary value problem [16, Chapter 5, Theorem 5.2], [60, Chapter 4, Theorem 9.1], [72, Theorem 7.32]) *Let conditions* **(B)** *and* **(D)** *hold, and Ω be of class C^2. Assume that compatibility condition*

$$\varphi(x) = g(x,0) \quad on \ \partial\Omega$$

is satisfied. Then problem (1.1) has a unique solution $u \in W_p^{2,1}(Q_T)$, and

$$\|u\|_{W_p^{2,1}(Q_T)} \leqslant C \left(\|f\|_{p,Q_T} + \|g\|_{W_p^{2,1}(Q_T)} + \|\varphi\|_{W_p^2(\Omega)} \right), \tag{1.4}$$

where constant C depends only on $\mathcal{A}$ and Ω.

Theorem 1.2 (L^p theory for the second initial-boundary value problem [60, Chapter 4, Section 9]) *Let conditions* (**B**) *and* (**D**) *be fulfilled, Ω be of class C^2, $b \in C^{1,1/2}(\overline{S}_T)$ and $b \geqslant 0$. Assume that compatibility condition*

$$\partial_{\boldsymbol{n}}\varphi(x) + b(x,0)\varphi(x) = g(x,0) \quad on \ \partial\Omega$$

holds. Then problem (1.2) has a unique solution $u \in W_p^{2,1}(Q_T)$, and the estimate (1.4) holds with constant C depending only on $\mathcal{A}$, Ω and $|b|_{1,\overline{S}_T}$.

For the L^p theory with $p > 2$, under more detailed conditions, in Theorem 1.1 and Theorem 1.2, the requirement $\varphi \in W_p^2(\Omega)$ in the condition (**D**) can be weakened to $\varphi \in W_p^{2-2/p}(\Omega)$ and the term $\|\varphi\|_{W_p^2(\Omega)}$ on the right hand side of (1.4) can be replaced by $\|\varphi\|_{W_p^{2-2/p}(\Omega)}$. The interested reader can refer to [60, Chapter 4, Section 9].

In order to study existence and uniqueness of $W_p^{2,1}(Q_T)$ solutions of problem (1.3), we first give the following L^p estimate.

Theorem 1.3 (Local L^p estimate) *Assume that Ω is of class C^2. For any given $\delta > 0$, we define $\Omega_i^\delta := \{x \in \Omega : d(x, \Gamma_i) \leqslant \delta\}$, $i = 1, 2$. Take $0 < \delta \ll 1$, so that $\overline{\Omega^\delta}_i \cap \partial\Omega = \Gamma_i$. Denote $D_i^\delta = \Omega_i^\delta \times (0, T]$. Assume that conditions* (**B**) *and* (**D**) *hold with Ω and Q_T replaced by Ω_i^δ and D_i^δ, respectively, and $b \in C^{1,1/2}(\Gamma_1 \times [0, T])$, $b \geqslant 0$. Let $u \in W_p^{2,1}(D_i^\delta)$ satisfy*

$$\begin{cases} \mathscr{L}u = f(x,t) & in \ D_i^\delta, \\ \mathscr{B}_i u = g_i & on \ \Gamma_i \times (0, T], \\ u(x,0) = \varphi(x) & in \ \Omega_i^\delta. \end{cases}$$

Then for any $0 < \sigma < \delta$ and $0 \leqslant \varepsilon < T$, we have the following local estimates:

$$\|u\|_{W_p^{2,1}(D_i^\sigma)} \leqslant C_1 \left(\|f\|_{p,D_i^\delta} + \|g_i\|_{W_p^{2,1}(D_i^\delta)} + \|\varphi\|_{W_p^2(\Omega_i^\delta)} + \|u\|_{p,D_i^\delta} \right) \quad when \ \varepsilon = 0,$$

$$\|u\|_{W_p^{2,1}(\Omega_i^\sigma \times (\varepsilon,T))} \leqslant C_2 \left(\|f\|_{p,D_i^\delta} + \|g_i\|_{W_p^{2,1}(D_i^\delta)} + \|u\|_{p,D_i^\delta} \right) \quad when \ \varepsilon > 0,$$

where constants C_1 and C_2 depend on $\mathcal{A}$ and $1/(\delta-\sigma)$, and also depend on $|b|_{1,\Gamma_1 \times [0,T]}$ when $i = 1$, and C_2 also depends on $1/\varepsilon$.

We state a lemma before giving the proof of Theorem 1.3.

Lemma 1.4 *Let $b > a \geqslant 0$ and $\varphi(s)$ be a bounded non-negative function defined in $[a, b]$. Assume that φ satisfies*

$$\varphi(t) \leqslant \theta\varphi(s) + \frac{A}{(s-t)^\beta} + B \quad \text{for all} \ \ a \leqslant t < s \leqslant b, \tag{1.5}$$

where constants θ, A, B, and β are non-negative with $\theta < 1$. Then

$$\varphi(\rho) \leqslant C\left(\frac{A}{(R-\rho)^\beta} + B\right) \quad \text{for all} \ \ a \leqslant \rho < R \leqslant b, \tag{1.6}$$

where C is a constant which depends only on β and θ.

Proof. Take $t_0 = \rho$ and $t_{i+1} = t_i + (1-\tau)\tau^i(R-\rho)$, $i = 0, 1, \ldots$, where $0 < \tau < 1$ is a constant to be determined. Then we have

$$\varphi(t_i) \leqslant \theta\varphi(t_{i+1}) + \frac{A}{[(1-\tau)\tau^i(R-\rho)]^\beta} + B, \quad i = 0, 1, \ldots$$

by (1.5), as a consequence

$$\varphi(t_0) \leqslant \theta^k\varphi(t_k) + \frac{A}{[(1-\tau)(R-\rho)]^\beta}\sum_{i=0}^{k-1}\theta^i\tau^{-i\beta} + B\sum_{i=0}^{k-1}\theta^i \quad \text{for all} \ \ k \geqslant 1.$$

Take τ to satisfy $\theta\tau^{-\beta} < 1$ and let $k \to \infty$, then we get inequality (1.6). $\qquad\square$

Proof of Theorem 1.3. Here we only consider the case $i = 1$ and $\varepsilon = 0$. For $\sigma \leqslant \rho < r \leqslant \delta$, let $\eta(x)$ be a smooth function satisfying $0 \leqslant \eta(x) \leqslant 1$, $\eta(x) \equiv 1$ in Ω_1^ρ, $\eta(x) \equiv 0$ in $\Omega \setminus \Omega_1^r$ and $|D^k\eta| \leqslant C/(r-\rho)^k$, $k = 1, 2$. Then function $v = \eta u$ satisfies

$$\begin{cases} \mathscr{L}v = \eta f - ua_{ij}D_{ij}\eta - 2a_{ij}D_iuD_j\eta + ub_iD_i\eta =: f^* & \text{in} \ \ Q_T, \\ \partial_{\boldsymbol{n}}v + bv = g & \text{on} \ \ S_T, \\ v(x, 0) = \eta(x)\varphi(x) & \text{in} \ \ \Omega, \end{cases}$$

where $g = g_1$ on $\Gamma_1 \times (0, T]$ and $b = g = 0$ on $\Gamma_2 \times (0, T]$. Compatibility condition

$$\partial_{\boldsymbol{n}}(\eta\varphi) + b(x, 0)\eta\varphi = g(x, 0) \ \ \text{on} \ \ \partial\Omega$$

is obvious. Denote $D_1^r = \Omega_1^r \times (0, T]$ and $B = \|f\|_{p, D_1^\delta} + \|g_1\|_{W_p^{2,1}(D_1^\delta)} + \|\varphi\|_{W_p^2(\Omega_1^\delta)}$. Applying Theorem 1.2 and Theorem A.9 (1), it yields

$$\|v\|_{W_p^{2,1}(Q_T)} \leqslant C\left(\|f^*\|_{p, Q_T} + \|g\|_{W_p^{2,1}(Q_T)} + \|\eta\varphi\|_{W_p^2(\Omega)}\right)$$

$$\leqslant C\left[\frac{1}{r-\rho}\|Du\|_{p, D_1^r} + \frac{1}{(r-\rho)^2}\left(\|u\|_{p, D_1^r} + \|\varphi\|_{W_p^1(\Omega_1^r)}\right) + B\right]$$

$$\leqslant C\varepsilon\sum_{i=0}^{2}\|D^iu\|_{p, D_1^r} + \frac{C(\varepsilon)}{(r-\rho)^2}\left(\|u\|_{p, D_1^r} + \|\varphi\|_{W_p^1(\Omega_1^r)}\right) + CB.$$

Noticing $\eta(x) \equiv 1$ in Ω_1^ρ and the above estimate, we have

$$\|u\|_{W_p^{2,1}(D_1^\rho)} \leqslant C\varepsilon \|u\|_{W_p^{2,1}(D_1^r)} + \frac{C(\varepsilon)}{(r-\rho)^2}\left(\|u\|_{p,\,D_1^\delta} + \|\varphi\|_{W_p^1(\Omega_1^\delta)}\right) + CB.$$

From Lemma 1.4, it follows that

$$\|u\|_{W_p^{2,1}(D_1^\rho)} \leqslant C\left[\frac{1}{(r-\rho)^2}\left(\|u\|_{p,\,D_1^\delta} + \|\varphi\|_{W_p^1(\Omega_1^\delta)}\right) + B\right].$$

The desired conclusion is obtained by taking $\rho = \sigma$ and $r = \delta$. □

Similar to the proof of Theorem 1.3, we can prove the following theorem.

Theorem 1.5 (Interior L^p estimate in x direction) *Assume that $u \in W_p^{2,1}(\Omega_0 \times (0,T)) \cap L^p(Q_T)$ for any $\Omega_0 \Subset \Omega$, and let u satisfy*

$$\mathscr{L}u = f(x,t) \quad in \ \ Q_T. \tag{1.7}$$

Then, for any given $\Omega_0 \Subset \Omega$, we have

$$\|u\|_{W_p^{2,1}(\Omega_0 \times (0,T))} \leqslant C\left(\|f\|_{p,\,Q_T} + \|u\|_{p,\,Q_T} + \|u(\cdot,0)\|_{W_p^2(\Omega)}\right),$$

where C is a constant depending only on $\mathcal{A}$ and $1/d(\Omega_0, \partial\Omega)$.

Theorem 1.6 (Global L^p estimate) *Let conditions $(\mathbf{B})$ and $(\mathbf{D})$ hold, Ω be of class C^2 and $b \in C^{1,1/2}(\Gamma_1 \times [0,T])$. Assume that $u \in W_p^{2,1}(Q_T)$ satisfies (1.3). Then we have global estimate*

$$\|u\|_{W_p^{2,1}(Q_T)} \leqslant C\left(\|f\|_{p,\,Q_T} + \|g_1, g_2\|_{W_p^{2,1}(Q_T)} + \|\varphi\|_{W_p^2(\Omega)}\right), \tag{1.8}$$

where C is a constant which depends on $\mathcal{A}$ and $|b|_{1,\,\Gamma_1 \times [0,T]}$.

Proof. Using Theorem 1.3 for $i = 1$ and $i = 2$, respectively, and interior L^p estimate in x direction (Theorem 1.5), we can obtain

$$\|u\|_{W_p^{2,1}(Q_T)} \leqslant C\left(\|f\|_{p,\,Q_T} + \|g_1, g_2\|_{W_p^{2,1}(Q_T)} + \|\varphi\|_{W_p^2(\Omega)} + \|u\|_{p,\,Q_T}\right). \tag{1.9}$$

Particularly,

$$\|u_t\|_{p,\,Q_s} \leqslant C\left(\|f\|_{p,\,Q_T} + \|g_1, g_2\|_{W_p^{2,1}(Q_T)} + \|\varphi\|_{W_p^2(\Omega)} + \|u\|_{p,\,Q_s}\right)$$

for all $0 \leqslant s \leqslant T$.

Denote

$$B = \|f\|_{p,\,Q_T} + \|g_1, g_2\|_{W_p^{2,1}(Q_T)} + \|\varphi\|_{W_p^2(\Omega)}.$$

By using the above estimate and Young's inequality,

$$
\begin{aligned}
\int_\Omega |u(x,s)|^p &= \int_\Omega |\varphi(x)|^p + \int_0^s \int_\Omega \frac{\mathrm{d}}{\mathrm{d}t} |u|^p \\
&\leqslant \|\varphi\|_{p,\Omega}^p + p \int_0^s \int_\Omega |u|^{p-1} |u_t| \\
&\leqslant \|\varphi\|_{p,\Omega}^p + (p-1) \int_0^s \int_\Omega |u|^p + \int_0^s \int_\Omega |u_t|^p \\
&\leqslant C \left(\int_0^s \int_\Omega |u|^p + B^p \right).
\end{aligned}
$$

Then the Gronwall inequality gives

$$
\int_0^T \int_\Omega |u|^p \leqslant C(T) B^p.
$$

This, together with (1.9), implies (1.8). □

Theorem 1.7 (L^p theory for the mixed initial-boundary value problem) *Set $g = g_i$ on $\Gamma_i \times (0,T]$. Assume that conditions* **(B)** *and* **(D)** *hold, Ω is of class C^2 and $b \in C^{1,1/2}(\Gamma_1 \times [0,T])$. If compatibility conditions*

$$
\partial_{\boldsymbol{n}} \varphi(x) + b(x,0)\varphi(x) = g_1(x,0) \quad \text{on} \quad \Gamma_1, \quad \text{and} \quad \varphi(x) = g_2(x,0) \quad \text{on} \quad \Gamma_2
$$

are satisfied, then problem (1.3) has a unique solution $u \in W_p^{2,1}(Q_T)$ and the estimate (1.8) holds.

Proof. Global estimate (1.8) implies that (1.3) has at most one solution in $W_p^{2,1}(Q_T)$. Proof of this theorem is similar to those of Theorem 1.1 and Theorem 1.2, hence we omit the details here. □

Theorem 1.8 (Interior L^p estimate in t direction) *Let Ω be of class C^2 and the condition* **(B)** *hold. Assume that, for any $0 < \delta < T$, $u \in W_p^{2,1}(\Omega \times (\delta, T)) \cap L^p(Q_T)$ satisfies the first three equations of (1.3) for $\delta \leqslant t \leqslant T$. Then for any $0 < \delta < T$ we have*

$$
\|u\|_{W_p^{2,1}(\Omega \times (\delta,T))} \leqslant C \left(\|f\|_{p,Q_T} + \|g_1, g_2\|_{W_p^{2,1}(Q_T)} + \|u\|_{p,Q_T} \right),
$$

where C is a constant depending on $\mathcal{A}$, $|b|_{1, \Gamma_1 \times [0,T]}$ and $1/\delta$.

Proof. Take a truncated function $\eta \in C^\infty([0,T])$ such that $0 \leqslant \eta \leqslant 1$, $\eta \equiv 1$ in $[\delta, T]$ and $\eta \equiv 0$ in $[0, \delta/2]$. Then there exists $C' > 0$ such that $|\eta'(t)| \leqslant C'/\delta$. Function $w = u\eta$ satisfies

$$
\begin{cases}
\mathscr{L}w = f(x,t)\eta(t) + u\eta'(t) & \text{in} \quad Q_T, \\
\partial_{\boldsymbol{n}} w + bw = g_1 \eta(t) & \text{on} \quad \Gamma_1 \times (0,T], \\
w = g_2 \eta(t) & \text{on} \quad \Gamma_2 \times (0,T], \\
w(x,0) = 0 & \text{in} \quad \Omega.
\end{cases}
$$

Making use of Theorem 1.7, we have

$$
\begin{aligned}
\|u\|_{W_p^{2,1}(\Omega \times (\delta, T))} &\leqslant \|w\|_{W_p^{2,1}(Q_T)} \\
&\leqslant C_1 \left(\|f\eta(t)\|_{p, Q_T} + \|u\eta'(t)\|_{p, Q_T} + \|g_1\eta(t), g_2\eta(t)\|_{W_p^{2,1}(Q_T)} \right) \\
&\leqslant C_1 \left(\|f\|_{p, Q_T} + C'\delta^{-1} \|g_1, g_2\|_{W_p^{2,1}(Q_T)} + C'\delta^{-1} \|u\|_{p, Q_T} \right) \\
&\leqslant C \left(\|f\|_{p, Q_T} + \|g_1, g_2\|_{W_p^{2,1}(Q_T)} + \|u\|_{p, Q_T} \right).
\end{aligned}
$$

The proof is complete. □

Theorem 1.9 (Interior L^p estimate) *Suppose that the condition* (**B**) *holds.*

(1) *Let* $u \in W_{p,\mathrm{loc}}^{2,1}(Q_T) \cap L^p(Q_T)$ *satisfy* (1.7). *Then, for any* $Q \Subset Q_T$,

$$
\|u\|_{W_p^{2,1}(Q)} \leqslant C \left(\|f\|_{p, Q_T} + \|u\|_{p, Q_T} \right),
$$

where C *is a constant which depends only on* $\mathcal{A}$ *and* $1/d(Q, \partial_p Q_T)$.

(2) *Let* $p \geqslant 2$, *and* $u \in W_2^{2,1}(Q_T)$ *satisfy* (1.7). *Then* $u \in W_{p,\mathrm{loc}}^{2,1}(Q_T)$ *and for any* $Q \Subset Q_T$,

$$
\|u\|_{W_p^{2,1}(Q)} \leqslant C \left(\|f\|_{p, Q_T} + \|u\|_{2, Q_T} \right), \tag{1.10}
$$

where C *is the same as above.*

Proof. (1) For $\rho \geqslant 0$, we define

$$
Q^\rho = \{(x, t) \in Q_T : d((x, t), Q) \leqslant \rho\}.
$$

Set $\delta = d(Q, \partial_p Q_T)$. For $0 \leqslant \rho < r \leqslant \delta/2$, let η be a smooth function with $0 \leqslant \eta \leqslant 1$, $\eta \equiv 1$ in Q^ρ and $\eta \equiv 0$ in $Q_T \setminus Q^r$. Then

$$
|\eta_t| \leqslant C/(r - \rho), \quad |D^k \eta| \leqslant C/(r - \rho)^k, \quad k = 1, 2,
$$

and the function $v = \eta u \in W_p^{2,1}(Q_T)$ satisfies

$$
\begin{cases}
\mathscr{L}v = \eta f + \eta_t u - u a_{ij} D_{ij} \eta - 2 a_{ij} D_i u D_j \eta + u b_i D_i \eta =: f^* & \text{in } Q_T, \\
v = 0 & \text{on } S_T, \quad (1.11) \\
v(x, 0) = 0 & \text{in } \Omega.
\end{cases}
$$

Applying Theorem 1.1 and Theorem A.9 (1), we have

$$
\begin{aligned}
\|v\|_{W_p^{2,1}(Q_T)} &\leqslant C \|f^*\|_{p, Q_T} \\
&\leqslant C \left(\frac{1}{r - \rho} \|Du\|_{p, Q^r} + \frac{1}{(r - \rho)^2} \|u\|_{p, Q^r} + \|f\|_{p, Q_T} \right) \\
&\leqslant C\varepsilon \sum_{i=0}^{2} \|D^i u\|_{p, Q^r} + \frac{C(\varepsilon)}{(r - \rho)^2} \|u\|_{p, Q^r} + C \|f\|_{p, Q_T}.
\end{aligned}
$$

Note that $\eta(x) \equiv 1$ in Q^ρ. From the above estimate it can be deduced that

$$\|u\|_{W_p^{2,1}(Q^\rho)} \leqslant C\varepsilon\|u\|_{W_p^{2,1}(Q^r)} + \frac{C(\varepsilon)}{(r-\rho)^2}\|u\|_{p,Q_T} + C\|f\|_{p,Q_T}.$$

Then, by the use of Lemma 1.4,

$$\|u\|_{W_p^{2,1}(Q^\rho)} \leqslant C\left(\frac{1}{(r-\rho)^2}\|u\|_{p,Q_T} + \|f\|_{p,Q_T}\right).$$

Therefore,

$$\|u\|_{W_p^{2,1}(Q)} \leqslant C\left(\frac{1}{(r-\rho)^2}\|u\|_{p,Q_T} + \|f\|_{p,Q_T}\right).$$

The conclusion can be obtained by taking $\rho = 0$ and $r = \delta/2$.

(2) In this case, taking advantage of Theorem 1.1 and Theorem A.9 (1), we have

$$\|v\|_{W_p^{2,1}(Q_T)} \leqslant C\left(\frac{1}{r-\rho}\|Du\|_{p,Q^r} + \frac{1}{(r-\rho)^2}\|u\|_{p,Q^r} + \|f\|_{p,Q_T}\right)$$

$$\leqslant C\varepsilon\sum_{i=0}^{2}\|D^i u\|_{p,Q^r} + \frac{C(\varepsilon)}{(r-\rho)^2}\|u\|_{p,Q^r} + C\|f\|_{p,Q_T}$$

$$\leqslant C\varepsilon\sum_{i=0}^{2}\|D^i u\|_{p,Q^r} + \frac{C(\varepsilon)}{(r-\rho)^q}\|u\|_{2,Q_T} + C\|f\|_{p,Q_T},$$

where $q = 2 + n(p-2)/(2p)$. The remaining proof is the same as above. □

Theorem 1.10 (Local L^p estimate) *Let the condition* **(B)** *hold. Assume that $\Sigma \subset S_T$ with C^2 boundary $\partial\Sigma$, $u \in W_p^{2,1}(Q_T)$ satisfies (1.7) and $u = 0$ on Σ. Then, for any $Q \subset Q_T$ with $d(Q, \partial_p Q_T \setminus \Sigma) = \delta > 0$, there is a constant C determined only by $\mathcal{A}$ and $1/\delta$ such that*

$$\|u\|_{W_p^{2,1}(Q)} \leqslant C\left(\|f\|_{p,Q_T} + \|u\|_{p,Q_T}\right).$$

Proof. For $0 \leqslant \rho < r \leqslant \delta/2$, we let η be a smooth function with $0 \leqslant \eta \leqslant 1$, $\eta \equiv 1$ in Q^ρ and $\eta \equiv 0$ in $Q_T \setminus Q^r$ as above. Then

$$|\eta_t| \leqslant C/(r-\rho)^2, \quad |D^k\eta| \leqslant C/(r-\rho)^k, \quad k = 1, 2,$$

and function $v = \eta u \in W_p^{2,1}(Q_T)$ which satisfies (1.11). The remaining proof is the same as that of Theorem 1.9. □

Before ending this subsection, we study existence and uniqueness of "weak" solutions. As an example, here we only consider the following special case of (1.1):

$$\begin{cases} u_t - \Delta u = f(x,t) & \text{in } Q_T, \\ u = 0 & \text{on } S_T, \\ u(x,0) = 1 & \text{in } \Omega \end{cases} \tag{1.12}$$

with corresponding functions $g = 0$ and $\varphi = 1$. In this case, the compatibility condition $\varphi(x) = g(x,0)$ on $\partial\Omega$ does not hold.

Theorem 1.11 *Assume that Ω is of class C^2 and $f \in L^p(Q_T)$. Then problem (1.12) has a unique solution u in the sense that $u \in W_p^{2,1}(\Omega \times [\tau, T])$ for any $0 < \tau < T$, and moreover, u satisfies the first two equations of (1.12), and*

$$\lim_{t \to 0} \|u(\cdot, t) - 1\|_{p,\Omega} = 0. \tag{1.13}$$

Proof. Take $\varphi^k \in \overset{\circ}{W}{}_p^1(\Omega) \cap W_p^2(\Omega)$ satisfying $\varphi^k \to 1$ in $L^p(\Omega)$ as $k \to \infty$. By Theorem 1.1, the problem

$$\begin{cases} u_t^k - \Delta u^k = f(x,t) & \text{in } Q_T, \\ u^k = 0 & \text{on } S_T, \\ u^k(x,0) = \varphi^k(x) & \text{in } \Omega \end{cases}$$

has a unique solution $u^k \in \overset{\circ}{W}{}_p^{2,1}(Q_T)$, where $\overset{\circ}{W}{}_p^{2,1}(Q_T)$ is the closure of $\overset{\circ}{C}{}^\infty(\overline{Q}_T)$ in $W_p^{2,1}(Q_T)$, and

$$\overset{\circ}{C}{}^\infty(\overline{Q}_T) := \{ u \in C^\infty(\overline{Q}_T) : u|_{S_T} = 0 \}.$$

Clearly, $u^k - u^l$ satisfies

$$\begin{cases} (u^k - u^l)_t - \Delta(u^k - u^l) = 0 & \text{in } Q_T, \\ u^k - u^l = 0 & \text{on } S_T, \\ (u^k - u^l)(x,0) = \varphi^k(x) - \varphi^l(x) & \text{in } \Omega. \end{cases}$$

Thus, for all $0 < t \leqslant T$,

$$\frac{1}{p}\frac{d}{dt}\int_\Omega |u^k - u^l|^p = -(p-1)\int_\Omega |u^k - u^l|^{p-2}|\nabla(u^k - u^l)|^2 \leqslant 0,$$

which implies

$$\int_\Omega |u^k - u^l|^p \leqslant \int_\Omega |\varphi^k - \varphi^l|^p \to 0 \tag{1.14}$$

as well as

$$\int_{Q_T} |u^k - u^l|^p dt \leqslant T \int_\Omega |\varphi^k - \varphi^l|^p \to 0$$

as $k, l \to \infty$. Applying Theorem 1.8, for any $0 < \tau < T$, we have

$$\|u^k - u^l\|_{W_p^{2,1}(\Omega \times [\tau, T])} \leqslant C\|u^k - u^l\|_{p, Q_T} \to 0$$

as $k, l \to \infty$. Therefore, $\{u^k\}$ is a Cauchy sequence in $\overset{\circ}{W}{}_p^{2,1}(\Omega \times [\tau, T])$. Consequently, $u^k \to u^\tau$ in $\overset{\circ}{W}{}_p^{2,1}(\Omega \times [\tau, T])$ for some $u^\tau \in \overset{\circ}{W}{}_p^{2,1}(\Omega \times [\tau, T])$. By the uniqueness of limit, $u^\tau = u^\varepsilon$ in $\Omega \times [\tau, T]$ when $\varepsilon < \tau$. Thus we get a function u having properties:

$$u = u^\tau \text{ in } \Omega \times [\tau, T], \quad u \in \overset{\circ}{W}{}_p^{2,1}(\Omega \times [\tau, T]),$$

and

$$\lim_{k \to \infty} \|u^k - u\|_{W_p^{2,1}(\Omega \times [\tau, T])} = 0$$

for any $0 < \tau < T$. This implies $u^k \to u$ in $C([\tau, T], L^p(\Omega))$ (cf. Exercise 1.2). Certainly, $u^k(\cdot, t) \to u(\cdot, t)$ in $L^p(\Omega)$ for any $0 < t \leqslant T$.

It is easy to see that u satisfies the first two equations of (1.12). We shall show that u satisfies (1.13). In fact, taking $l \to \infty$ in the inequality of (1.14), we have

$$\int_\Omega |u^k - u|^p \leqslant \int_\Omega |\varphi^k - 1|^p,$$

which implies

$$\lim_{k \to \infty} \int_\Omega |u^k - u|^p = 0 \quad \text{uniformly in } t \in (0, T].$$

As a consequence, there exists $k \gg 1$ such that

$$\int_\Omega |u^k - u|^p + \int_\Omega |\varphi^k - 1|^p < \varepsilon \quad \text{for all } t \in (0, T].$$

Since

$$\lim_{t \to 0} \int_\Omega |u^k - \varphi^k|^p = 0,$$

making use of

$$\int_\Omega |u - 1|^p \leqslant C \left(\int_\Omega |u^k - u|^p + \int_\Omega |\varphi^k - 1|^p + \int_\Omega |u^k - \varphi^k|^p \right),$$

we can prove (1.13). Existence is proved.

We now prove the uniqueness of such solutions. Let u_1 and u_2 be two such solutions. Then $u_1 - u_2$ satisfies

$$\begin{cases} (u_1 - u_2)_t - \Delta(u_1 - u_2) = 0 & \text{in } Q_T, \\ u_1 - u_2 = 0 & \text{on } S_T. \end{cases}$$

Similar to the above arguments,

$$\int_\Omega |u_1(\cdot, t) - u_2(\cdot, t)|^p \leqslant \int_\Omega |u_1(\cdot, \tau) - u_2(\cdot, \tau)|^p, \quad 0 < \tau < t \leqslant T.$$

As both u_1 and u_2 satisfy (1.13), letting $\tau \to 0$, we have

$$\int_\Omega |u_1(\cdot, t) - u_2(\cdot, t)|^p = 0, \quad 0 < t \leqslant T.$$

Thus $u_1 = u_2$, and the proof is finished. □

1.2.2 Schauder theory

For convenience, let us denote

$$\mathcal{B} = \{n, \lambda, \Lambda, T\}.$$

Let $\Omega' \subset \Omega'' \subset \Omega$ with $\text{dist}(\Omega', \Omega \backslash \Omega'') > 0$. Denote $S_1 = \partial\Omega \cap \overline{\Omega'}$ and $S_2 = \partial\Omega \cap \overline{\Omega''}$. If $S_1 = \emptyset$, then Ω'' will be chosen so that $S_2 = \emptyset$. Let $0 \leqslant t_1 < t_2 \leqslant T$. We take $0 < t_0 < t_1$ if $t_1 > 0$, and $t_0 = 0$ if $t_1 = 0$. Set $Q_1 = \Omega' \times (t_1, t_2)$, $Q_2 = \Omega'' \times (t_0, t_2)$ and $S_{02} = S_2 \times (t_0, t_2)$. We first state the local *Schauder estimate*.

Theorem 1.12 (Local Schauder estimate [60, Chapter 4, Theorem 10.1]) *Assume that the condition* **(A)** *holds. Let* $u \in C^{2+\alpha, 1+\alpha/2}(\overline{Q}_2)$ *and* $\mathscr{L}u = f$ *in* Q_2. *Suppose* $a_{ij}, b_i, c, f \in C^{\alpha, \alpha/2}(\overline{Q}_2)$. *In addition, if* $t_0 = 0$ *we assume* $u|_{t=0} = \varphi(x)$ *in* Ω'', *and if* $S_2 \neq \emptyset$ *we assume either*

$$u = g \quad on \ \ S_{02}, \tag{1.15}$$

or

$$\partial_{\boldsymbol{n}} u + bu = g \quad on \ \ S_{02} \quad and \ \ b \in C^{1+\alpha, (1+\alpha)/2}(\overline{S}_{02}). \tag{1.16}$$

Then there exists a constant C *depending on* $\mathcal{B}$, $|a_{ij}, b_i, c|_{\alpha, \overline{Q}_2}$ *and* $|b|_{1+\alpha, \overline{S}_{02}}$ *so that*

$$|u|_{2+\alpha, \overline{Q}_1} \leqslant C \left(|f|_{\alpha, \overline{Q}_2} + |g|_{2+\alpha, \overline{S}_{02}} + |\varphi|_{2+\alpha, \overline{\Omega''}} + \|u\|_{\infty, Q_2} \right) \tag{1.17}$$

when the boundary condition is (1.15), *and*

$$|u|_{2+\alpha, \overline{Q}_1} \leqslant C \left(|f|_{\alpha, \overline{Q}_2} + |g|_{1+\alpha, \overline{S}_{02}} + |\varphi|_{2+\alpha, \overline{\Omega''}} + \|u\|_{\infty, Q_2} \right) \tag{1.18}$$

when the boundary condition is (1.16).

Moreover, if $S_2 = \emptyset$ *then the right hand sides of* (1.17) *and* (1.18) *will not contain norm of* g, *and the constant* C *also depends on* $1/d(\Omega', \partial\Omega'')$; *if* $t_0 > 0$ *then they will not contain* $|\varphi|_{2+\alpha, \overline{\Omega''}}$, *and the constant* C *also depends on* t_0 *and* t_1.

As a consequence of Theorem 1.12, we have the following results.

Remark 1.13 *Assume that* $\mathscr{L}u = f$ *in* Q_T.

(1) Interior Schauder estimate in t direction: *Suppose* $u \in C^{2+\alpha, 1+\alpha/2}(\overline{\Omega} \times (0, T])$. *Taking into account that* Ω *is connected, we can choose* Ω_i' *and* Ω_i'' *to have the above properties such that* $\Omega_1' \cup \Omega_2' = \Omega$. *Take* $0 < t_0 < t_1 < T$ *and* $t_2 = T$. *Using the estimates* (1.17) *and* (1.18), *it follows that*

$$|u|_{2+\alpha, \overline{\Omega} \times [t_1, T]} \leqslant C \left(|f|_{\alpha, \overline{\Omega} \times [t_0, T]} + |g|_{2+\alpha, \overline{\Omega} \times [t_0, T]} + \|u\|_{\infty, \Omega \times (t_0, T)} \right)$$

when the boundary condition is $u = g$, *here constant* C *depends on* $\mathcal{B}$, t_0, t_1, $|a_{ij}, b_i, c|_{\alpha, \overline{\Omega} \times [t_0, T]}$; *while*

$$|u|_{2+\alpha, \overline{\Omega} \times [t_1, T]} \leqslant C \left(|f|_{\alpha, \overline{\Omega} \times [t_0, T]} + |g|_{1+\alpha, \overline{\Omega} \times [t_0, T]} + \|u\|_{\infty, \Omega \times (t_0, T)} \right) \tag{1.19}$$

when the boundary condition is $\partial_{\boldsymbol{n}} u + bu = g$ *with* $b \in C^{1+\alpha, (1+\alpha)/2}(\overline{S}_T)$ *and* $b \geqslant 0$. *Here constant* C *also depends on* $|b|_{1+\alpha, \partial\Omega \times [t_0, T]}$.

(2) Interior Schauder estimate in x direction: *Suppose $u \in C^{2+\alpha, 1+\alpha/2}(\Omega \times [0, T])$. Take $\Omega' \Subset \Omega'' \Subset \Omega$, $t_0 = t_1 = 0$ and $t_2 = T$. Then estimates (1.17) and (1.18) coincide and are reduced to*

$$|u|_{2+\alpha, \overline{\Omega'} \times [0,T]} \leqslant C \left(|f|_{\alpha, \overline{\Omega''} \times [0,T]} + |\varphi|_{2+\alpha, \overline{\Omega''}} + \|u\|_{\infty, \Omega'' \times (0,T)} \right),$$

where constant C depends on $\mathcal{B}, |a_{ij}, b_i, c|_{\alpha, \overline{\Omega''} \times [0,T]}$ and $1/d(\Omega', \partial\Omega'')$.

Theorem 1.14 *Assume that Ω is of class $C^{2+\alpha}$, $b \in C^{1+\alpha, (1+\alpha)/2}(\Gamma_1 \times [0, T])$.*

(1) Local Schauder estimate in x direction: *Let $\Omega_l \subset \Omega'_l \subset \Omega$ with $d(\Omega_l, \Omega \setminus \Omega'_l) = \varepsilon > 0$, and $\overline{\Omega}_l \cap \partial\Omega = \overline{\Omega'_l} \cap \partial\Omega = \Gamma_l$ for $l = 1$ or $l = 2$. Denote $D'_l = \Omega'_l \times (0, T]$. Assume that $a_{ij}, b_i, c \in C^{\alpha, \alpha/2}(\overline{D'_l})$ and $u \in C^{2+\alpha, 1+\alpha/2}(\overline{D'_l})$ satisfies*

$$\begin{cases} \mathscr{L}u = f(x,t) & \text{in } D'_l, \\ \mathscr{B}_l u = g_l & \text{on } \Gamma_l \times (0, T]. \end{cases}$$

Then

$$|u|_{2+\alpha, \overline{\Omega}_l \times [0,T]} \leqslant C \left(|f|_{\alpha, \overline{D'_l}} + |g_l|_{l+\alpha, \Gamma_l \times [0,T]} + |u(\cdot, 0)|_{2+\alpha, \overline{\Omega'_l}} + \|u\|_{\infty, D'_l} \right),$$

where the constant C depends on $\mathcal{B}$, $|a_{ij}, b_i, c|_{\alpha, \overline{Q'_l}}$ and $1/\varepsilon$, and also depends on $|b|_{1+\alpha, \Gamma_1 \times [0,T]}$ when $l = 1$.

(2) Interior Schauder estimate in t direction: *Let $a_{ij}, b_i, c, f \in C^{\alpha, \alpha/2}(\overline{\Omega} \times (0, T])$, $u \in C^{2+\alpha, 1+\alpha/2}(\overline{\Omega} \times (0, T])$ satisfy (1.3). Take Ω_l and Ω'_l as in conclusion (1). Then, for any $0 < \delta < \tau < T$, we have*

$$|u|_{2+\alpha, \overline{\Omega}_l \times [\tau,T]} \leqslant C \left(|f|_{\alpha, \overline{\Omega'_l} \times [\delta,T]} + |g_l|_{l+\alpha, \Gamma_l \times [\delta,T]} + \|u\|_{\infty, \Omega'_l \times [\delta,T]} \right)$$

for $l = 1, 2$. Especially, if we take Ω_l and Ω'_l such that $\Omega_1 \cup \Omega_2 = \Omega$, then

$$|u|_{2+\alpha, \overline{\Omega} \times [\tau,T]} \leqslant C \left(|f|_{\alpha, \overline{\Omega} \times [\delta,T]} + \sum_{l=1}^{2} |g_l|_{l+\alpha, \Gamma_l \times [\delta,T]} + \|u\|_{\infty, \Omega \times (\delta,T)} \right),$$

where the constant C depends only on $\mathcal{B}$, $|a_{ij}, b_i, c|_{\alpha, \overline{\Omega} \times [\delta,T]}$, $1/(\tau - \delta)$ and $|b|_{1+\alpha, \Gamma_1 \times [\delta,T]}$.

Theorem 1.15 (Global Schauder estimate) *Define $g = g_l$ on $\Gamma_l \times (0, T]$. Let conditions (**C**) and (**E**) hold, Ω be of class $C^{2+\alpha}$ and $b \in C^{1+\alpha, (1+\alpha)/2}(\Gamma_1 \times [0, T])$. If $u \in C^{2+\alpha, 1+\alpha/2}(\overline{Q}_T)$ solves (1.3), then we have the global estimate:*

$$|u|_{2+\alpha, \overline{Q}_T} \leqslant C \left(|f|_{\alpha, \overline{Q}_T} + \sum_{l=1}^{2} |g_l|_{l+\alpha, \Gamma_l \times [0,T]} + |\varphi|_{2+\alpha, \overline{\Omega}} \right), \tag{1.20}$$

where the constant C depends on $\mathcal{B}$, $|a_{ij}, b_i, c|_{\alpha, \overline{Q}_T}$, Ω and $|b|_{1+\alpha, \Gamma_1 \times [0,T]}$.

Proof. Taking advantage of Theorem 1.14 (1), we have

$$|u|_{2+\alpha,\overline{Q}_s} \leqslant C \left(|f|_{\alpha,\overline{Q}_T} + \sum_{l=1}^{2} |g_l|_{l+\alpha,\, \Gamma_l \times [0,T]} + |\varphi|_{2+\alpha,\overline{\Omega}} + \|u\|_{\infty,Q_s} \right) \qquad (1.21)$$

for all $0 \leqslant s \leqslant T$, which implies

$$\|u_t\|_{\infty,Q_s} \leqslant C \left(|f|_{\alpha,\overline{Q}_T} + \sum_{l=1}^{2} |g_l|_{l+\alpha,\, \Gamma_l \times [0,T]} + |\varphi|_{2+\alpha,\overline{\Omega}} + \|u\|_{\infty,Q_s} \right). \qquad (1.22)$$

Denote

$$A = |f|_{\alpha,\overline{Q}_T} + \sum_{l=1}^{2} |g_l|_{l+\alpha,\, \Gamma_l \times [0,T]} + |\varphi|_{2+\alpha,\overline{\Omega}}.$$

It then follows from (1.22) that, for all $x \in \overline{\Omega}$ and $0 \leqslant t \leqslant \tau \leqslant T$,

$$\begin{aligned}
|u(x,t)| &= |\varphi(x)| + \int_0^t \frac{\mathrm{d}}{\mathrm{d}s} |u(x,s)| \\
&\leqslant \|\varphi\|_{\infty,\Omega} + \int_0^t |u_s(x,s)| \\
&\leqslant CA + C \int_0^t \|u\|_{\infty,Q_s} \\
&\leqslant CA + C \int_0^\tau \|u\|_{\infty,Q_s}.
\end{aligned}$$

Taking the maximum with respect to $x \in \overline{\Omega}$ and $0 \leqslant t \leqslant \tau$, we have

$$\|u\|_{\infty,Q_\tau} \leqslant CA + C \int_0^\tau \|u\|_{\infty,Q_s} \quad \text{for all } 0 < \tau \leqslant T.$$

Set $v(\tau) = \|u\|_{\infty,Q_\tau}$, then the Gronwall inequality gives $v(T) \leqslant C(T)A$. This combines with (1.21) enables us to derive the estimate (1.20). $\qquad \square$

Theorem 1.16 (Schauder theory for the first initial-boundary value problem [72, Theorem 5.14], [60, Chapter 7, Theorem 10.1]) *Let conditions* **(C)** *and* **(E)** *hold, and Ω be of class $C^{2+\alpha}$. Then problem (1.1) has a unique solution $u \in C(\overline{Q}_T) \cap C^{2,1}(Q_T)$. Moreover, if φ and g satisfy compatibility conditions:*

$$\varphi(x) = g(x,0) \quad \text{for } x \in \partial\Omega,$$

$$g_t - a_{ij} D_{ij}\varphi + b_i D_i\varphi + c\varphi = f \quad \text{for } x \in \partial\Omega, \ t = 0,$$

then $u \in C^{2+\alpha,\, 1+\alpha/2}(\overline{Q}_T)$ and

$$|u|_{2+\alpha,\overline{Q}_T} \leqslant C \left(|f|_{\alpha,\overline{Q}_T} + |g|_{2+\alpha,\overline{Q}_T} + |\varphi|_{2+\alpha,\overline{\Omega}} \right),$$

where C is a constant which depends only on $\mathcal{B}$, $|a_{ij}, b_i, c|_{\alpha,\overline{Q}_T}$ and Ω.

Theorem 1.17 (Schauder theory for the second initial-boundary value problem [72, Theorem 5.18]) *Suppose that conditions* **(C)** *and* **(E)** *hold,* Ω *is of class* $C^{2+\alpha}$ *and* $b \in C^{1+\alpha,\,(1+\alpha)/2}(\overline{S}_T)$. *If* φ *and* g *satisfy the compatibility condition*

$$\partial_{\boldsymbol{n}}\varphi + b\varphi = g \quad \text{for } t = 0, \; x \in \partial\Omega,$$

then problem (1.2) has a unique solution $u \in C^{2+\alpha,1+\alpha/2}(\overline{Q}_T)$, *and*

$$|u|_{2+\alpha,\overline{Q}_T} \leqslant C \left(|f|_{\alpha,\overline{Q}_T} + |g|_{1+\alpha,\overline{S}_T} + |\varphi|_{2+\alpha,\overline{\Omega}} \right)$$

with a constant C *relying on* $\mathcal{B}$, $|a_{ij}, b_i, c|_{\alpha,\overline{Q}_T}$, $|b|_{1+\alpha,\overline{S}_T}$ *and* Ω.

Theorem 1.18 (Schauder theory for the mixed initial-boundary value problem) *Set* $g = g_l$ *on* $\Gamma_l \times (0,T]$. *Let conditions* **(C)** *and* **(E)** *hold,* Ω *be of class* $C^{2+\alpha}$ *and* $b \in C^{1+\alpha,\,(1+\alpha)/2}(\Gamma_1 \times [0,T])$. *Assume that compatibility conditions*

$$\partial_{\boldsymbol{n}}\varphi + b\varphi = g_1 \quad \text{for } t = 0, \; x \in \Gamma_1, \quad \text{and}$$

$$\varphi = g_2, \quad g_{2t} - a_{ij}D_{ij}\varphi + b_iD_i\varphi + c\varphi = f \quad \text{for } t = 0, \; x \in \Gamma_2$$

are satisfied. Then problem (1.3) has a unique solution $u \in C^{2+\alpha,\,1+\alpha/2}(\overline{Q}_T)$, *and*

$$|u|_{2+\alpha,\overline{Q}_T} \leqslant C \left(|f|_{\alpha,\overline{Q}_T} + \sum_{l=1}^{2} |g_l|_{l+\alpha,\,\Gamma_l\times[0,T]} + |\varphi|_{2+\alpha,\overline{\Omega}} \right),$$

where C *is a constant depending on* $\mathcal{B}$, $|a_{ij}, b_i, c|_{\alpha,\overline{Q}_T}$, Ω *and* $|b|_{1+\alpha,\,\Gamma_1\times[0,T]}$.

Proof. The global estimate (1.20) implies that problem (1.3) has at most one solution in the class of $C^{2+\alpha,\,1+\alpha/2}(\overline{Q}_T)$. The remaining proof of this theorem is similar to those of Theorems 1.16 and 1.17, and therefore is omitted. □

A conclusion similar to Theorem 1.18 is given in [71].

Theorem 1.19 (Interior Schauder estimate) *Let the condition* **(C)** *hold, and* $u \in W_2^{2,1}(Q_T)$ *be a solution of (1.7). Then* $u \in C^{2+\alpha,\,1+\alpha/2}(Q_T)$, *and for any* $Q \Subset Q_T$,

$$|u|_{2+\alpha,\overline{Q}} \leqslant C \left(|f|_{\alpha,\overline{Q}_T} + \|u\|_{2,Q_T} \right), \tag{1.23}$$

where C *is a constant which relies on* $\mathcal{B}$, $|a_{ij}, b_i, c|_{\alpha,\overline{Q}_T}$ *and* $1/d(Q, \partial_p Q_T)$.

This theorem can be proved in the same way as that of Theorem 1.9 (2), and we leave the proof to readers as an exercise.

1.2.3 Existence and uniqueness of solutions of linear system

Give m parabolic operators

$$\mathscr{L}_k = \partial_t - a_{ij}^k(x,t)D_{ij} + b_i^k(x,t)D_i, \quad k = 1,\dots,m. \tag{1.24}$$

Assume that a_{ij}^k satisfies the condition **(A)** with the same λ and Λ for all $1 \leqslant k \leqslant m$, and the modulus of continuity $\omega_k(R)$ of a_{ij}^k are the same as $\omega(R)$ defined in **(A)**. Let $b_i^k, h_{kl} \in L^\infty(Q_T)$ with $\|b_i^k, h_{kl}\|_\infty \leqslant \Lambda$, g_k and φ_k satisfy the condition **(D)**.

Theorem 1.20 (L^p theory) *Consider the following problem*

$$
\begin{cases}
\mathscr{L}_k u_k + \sum_{l=1}^{m} h_{kl}(x,t)u_l = f_k(x,t) & in \ \ Q_T, \\
\mathscr{B}_k u_k = g_k & on \ \ S_T, \\
u_k(x,0) = \varphi_k(x) & in \ \ \Omega, \\
k = 1, \ldots, m,
\end{cases}
\tag{1.25}
$$

where either $\mathscr{B}_k u_k = u_k$, or $\mathscr{B}_k u_k = \partial_{\boldsymbol{n}} u_k + b^k u_k$ with $b^k \geqslant 0$.

Assume that Ω is of class C^2, g_k and φ_k satisfy requirements of Theorem 1.1 when $\mathscr{B}_k u_k = u_k$, while b^k, g_k and φ_k satisfy requirements of Theorem 1.2 when $\mathscr{B}_k u_k = \partial_{\boldsymbol{n}} u_k + b^k u_k$. If $f_k \in L^p(Q_T)$, then problem (1.25) has a unique solution $u \in [W_p^{2,1}(Q_T)]^m$, and

$$
\|u_k\|_{W_p^{2,1}(Q_T)} \leqslant C \sum_{k=1}^{m} \left(\|f_k\|_{p, Q_T} + \|g_k\|_{W_p^{2,1}(Q_T)} + \|\varphi_k\|_{W_p^2(\Omega)} \right),
$$

$k = 1, \ldots, m$, where C is a constant depending on n, Q_T, λ, Λ, $\max_{1 \leqslant l \leqslant m} |b^l|_{1, \overline{S}_T}$ and $\max_{1 \leqslant l \leqslant m} \omega_l(R)$.

This theorem can refer to Theorem 10.4 of [60, Chapter 7].

1.3   MAXIMUM PRINCIPLE OF CLASSICAL SOLUTIONS

An important tool in theory of second order parabolic equations is the *maximum principle*, which avers that a solution of a given homogeneous linear parabolic equation in a domain must achieve its maximum on parabolic boundary of that domain. The maximum principle will be used to prove uniqueness results for various initial-boundary value problems, to obtain L^∞ bounds for solutions and their derivatives, and to derive various continuity estimates as well.

1.3.1   Initial-boundary value problems

An available result related to the strong maximum principle is the so-called boundary point lemma, which states that if a non-constant solution takes local strict maximum at some point on boundary, then its outward normal derivative at this point must be positive.

Lemma 1.21 (Hopf boundary lemma [91] or [110, Lemmma 2.4.5]) *Let $a_{ii}, b_i \in L^\infty(Q)$, $\mathscr{L}$ be parabolic and $c \geqslant 0$ in Q. Let $u \in C(\overline{Q}) \cap C^{2,1}(Q)$ satisfy $\mathscr{L}u \leqslant 0 \, (\geqslant 0)$ in Q.*

Suppose $u(\mathbf{x}_0) = M = \max_{\overline{Q}} u \, (u(\mathbf{x}_0) = m = \min_{\overline{Q}} u)$ for some $\mathbf{x}_0 = (x_0, t_0) \in \partial Q$, and there is a neighbourhood V of $\mathbf{x}_0$ such that $u < M \, (u > m)$ in $Q \cap V$, where $M \geqslant 0 \, (m \leqslant 0)$ when $c \not\equiv 0$. We further assume that there exists a ball $B_r(\mathbf{x}^*)$ centred at $\mathbf{x}^* = (x^*, t^*)$ with radius $r > 0$ and $x^* \neq x_0$, such that $\overline{B_r(\mathbf{x}^*)} \subset \overline{Q}$ and $\overline{B_r(\mathbf{x}^*)} \cap \partial Q = \{\mathbf{x}_0\}$. Then the derivative $\partial_{\boldsymbol{\nu}} u(\mathbf{x}_0) > 0 \, (< 0)$ when it exists, where $\boldsymbol{\nu}$ is the outward normal vector of ∂Q at point $\mathbf{x}_0$.

Lemma 1.22 (Hopf boundary lemma [72, Lemmma 2.8]) *Let R and η be positive constants. We define*

$$\Gamma = \Gamma(R, \eta, (x_0, t_0)) = \{|x - x_0|^2 + \eta(t_0 - t) < R^2, \ t < t_0\}.$$

Assume that $a_{ii}, b_i \in L^\infty(\Gamma)$, $\mathscr{L}$ is parabolic in Γ, $u \in C(\overline{\Gamma}) \cap C^{2,1}(\Gamma)$ satisfies $\mathscr{L}u \leqslant 0 (\geqslant 0)$ in Γ. Also suppose that there exists (x^*, t_0) with $|x^* - x_0| = R$ such that

$$u(x^*, t_0) \geqslant (\leqslant) \, u(x, t), \quad c(x, t)u(x^*, t_0) \geqslant (\leqslant) \, 0 \quad \text{for all } (x, t) \in \Gamma,$$

$$u(x^*, t_0) > (<) \, u(x, t) \quad \text{for all } (x, t) \in \Gamma \text{ with } |x - x_0| \leqslant R/2.$$

Then the derivative $\partial_{\boldsymbol{\nu}} u(x^*, t_0) > 0 \ (< 0)$ when it exists.

Lemma 1.23 (Hopf boundary lemma) *Let $a_{ii}, b_i \in L^\infty(Q_T)$, $\mathscr{L}$ be parabolic and $c \geqslant 0$ in Q_T. Let $u \in C(\overline{Q}_T) \cap C^{2,1}(Q_T)$ satisfy $\mathscr{L}u \leqslant 0 (\geqslant 0)$ in Q_T. Suppose that $u(x_0, t_0) = M = \max_{\overline{Q}_T} u$ $(u(x_0, t_0) = m = \min_{\overline{Q}_T} u)$ for some $(x_0, t_0) \in S_T$, and there is a neighbourhood V of (x_0, t_0) such that $u < M$ $(u > m)$ in $Q_T \cap V$. We further assume that Ω has the interior ball property at x_0 and $M \geqslant 0$ $(m \leqslant 0)$ when $c \not\equiv 0$. Then the derivative $\partial_{\boldsymbol{n}} u(x_0, t_0) > 0 \ (< 0)$ when it exists.*

Theorem 1.24 (Strong maximum principle [91] or [72, Theorem 2.7]) *Let $a_{ii}, b_i \in L^\infty(Q)$, $\mathscr{L}$ be parabolic and $c \geqslant 0$ in Q. Assume that $u \in C(\overline{Q}) \cap C^{2,1}(Q)$ satisfies $\mathscr{L}u \leqslant 0 (\geqslant 0)$ in Q. If $u(\mathbf{x}_0) = \max_{\overline{Q}} u = M$ $(u(\mathbf{x}_0) = \min_{\overline{Q}} u = m)$ for some $\mathbf{x}_0 \in \overline{Q} \setminus \partial_p Q$, and $M \geqslant 0 \, (m \leqslant 0)$ when $c \not\equiv 0$, then $u \equiv u(\mathbf{x}_0)$ in $S(\mathbf{x}_0)$.*

Corollary 1.25 (Strong maximum principle) *Let $a_{ii}, b_i \in L^\infty(Q_T)$, $\mathscr{L}$ be parabolic and $c \geqslant 0$ in Q_T. Assume that $u \in C(\overline{Q}_T) \cap C^{2,1}(Q_T)$ satisfies $\mathscr{L}u \leqslant 0 (\geqslant 0)$ in Q_T. If $u(x_0, t_0) = \max_{\overline{Q}_T} u = M$ $(u(x_0, t_0) = \min_{\overline{Q}_T} u = m)$ for some $x_0 \in \Omega$, $0 < t_0 \leqslant T$, and $M \geqslant 0 \, (m \leqslant 0)$ when $c \not\equiv 0$, then $u \equiv u(x_0, t_0)$ in Q_{t_0}.*

Lemma 1.26 (Positivity lemma) *Let $a_{ii}, b_i \in L^\infty(Q_T)$, $\mathscr{L}$ be parabolic and c be bounded below in Q_T. Let $a, b \geqslant 0$ with $a + b > 0$ on S_T. For any $(x, t) \in S_T$, if $a(x, t) \neq 0$, then we further assume that Ω has the interior ball property at point x. Suppose $u \in C(\overline{Q}_T) \cap C^{2,1}(Q_T)$, and when $a \not\equiv 0$ we further assume $u \in C^{1,0}(\overline{\Omega} \times (0, T])$. If u satisfies*

$$\begin{cases} \mathscr{L}u \geqslant 0 & \text{in } Q_T, \\ a\partial_{\boldsymbol{n}} u + bu \geqslant 0 & \text{on } S_T, \\ u(x, 0) \geqslant 0 & \text{in } \Omega, \end{cases} \tag{1.26}$$

then $u \geqslant 0$ in Q_T. If in addition $u(x, 0) \not\equiv 0$, then $u > 0$ in Q_T, and $u > 0$ in $\overline{\Omega} \times (0, T]$ when $a > 0$ on S_T.

Proof. Take a constant k large enough so that $k + c > 0$ in Q_T, and let $w = ue^{-kt}$. Then $w(x, 0) \geqslant 0$ in $\overline{\Omega}$, $a\partial_n w + bw \geqslant 0$ on S_T, and

$$w_t - a_{ij} D_{ij}w + b_i D_i w + (k+c)w \geqslant 0 \quad \text{in} \quad Q_T.$$

Assume on the contrary that $w(x_0, t_0) = \min_{\overline{Q}_T} w < 0$, then $t_0 > 0$. Moreover, $x_0 \notin \Omega$ by Corollary 1.25. Thus $x_0 \in \partial\Omega$ and $w > w(x_0, t_0)$ in Q_T. Obviously $a(x_0, t_0) \neq 0$. Applying Lemma 1.23 we have $\partial_n w(x_0, t_0) < 0$, which leads to $a\partial_n w(x_0, t_0) + bw(x_0, t_0) < 0$. It is a contradiction.

In addition, if $w(x, 0) \not\equiv 0$, then $w > 0$ in Q_T by Corollary 1.25. Furthermore, if $a > 0$ on S_T, then $w > 0$ in $\overline{\Omega} \times (0, T]$ by Lemma 1.23. □

Here we should remark that, in Lemma 1.26, if it is only assumed that $u \not\equiv 0$ then we cannot conclude $u > 0$ in Q_T. For example, take $\mathscr{L}u = u_t - u_{xx}$, $\Omega = (0, 1)$ and $T = 2$. Let

$$u(t) = e^{-1/(t-1)} \quad \text{for} \quad 1 < t \leqslant 2, \quad u(t) \equiv 0 \quad \text{for} \quad 0 \leqslant t \leqslant 1.$$

Then u satisfies (1.26) and $u \not\equiv 0$. However, $u > 0$ does not hold in $Q_T = (0, 1) \times (0, 2]$.

1.3.2 Cauchy problems

Let us consider the following Cauchy problem

$$\begin{cases} \mathscr{L}u = f & \text{in} \quad \mathbb{R}^n \times (0, T], \\ u(x, 0) = \varphi(x) & \text{in} \quad \mathbb{R}^n, \end{cases} \tag{1.27}$$

where the coefficient c of $\mathscr{L}$ may be unbounded in $\mathbb{R}^n \times (0, T]$.

Lemma 1.27 *Assume that $a_{ii}, b_i \in L^\infty(\mathbb{R}^n \times (0, T])$, $\mathscr{L}$ is parabolic in $\mathbb{R}^n \times (0, T]$ and there exists a constant $c_0 > 0$ such that $c + c_0 > 0$ in $\mathbb{R}^n \times (0, T]$. Let $u \in C^{2,1}(\mathbb{R}^n \times (0, T]) \cap C(\mathbb{R}^n \times [0, T])$ be a solution of (1.27). If $f \geqslant 0$ in $\mathbb{R}^n \times (0, T]$, $\varphi \geqslant 0$ in $\mathbb{R}^n$ and*

$$\liminf_{|x| \to \infty} u(x, t) \geqslant 0$$

uniformly in $t \in [0, T]$. Then $u \geqslant 0$ in $\mathbb{R}^n \times (0, T]$.

Proof. Invoking a transform $v = ue^{-c_0 t}$ we assume without loss of generality that $c > 0$. For any given $\varepsilon > 0$, there holds that $\varphi(x) + \varepsilon > 0$ in $\mathbb{R}^n$, and

$$\mathscr{L}[u + \varepsilon] = \mathscr{L}u + c\varepsilon > 0 \quad \text{in} \quad \mathbb{R}^n \times (0, T].$$

By our assumption, there exists $R_0 \gg 1$ such that $u + \varepsilon > 0$ on $\{|x| = R, \, 0 \leqslant t \leqslant T\}$ for all $R > R_0$. Then we can apply Corollary 1.25 to deduce $u + \varepsilon > 0$ in $Q_T^R := \{|x| < R\} \times [0, T]$. Hence $u \geqslant 0$ in Q_T^R by letting $\varepsilon \to 0$. Arbitrariness of R implies $u \geqslant 0$ in $\mathbb{R}^n \times (0, T]$. □

Theorem 1.28 (Maximum principle [36, 110]) *Let the operator $\mathscr{L}$ be parabolic in $\mathbb{R}^n \times (0, T]$, and*

(F) $\quad |a_{ij}| \leqslant M, \quad |b_i| \leqslant M(1 + |x|) \quad \text{and} \quad c \geqslant -M(1 + |x|^2).$

Assume that $u \in C^{2,1}(\mathbb{R}^n \times (0,T]) \cap C(\mathbb{R}^n \times [0,T])$ satisfies $\mathscr{L}u \geqslant 0$ in $\mathbb{R}^n \times (0,T]$, and there exist positive constants B and k such that

$$u \geqslant -Be^{k|x|^2} \quad \text{in } \mathbb{R}^n \times (0,T]. \tag{1.28}$$

If $u(x,0) \geqslant 0$ in $\mathbb{R}^n$, then $u \geqslant 0$ in $\mathbb{R}^n \times (0,T]$.

Proof. Take $\gamma, \ell > 0$ which will be determined later. Define a function

$$w = \exp\left\{\frac{2k|x|^2}{1 - \ell t} + \gamma t\right\}, \quad 0 \leqslant t \leqslant \frac{1}{2\ell}.$$

A straightforward computation shows that

$$\frac{\mathscr{L}w}{w} = \gamma - \frac{16k^2}{(1 - \ell t)^2}a_{ij}x_i x_j - \frac{4k}{1 - \ell t}a_{ii} + \frac{4k}{1 - \ell t}b_i x_i + c + \frac{2\ell k|x|^2}{(1 - \ell t)^2}.$$

Making use of the condition **(F)**,

$$\frac{\mathscr{L}w}{w} \geqslant [2k\ell - (64k^2 n^2 + 12kn + 1)M]|x|^2 + \gamma - (12kn + 1)M.$$

We can take $\ell, \gamma > 0$ so large that

$$\frac{\mathscr{L}w}{w} \geqslant 0. \tag{1.29}$$

For such fixed ℓ, γ, we denote $T_1 = 1/(2\ell)$ and investigate function $v = u/w$ in D_{T_1}. Condition (1.28) implies $\liminf_{|x| \to \infty} v(x,t) \geqslant 0$ uniformly in $t \in [0, T_1]$. A straightforward computation leads to

$$\bar{\mathscr{L}}v := v_t - a_{ij}D_{ij}v + \bar{b}_i D_i v + \bar{c}v = \bar{f},$$

where

$$\bar{b}_i = b_i - 2\sum_j a_{ij}\frac{D_j w}{w}, \quad \bar{c} = \frac{\mathscr{L}w}{w} \quad \text{and} \quad \bar{f} = \frac{\mathscr{L}u}{w} \geqslant 0.$$

According to (1.29), $\bar{c} \geqslant 0$. By a similar argument in the proof of Lemma 1.27 we have $v \geqslant 0$. As a result, $u \geqslant 0$ in $\mathbb{R}^n \times (0, T_1]$. Since $T_1 > 0$ is a fixed constant, repeating the above procedure we can deduce $u \geqslant 0$ in $\mathbb{R}^n \times (0,T]$. $\quad\square$

Using Theorem 1.28 we can prove the following theorem, and its proof will be omitted and left to readers.

Theorem 1.29 *Let $\mathscr{L}$ be a parabolic operator in $\mathbb{R}^n \times (0,T]$ and the condition **(F)** be fulfilled. Then problem (1.27) has at most one solution in the sense that $|u| \leqslant Be^{\beta|x|^2}$ with some positive constants B and β.*

1.3.3 Weakly coupled systems

Contents of this part are taken from [91, 124]. Let $\mathscr{L}_k$ be strongly parabolic operators given by (1.24), and let

$$H(x,t) = (h_{kl}(x,t))_{m \times m}$$

be a function matrix of order m. We will establish the maximum principle for the following weakly coupled linear parabolic inequalities

$$\mathscr{L}_k u_k + \sum_{l=1}^{m} h_{kl} u_l \leqslant 0 \ (\geqslant 0), \quad k = 1, \ldots, m.$$

It is assumed that a_{ij}^k, b_i^k, h_{kl} are bounded in Q_T, and functions a^k, b^k appearing in boundary conditions are non-negative and $a^k + b^k > 0$. Denote $u = (u_1, \ldots, u_m)$.

Lemma 1.30 *Let Ω be of class C^2 and $u_k \in C(\overline{Q}_T) \cap C^{2,1}(Q_T)$ for all $1 \leqslant k \leqslant m$. And when $a^k \not\equiv 0$ for some $1 \leqslant k \leqslant m$, we further assume $u_k \in C^{1,0}(\overline{\Omega} \times (0,T])$. Suppose that*

(i) $h_{kl} \leqslant 0$ for $k \neq l, k, l = 1, \ldots, m$;

(ii) $\mathscr{L}_k u_k + \sum_{l=1}^{m} h_{kl} u_l < 0 \ (> 0)$ in Q_T, $k = 1, \ldots, m$;

(iii) $u(x,0) < 0 \ (> 0)$ in $\overline{\Omega}$;

(iv) *for each $1 \leqslant k \leqslant m$ and each point $(x,t) \in S_T$, either $u_k < 0 \ (> 0)$, or $a^k \partial_{\boldsymbol{n}} u_k + b^k u_k \leqslant 0 \ (\geqslant 0)$ and $a^k > 0$ at (x,t).*

Then $u < 0 \ (> 0)$ in $\overline{Q}_T$.

Proof. Take a constant $\alpha > 0$ so that $h_{kk} + \alpha > 0$ in $\overline{Q}_T$ for all k, and let $u_k = v_k e^{\alpha t}$. In view of the condition (ii),

$$\mathscr{L}_k v_k + \sum_{l=1}^{m} \tilde{h}_{kl} v_l < 0 \ \text{ in } Q_T, \quad k = 1, \ldots, m, \tag{1.30}$$

where

$$\begin{cases} \tilde{h}_{kk} = h_{kk} + \alpha > 0, & k = 1, \ldots, m, \\ \tilde{h}_{kl} = h_{kl} \leqslant 0, & k \neq l, \ k, l = 1, \ldots, m. \end{cases} \tag{1.31}$$

Set $v = (v_1, \ldots, v_m)$. Thanks to the fact that $v(x,0) < 0$ in $\overline{\Omega}$, there exists $\delta > 0$ such that $v(x,t) < 0$ for $x \in \overline{\Omega}$ and $0 \leqslant t \leqslant \delta$. Define

$$A = \{t : t \leqslant T, \ v(x,s) < 0 \text{ for all } x \in \overline{\Omega}, \ 0 \leqslant s \leqslant t\}.$$

Then $t_0 = \sup A$ exists and $0 < t_0 \leqslant T$.

Assume, on the contrary, that our conclusion is not valid, then $v < 0$ in $\overline{\Omega} \times [0, t_0)$ and there exists $x_0 \in \overline{\Omega}$ such that $v_k(x_0, t_0) = 0$ for some k. If $x_0 \in \Omega$, then v_k achieves its maximum over $\overline{Q}_{t_0}$ at (x_0, t_0), and

$$\partial_t v_k(x_0, t_0) \geqslant 0, \quad D_i v_k(x_0, t_0) = 0, \quad a_{ij}^k D_{ij} v_k(x_0, t_0) \leqslant 0.$$

These imply $\mathscr{L}_k v_k|_{(x_0,t_0)} \geqslant 0$. On the other hand, thanks to $v_k(x_0, t_0) = 0$ and $v_l(x_0, t_0) \leqslant 0$ for all $l \neq k$, it follows from (1.31) that

$$\mathscr{L}_k v_k(x_0, t_0) + \sum_{l=1}^{m} \tilde{h}_{kl} v_l(x_0, t_0) \geqslant 0.$$

This contradicts (1.30). Thus, $x_0 \in \partial\Omega$ and $v_k(x, t_0) < 0$ in Ω, and subsequently, $v_k < 0$ in Q_{t_0}. Moreover, $a^k(x_0, t_0) > 0$ by the condition (iv). Due to $v \leqslant 0$ in Q_{t_0}, we have

$$0 > \mathscr{L}_k v_k + \tilde{h}_{kk} v_k + \sum_{l \neq k} h_{kl} v_l \geqslant \mathscr{L}_k v_k + \tilde{h}_{kk} v_k \quad \text{in } Q_{t_0}.$$

According to the Hopf boundary lemma, we have $\partial_{\boldsymbol{n}} v_k(x_0, t_0) > 0$, and so

$$0 < a^k \partial_{\boldsymbol{n}} v_k|_{(x_0,t_0)} = (a^k \partial_{\boldsymbol{n}} v_k + b^k v_k)|_{(x_0,t_0)} \leqslant 0.$$

This contradiction shows that $v < 0$, i.e., $u < 0$ in Q_T. $\qquad\square$

Theorem 1.31 (Maximum principle) *Let Ω be of class C^2 and*

$$h_{kl} \leqslant 0 \quad \text{in } Q_T \quad \text{for } k \neq l, \ k, l = 1, \ldots, m. \tag{1.32}$$

Assume $u_k \in C(\overline{Q}_T) \cap C^{2,1}(Q_T)$ for all $1 \leqslant k \leqslant m$. And when $a^k \not\equiv 0$ for some $1 \leqslant k \leqslant m$, we further assume $u_k \in C^{1,0}(\overline{\Omega} \times (0, T])$. Suppose that the followings hold:

(i) *$\mathscr{L}_k u_k + \sum_{l=1}^{m} h_{kl} u_l \leqslant 0 \ (\geqslant 0)$ in Q_T for each $1 \leqslant k \leqslant m$;*

(ii) *$u(x, 0) \leqslant 0 \ (\geqslant 0)$ in $\overline{\Omega}$;*

(iii) *$a^k \partial_{\boldsymbol{n}} u_k + b^k u_k \leqslant 0 \ (\geqslant 0)$ on S_T for each $1 \leqslant k \leqslant m$.*

Then we have

(1) *$u \leqslant 0 \ (\geqslant 0)$ in Q_T;*

(2) *if there exist $(x_0, t_0) \in Q_T$ and $1 \leqslant k \leqslant m$ such that $u_k(x_0, t_0) = 0$, then $u_k \equiv 0$ in Q_{t_0}.*

Proof. We first prove (1). Since h_{kl} are bounded, there is $\beta > 0$ so that

$$\beta + \sum_{l=1}^{m} h_{kl} > 0 \quad \text{in } Q_T \quad \text{for all } 1 \leqslant k \leqslant m.$$

Then function $v_k = u_k - \varepsilon e^{\beta t}$ satisfies

$$\mathscr{L}_k v_k + \sum_{l=1}^{m} h_{kl} v_l = \mathscr{L}_k u_k + \sum_{l=1}^{m} h_{kl} u_l - \varepsilon e^{\beta t} \left(\beta + \sum_{l=1}^{m} h_{kl} \right) < 0,$$

$v_k(x, 0) < 0$ in $\overline{\Omega}$, and

$$a^k \partial_{\boldsymbol{n}} v_k + b^k v_k = a^k \partial_{\boldsymbol{n}} u_k + b^k u_k - \varepsilon b^k e^{-\beta t} \leqslant 0 \quad \text{on } S_T.$$

By means of Lemma 1.30, $v_k < 0$ in Q_T. Letting $\varepsilon \to 0$ we derive $u_k \leqslant 0$ in Q_T.

Now we prove the conclusion (2). Note that

$$0 \geqslant \mathscr{L}_k u_k + \sum_{l=1}^{m} h_{kl} u_l = \mathscr{L}_k u_k + h_{kk} u_k + \sum_{l \neq k} h_{kl} u_l \geqslant \mathscr{L}_k u_k + h_{kk} u_k.$$

Take $\gamma > 0$ to satisfy $\gamma + h_{kk} > 0$, and set $u_k = v_k e^{\gamma t}$. We have

$$\mathscr{L}_k v_k + (\gamma + h_{kk}) v_k \leqslant 0.$$

The desired result can be deduced by the maximum principle. $\qquad\square$

Similarly, we can prove the Hopf boundary lemma for weakly coupled systems.

Theorem 1.32 (Hopf boundary lemma) *Let $u \in [C(\overline{Q_T}) \cap C^{2,1}(Q_T)]^m$ and $u \leqslant 0 \, (\geqslant 0)$ satisfy the condition (i) of Theorem 1.31. Let H satisfy (1.32). Assume that there exist $1 \leqslant k \leqslant m$ and $(x_0, t_0) \in S_T$ such that $u_k(x_0, t_0) = 0$. If Ω has the interior ball property at $x_0 \in \partial\Omega$, and there is a neighbourhood V of (x_0, t_0) so that $u_k < 0 \, (u_k > 0)$ in $Q_T \cap V$, then $\partial_{\boldsymbol{n}} u_k(x_0, t_0) > 0 \, (< 0)$ when it exists.*

We should mention that, in general, the maximum principle does not hold for strongly coupled parabolic systems. In fact, if we take $u = -e^{x+t}$ and $v = t - 4(x - 1/2)^2$, then it is easily verified that (u, v) satisfies

$$\begin{cases} u_t - u_{xx} \leqslant 0 & \text{in } (0, 1] \times (0, 1], \\ v_t - v_{xx} + 9u_x \leqslant 0 & \text{in } (0, 1] \times (0, 1], \end{cases}$$

and

$$(u, v)|_{t=0} \leqslant 0 \text{ for } x \in [0, 1], \quad (u, v)|_{x=0,1} \leqslant 0 \text{ for } t \in (0, 1].$$

However, $v|_{x=\frac{1}{2}} = t \geqslant 0$.

In addition, such (u, v) also satisfies

$$u_t - u_{xx} \leqslant 0, \quad v_t - v_{xx} + 9u \leqslant 0.$$

These discussions show that the maximum principle is not valid for such a system of inequalities. The reason is that $h_{21} = 9 > 0$, and the requirement (1.32) is not satisfied.

1.3.4 Equations with non-classical boundary conditions

This subsection is concerned with the maximum principle for initial-boundary value problems of parabolic equations with *non-classical boundary conditions*. The content of this part is due to [105].

Theorem 1.33 *Let Ω be of class C^2, $b \in C(\overline{S}_T)$, and $a_{ij}, b_i, c \in L^\infty(Q_T)$. Assume that there exists $\lambda > 0$ such that*

$$a_{ij}(x, t)\xi_i \xi_j \geqslant \lambda |\xi|^2 \quad \text{for all } (x, t) \in \overline{Q}_T, \ \xi \in \mathbb{R}^n.$$

If $u \in C(\overline{Q}_T) \cap C^{2,1}(Q_T) \cap C^{1,0}(\overline{\Omega} \times (0,T])$ satisfies

$$\begin{cases} u_t \geqslant a_{ij} D_{ij} u + b_i D_i u + cu & \text{in } Q_T, \\ \partial_{\boldsymbol{n}} u \geqslant b(x,t) u & \text{on } S_T, \\ u(x,0) \geqslant 0 & \text{in } \Omega, \end{cases}$$

then $u \geqslant 0$ in $\overline{Q}_T$. Additionally, if $u(x,0) \not\equiv 0$, then $u > 0$ in Q_T.

Proof. Take a constant $M > 0$ so that $\max_{\overline{S}_T} |b| \leqslant M$. Let λ be the first eigenvalue of the eigenvalue problem

$$\begin{cases} -\Delta \phi = \lambda \phi & \text{in } \Omega, \\ \phi = 0 & \text{on } \partial\Omega, \end{cases} \tag{1.33}$$

and ϕ be the corresponding positive eigenfunction. Then $\phi > 0$ in Ω and $\partial_{\boldsymbol{n}} \phi < 0$ on $\partial\Omega$. We call (λ, ϕ) the *first eigen-pair* of (1.33).

There exists $k > 0$ such that $\partial_{\boldsymbol{n}} \phi \leqslant -k$ on $\partial\Omega$. Let us define

$$h(x) = \exp\{\ell(1 + \phi(x))\}, \quad w(x,t) = u(x,t) h(x),$$

where ℓ is a positive constant to be determined. A direct calculation yields

$$\begin{cases} w_t \geqslant a_{ij} D_{ij} w - 2a_{ij} \dfrac{D_i h}{h} D_j w + b_i D_i w \\ \qquad + \left(a_{ij} \dfrac{2D_i h D_j h - h D_{ij} h}{h^2} - b_i \dfrac{D_i h}{h} + c \right) w & \text{in } Q_T, \\ \partial_{\boldsymbol{n}} w + \left(-b - \dfrac{1}{h} \partial_{\boldsymbol{n}} h \right) w \geqslant 0 & \text{on } S_T, \\ w(x,0) = u(x,0) h(x) \geqslant 0 & \text{in } \Omega. \end{cases} \tag{1.34}$$

As

$$\frac{1}{h} \partial_{\boldsymbol{n}} h = \ell \partial_{\boldsymbol{n}} \phi \leqslant -\ell k,$$

we have

$$-b - \frac{1}{h} \partial_{\boldsymbol{n}} h \geqslant -b + \ell k > 0 \quad \text{when } \ell k > M.$$

Applying Lemma 1.26 to (1.34) we get $w \geqslant 0$, and $w > 0$ in Q_T when $u(x,0) \not\equiv 0$. $\square$

1.3.5 Equations with non-local terms

In this subsection we study the maximum principle of initial-boundary value problems with *non-local terms*. The content in this part is taken from [110].

We first discuss the case of non-local terms appearing in differential equations.

Theorem 1.34 *Let Ω be of class C^2, a and b be two non-negative continuous functions with $a + b > 0$ on S_T, and let c_1 and c_2 be two bounded continuous functions with $c_2 \geqslant 0$ in $\overline{Q}_T$. If $u \in C^{2,1}(Q_T) \cap C^{1,0}(\overline{Q}_T)$ verifies*

$$\begin{cases} u_t - \Delta u \geqslant c_1 u + \displaystyle\int_\Omega c_2(y,t)u(y,t)\mathrm{d}y & \text{in } Q_T, \\ a\partial_{\boldsymbol{n}} u + bu \geqslant 0 & \text{on } S_T, \\ u(x,0) \geqslant 0 & \text{in } \Omega, \end{cases}$$

then $u \geqslant 0$ in $\overline{Q}_T$. In addition, if $u(x,0) \not\equiv 0$, then $u > 0$ in Q_T.

Proof. Take $\bar{c}_i = \sup_{Q_T} |c_i|$ for $i = 1, 2$, and $\gamma > \bar{c}_1 + 2\bar{c}_2|\Omega|$. Set $v = \mathrm{e}^{-\gamma t}u$, then we have

$$\begin{cases} v_t - \Delta v \geqslant (c_1 - \gamma)v + \displaystyle\int_\Omega c_2(y,t)v(y,t)\mathrm{d}y & \text{in } Q_T, \\ a\partial_{\boldsymbol{n}} v + bv \geqslant 0 & \text{on } S_T, \\ v(x,0) \geqslant 0 & \text{in } \Omega. \end{cases}$$

Assume that v acquires its negative minimum at some point (x_0, t_0). In the case of $x_0 \in \Omega$, we have $v_t(x_0, t_0) \leqslant 0$ and $\Delta v(x_0, t_0) \geqslant 0$, and hence

$$\begin{aligned} [\gamma - c_1(x_0, t_0)]v(x_0, t_0) &\geqslant \int_\Omega c_2(y, t_0)v(y, t_0)\mathrm{d}y \\ &\geqslant \int_\Omega c_2(y, t_0)v(x_0, t_0)\mathrm{d}y \\ &\geqslant \bar{c}_2 v(x_0, t_0)|\Omega|, \end{aligned}$$

where we have used $0 \leqslant c_2(y, t_0) \leqslant \bar{c}_2$ and $v(x_0, t_0) < 0$. The above inequality implies $\gamma - c_1(x_0, t_0) \leqslant \bar{c}_2|\Omega|$, and thereupon

$$\gamma \leqslant c_1(x_0, t_0) + \bar{c}_2|\Omega| \leqslant \bar{c}_1 + \bar{c}_2|\Omega|.$$

This contradicts the fact $\gamma > \bar{c}_1 + 2\bar{c}_2|\Omega|$.

In the case of $x_0 \in \partial\Omega$, we have $\partial_{\boldsymbol{n}} v(x_0, t_0) \leqslant 0$. Then the boundary condition implies that $a(x_0, t_0) > 0, b(x_0, t_0) = 0$ and $\partial_{\boldsymbol{n}} v(x_0, t_0) = 0$. On the other hand, as $v(x_0, t_0) < 0$, there exists a neighbourhood V of (x_0, t_0) such that $v \leqslant \frac{1}{2}v(x_0, t_0)$ in $Q_T \cap V$. Noticing

$$c_1 - \gamma < 0, \quad c_2 \geqslant 0, \quad \text{and} \quad v(x_0, t_0) < 0,$$

we see that in $Q_T \cap V$,

$$\begin{aligned} v_t - \Delta v &\geqslant (c_1 - \gamma)v + \int_\Omega c_2(y, t)v(y, t)\mathrm{d}y \\ &\geqslant \frac{c_1 - \gamma}{2}v(x_0, t_0) + \int_\Omega c_2(y, t)v(x_0, t_0)\mathrm{d}y \\ &\geqslant \frac{c_1 - \gamma}{2}v(x_0, t_0) + \bar{c}_2|\Omega|v(x_0, t_0) \\ &= \frac{c_1 - \gamma + 2\bar{c}_2|\Omega|}{2}v(x_0, t_0) > 0. \end{aligned}$$

By the Hopf boundary lemma, $\partial_n v(x_0, t_0) < 0$. This contradicts the fact $\partial_n v(x_0, t_0) = 0$. Thus $v \geqslant 0$, and then $u \geqslant 0$.

Since $c_2 \geqslant 0$, we have

$$u_t - \Delta u \geqslant c_1 u + \int_\Omega c_2(y, t)u(y, t)dy \geqslant c_1 u \quad \text{in } Q_T.$$

Therefore, $u > 0$ in Q_T so long as $u(x, 0) \not\equiv 0$. □

Now we study the case that the boundary conditions have non-local terms. The following theorem can be proved in the same way as that of Theorem 1.33, and therefore it is omitted.

Theorem 1.35 *Let Ω be of class C^2, $d > 0$ be a constant and $u \in C(\overline{Q}_T) \cap C^{2,1}(Q_T) \cap C^{1,0}(\overline{\Omega} \times (0, T])$, satisfy*

$$\begin{cases} u_t - d\Delta u \geqslant 0 & \text{in } Q_T, \\ \partial_n u \geqslant c_1 u + \int_{\partial\Omega} c_2(y, t)u(y, t)dS_y & \text{on } S_T, \\ u(x, 0) \geqslant 0 & \text{in } \Omega. \end{cases}$$

Assume that c_i is continuous and bounded for $i = 1, 2$, and $c_2 \geqslant 0$ on S_T. Then $u \geqslant 0$ in $\overline{Q}_T$. Furthermore, $u > 0$ in Q_T when $u(x, 0) \not\equiv 0$.

In the sequel we investigate the case that non-local terms appear in both the differential equation and boundary condition.

Lemma 1.36 *Let $b_i, b \in L^\infty(Q_T)$, $c, f \geqslant 0$ in Q_T and $g \geqslant 0$ in $\Omega \times S_T$. Assume that $u \in C(\overline{Q}_T) \cap C^{2,1}(Q_T)$ satisfies*

$$\begin{cases} u_t - \Delta u - b_i D_i u - bu \geqslant f(x, t) \int_\Omega c(y, t)u dy & \text{in } Q_T, \\ u > \int_\Omega g(y, x, t)u(y, t)dy & \text{on } S_T, \\ u(x, 0) = u_0(x) > 0 & \text{in } \overline{\Omega}. \end{cases}$$

Then $u > 0$ in $\overline{Q}_T$.

Proof. Take $\bar{b} = \sup_{\overline{Q}_T} |b|$ and $\gamma > \bar{b}$. Then function $v = e^{-\gamma t}u$ satisfies

$$\begin{cases} v_t - \Delta v + (\gamma - b)v - b_i D_i v \geqslant f(x, t) \int_\Omega c(y, t)v dy & \text{in } Q_T, \\ v > \int_\Omega g(y, x, t)v(y, t)dy & \text{on } S_T, \\ v(x, 0) = u_0(x) > 0 & \text{in } \overline{\Omega}. \end{cases} \quad (1.35)$$

As $v(x, 0) > 0$, there exists $\delta > 0$ so that $v > 0$ in $\overline{\Omega} \times [0, \delta]$. Set

$$A := \{t \leqslant T : v > 0 \text{ in } \overline{\Omega} \times [0, t]\}, \quad \text{and} \quad t_0 := \sup A,$$

then $0 < t_0 \leqslant T$. If $t_0 < T$, then $v > 0$ in $\overline{\Omega} \times (0, t_0)$, and $v(x_0, t_0) = 0$ for some $x_0 \in \Omega$. The boundary condition implies $v > 0$ on $\partial\Omega \times (0, t_0]$. Consequently v achieves its minimum over $\overline{Q}_{t_0}$ at (x_0, t_0). It then follows from the first inequality of (1.35) that

$$v_t - \Delta v + (\gamma - b)v - b_i D_i v \geqslant 0 \quad \text{in} \ \ Q_{t_0}.$$

Corollary 1.25 asserts $v \equiv 0$ in Q_{t_0}. We get a contradiction, and therefore $v > 0$, i.e., $u > 0$ in $\overline{Q}_T$. □

Lemma 1.37 *Assume that $b_i, b, f, c \in L^\infty(Q_T)$ and $f, c \geqslant 0$. Let $g \geqslant 0$ in $\Omega \times S_T$ with $\int_\Omega g(y, x, t)\mathrm{d}y < 1$ for all $(x, t) \in S_T$. Let $u \in C(\overline{Q}_T) \cap C^{2,1}(Q_T)$ satisfy*

$$\begin{cases} u_t - \Delta u - b_i D_i u - bu \geqslant f(x, t) \displaystyle\int_\Omega c(y, t)u\mathrm{d}y & \text{in} \ \ Q_T, \\ u \geqslant \displaystyle\int_\Omega g(y, x, t)u(y, t)\mathrm{d}y & \text{on} \ \ S_T, \\ u(x, 0) = u_0(x) \geqslant 0 & \text{in} \ \ \Omega. \end{cases}$$

Then $u \geqslant 0$ in Q_T.

Proof. Define $\bar{b} = \sup_{Q_T} |b|$ and $\beta = \sup_{Q_T} f(x, t) \int_\Omega c(y, t)\mathrm{d}y$. Take $\gamma > \bar{b} + \beta$, and set $w = u + \varepsilon e^{\gamma t}$ with $\varepsilon > 0$. Then we have

$$\begin{cases} w_t - \Delta w - b_i D_i w - bw \geqslant f(x, t) \displaystyle\int_\Omega c(y, t)w\mathrm{d}y & \text{in} \ \ Q_T, \\ w > \displaystyle\int_\Omega g(y, x, t)w(y, t)\mathrm{d}y & \text{on} \ \ S_T, \\ w(x, 0) = u_0(x) + \varepsilon > 0 & \text{in} \ \ \overline{\Omega}. \end{cases}$$

As a consequence, $w > 0$ by Lemma 1.36, i.e., $u + \varepsilon e^{\gamma t} > 0$ in $\overline{Q}_T$. The desired result is obtained by sending $\varepsilon \to 0^+$. □

Theorem 1.38 *Assume $b_i, b, f, c \in L^\infty(Q_T)$, $g \in L^\infty(\Omega \times S_T)$ and $f, c, g \geqslant 0$. If $u \in C(\overline{Q}_T) \cap C^{2,1}(Q_T)$ satisfies*

$$\begin{cases} u_t - \Delta u - b_i D_i u - bu \geqslant f(x, t) \displaystyle\int_\Omega c(y, t)u\mathrm{d}y & \text{in} \ \ Q_T, \\ u \geqslant \displaystyle\int_\Omega g(y, x, t)u(y, t)\mathrm{d}y & \text{on} \ \ S_T, \\ u(x, 0) = u_0(x) \geqslant 0 & \text{in} \ \ \Omega, \end{cases}$$

then $u \geqslant 0$ in Q_T.

Proof. As $g \in L^\infty(\Omega \times S_T)$, we can take a function $z \in C^2(\overline{\Omega})$ with properties: $z > 0$ in $\overline{\Omega}$, $z \equiv 1$ on $\partial\Omega$, and

$$\int_\Omega z(y)g(y, x, t)\mathrm{d}y < 1 \quad \text{for} \ \ (x, t) \in \partial\Omega \times [0, T].$$

Let $u = z(x)w$. Then a straightforward calculation yields

$$w_t - \Delta w \geqslant \left(b_i + \frac{2D_i z}{z}\right) D_i w + \left(b + \frac{\Delta z}{z} + b_i \frac{D_i z}{z}\right) w + \frac{f(x,t)}{z(x)} \int_\Omega z(y)c(y,t)wdy$$

in Q_T, and

$$w(x,t) \geqslant \int_\Omega g(y,x,t)z(y)wdy \quad \text{on } S_T,$$

$$w(x,0) = \frac{u_0(x)}{z(x)} \geqslant 0 \qquad \qquad \text{in } \Omega.$$

It follows from our assumptions that

$$b_i + \frac{2D_i z}{z}, \quad b + \frac{\Delta z}{z} + b_i \frac{D_i z}{z}, \quad \frac{f}{z}, \quad zc \in L^\infty(Q_T),$$

$$zg \geqslant 0, \quad \int_\Omega g(y,x,t)z(y)dy < 1 \quad \text{for all } (x,t) \in S_T,$$

and $f/z \geqslant 0$, $zc \geqslant 0$. Lemma 1.37 leads to $w \geqslant 0$ in Q_T, and so $u \geqslant 0$ in Q_T. □

1.4 MAXIMUM PRINCIPLE OF STRONG SOLUTIONS

In this section we discuss the *maximum principle of strong solutions* (solutions in $W_p^{2,1}(Q_T)$) for non-divergence equations.

Lemma 1.39 (Hopf boundary lemma for strong solutions, [84, Theorem 3.4]) *Let $a_{ij}, b_i \in L^\infty(Q_T)$, Ω be of class C^2, $\mathscr{L}$ be parabolic and $c \geqslant 0$ in Q_T. Let $u \in C(\overline{Q}_T) \cap W_{n+1,\text{loc}}^{2,1}(Q_T)$ satisfy $\mathscr{L}u \leqslant 0 (\geqslant 0)$ in Q_T. Suppose that $u(x_0,t_0) = M = \max_{\overline{Q}_T} u$ $(u(x_0,t_0) = m = \min_{\overline{Q}_T} u)$ for some $(x_0,t_0) \in S_T$, and there is a neighborhood V of (x_0,t_0) such that $u < M$ $(u > m)$ in $Q_T \cap V$, and $M \geqslant 0$ $(m \leqslant 0)$ when $c \not\equiv 0$. Then the derivative $\partial_{\boldsymbol{n}} u(x_0,t_0) > 0$ (< 0) when it exists.*

Theorem 1.40 (1) (Strong maximum principle of strong solutions) *Let $a_{ij}, b_i \in L^\infty(Q_T)$, $\mathscr{L}$ be parabolic and $c \geqslant 0$ in Q_T. Assume that $u \in C(\overline{Q}_T) \cap C^{1,0}(Q_T) \cap W_{n+1,\text{loc}}^{2,1}(Q_T)$ satisfies $\mathscr{L}u \leqslant 0 (\geqslant 0)$ in Q_T. If $u(x_0,t_0) = \max_{\overline{Q}_T} u = M$ $(u(x_0,t_0) = \min_{\overline{Q}_T} u = m)$ for some $x_0 \in \Omega$, $0 < t_0 \leqslant T$, and $M \geqslant 0$ $(m \leqslant 0)$ when $c \not\equiv 0$, then $u \equiv u(x_0,t_0)$ in Q_{t_0}.*

(2) (Positivity lemma for strong solutions) *Let $a_{ij}, b_i \in L^\infty(Q_T)$, $\mathscr{L}$ be parabolic and c be bounded below in Q_T. Let $a, b \in C(S_T)$ with $a, b \geqslant 0$ and $a + b > 0$ on S_T. For any $(x,t) \in S_T$, if $a(x,t) \neq 0$, then we further assume that $\partial\Omega$ is of class C^2 near x. Suppose $u \in C(\overline{Q}_T) \cap C^{1,0}(Q_T) \cap W_{n+1,\text{loc}}^{2,1}(Q_T)$, and when $a \not\equiv 0$ we further assume $u \in C^{1,0}(\overline{\Omega} \times (0,T])$. If u satisfies*

$$\begin{cases} \mathscr{L}u \geqslant 0 & \text{in } Q_T, \\ a\partial_{\boldsymbol{n}} u + bu \geqslant 0 & \text{on } S_T, \\ u(x,0) \geqslant 0 & \text{in } \Omega, \end{cases}$$

then $u \geqslant 0$ in Q_T. If in addition $u(x,0) \not\equiv 0$, then $u > 0$ in Q_T, and $u > 0$ in $\overline{\Omega} \times (0,T]$ when $a > 0$ on S_T.

Proof. The conclusion (1) can be deduced by Lemma 1.39 directly. The conclusion (2) can be proved by the same way as that of Lemma 1.26 using the conclusion (1) and Lemma 1.39 instead of Corollary 1.25 and Lemma 1.23, respectively. The details are omitted here. $\qquad\square$

In the following we consider the general case $p > 1 + n/2$. We first consider the first initial-boundary value problem.

Theorem 1.41 *Let Ω be of class C^2, the condition* **(A)** *hold and $b_i, c \in C(\overline{Q}_T)$. Assume that $p > 1 + n/2$ and $u \in W_p^{2,1}(Q_T)$ satisfies*

$$
\begin{cases}
\mathscr{L}u = f(x,t) \geqslant 0 & in \ Q_T, \\
u = g(x,t) \geqslant 0 & on \ S_T, \\
u(x,0) = \varphi(x) \geqslant 0 & in \ \Omega.
\end{cases}
$$

Then $u \geqslant 0$ in Q_T.

Proof. Step 1. As $u \in W_p^{2,1}(Q_T)$, one has $f \in L^p(Q_T)$. Moreover, $u \in C^{\alpha, \alpha/2}(\overline{Q}_T)$ as $p > 1 + n/2$. We assume without loss of generality that $c \geqslant 0$. For any given $\sigma > 0$, let $v = u + \sigma e^t$. Then v satisfies

$$
\begin{cases}
\mathscr{L}v = (1+c)\sigma e^t + \mathscr{L}u = (1+c)\sigma e^t + f =: f_\sigma \geqslant \sigma & in \ Q_T, \\
v = \sigma e^t + g =: g_\sigma \geqslant \sigma & on \ S_T, \\
v(x,0) = \sigma + \varphi(x) =: \varphi_\sigma \geqslant \sigma & in \ \Omega.
\end{cases}
$$

Step 2. Take $\tau_k \searrow 0$ satisfying $\tau_{k+1} < \tau_k$, and subdomain $\Omega_k \Subset \Omega$ being of class $C^{2+\alpha}$ with $\Omega_k \Subset \Omega_{k+1}$, $\Omega_k \to \Omega$. Define

$$ Q^k = \Omega_k \times (\tau_k, T]. $$

Since v is continuous in $\overline{Q}_T$, there exists $k_0 \gg 1$ such that, for all $k \geqslant k_0$,

$$ v \geqslant \sigma/2 \quad in \ Q_T \setminus Q^{k-1}. $$

For such fixed k, take a truncated function $\eta \in C^\infty(\overline{Q}_T)$ so that $0 \leqslant \eta \leqslant 1$, $\eta \equiv 1$ in Q^k and $\eta \equiv 0$ in $Q_T \setminus Q^{k+1}$. We will see that the following proof is still valid with Q^{k+3} instead of Q_T, and hence we may assume that Ω is of class $C^{2+\alpha}$. Let $w = v\eta$. Then w satisfies

$$
\begin{cases}
\mathscr{L}w = v\eta_t + \eta f_\sigma - va_{ij}D_{ij}\eta - 2a_{ij}D_iv D_j\eta + vb_iD_i\eta =: f^* & in \ Q_T, \\
w = 0 & on \ \partial_p Q_T.
\end{cases}
$$

Obviously, $f^* = f_\sigma$ in Q^k and $f^* = 0$ in $Q_T \setminus Q^{k+1}$.

Step 3. Take $a_{ij}^\varepsilon, b_i^\varepsilon, c^\varepsilon, f^\varepsilon \in C^{\alpha, \alpha/2}(\overline{Q}_T)$ such that

(a) $f^\varepsilon = 0$ in $Q_T \setminus Q^{k+2}$ and $f^\varepsilon \geqslant \sigma/2$ in Q^k;

(b) $a_{ij}^\varepsilon \to a_{ij}, b_i^\varepsilon \to b_i, c^\varepsilon \to c$ in $C(\overline{Q}_T)$, and $f^\varepsilon \to f^*$ in $L^p(Q_T)$ as $\varepsilon \to 0^+$.

Consider problem

$$\begin{cases} w_t^\varepsilon - a_{ij}^\varepsilon D_{ij} w^\varepsilon + b_i^\varepsilon D_i w^\varepsilon + c^\varepsilon w^\varepsilon = f^\varepsilon & \text{in } Q_T, \\ w^\varepsilon = 0 & \text{on } \partial_p Q_T. \end{cases} \tag{1.36}$$

By Theorem 1.16, this problem has a unique solution $w^\varepsilon \in C^{2+\alpha, 1+\alpha/2}(\overline{Q}_T)$.

We claim that $w^\varepsilon \to w$ in $W_p^{2,1}(Q_T)$. In fact, let $z^\varepsilon = w - w^\varepsilon$ and

$$F^\varepsilon(x,t) = (a_{ij} - a_{ij}^\varepsilon) D_{ij} w^\varepsilon + (b_i^\varepsilon - b_i) D_i w^\varepsilon + (c^\varepsilon - c) w^\varepsilon + f^* - f^\varepsilon.$$

Then z^ε satisfies

$$\begin{cases} \mathscr{L} z^\varepsilon = F^\varepsilon & \text{in } Q_T, \\ z^\varepsilon = 0 & \text{on } \partial_p Q_T. \end{cases}$$

By use of (1.4),

$$\|z^\varepsilon\|_{W_p^{2,1}(Q_T)} \leqslant C \|F^\varepsilon\|_{p, Q_T}.$$

Applying the estimate (1.4) to (1.36), we may assume $\|w^\varepsilon\|_{W_p^{2,1}(Q_T)} \leqslant K$ for some K and all $0 < \varepsilon \ll 1$. By the choice of $(a_{ij}^\varepsilon, b_i^\varepsilon, c^\varepsilon, f^\varepsilon)$, it is easy to see that $\lim_{\varepsilon \to 0^+} \|F^\varepsilon\|_{p, Q_T} = 0$, and this implies $\lim_{\varepsilon \to 0^+} \|z^\varepsilon\|_{W_p^{2,1}(Q_T)} = 0$, i.e., $\lim_{\varepsilon \to 0^+} w^\varepsilon = w$ in $W_p^{2,1}(Q_T)$.

Step 4. Since $W_p^{2,1}(Q_T) \hookrightarrow C(\overline{Q}_T)$, we have $\lim_{\varepsilon \to 0^+} w^\varepsilon = w$ in $C(\overline{Q}_T)$. Recall $v \geqslant \sigma/2$ in $Q_T \setminus Q^{k-1}$ and $\eta \equiv 1$ in Q^k. It follows that $w \geqslant \sigma/2$ in $Q^k \setminus Q^{k-1}$. We may assume that $w^\varepsilon \geqslant \sigma/3$ in $Q^k \setminus Q^{k-1}$. Consequently, w^ε satisfies

$$\begin{cases} w_t^\varepsilon - a_{ij}^\varepsilon D_{ij} w^\varepsilon + b_i^\varepsilon D_i w^\varepsilon + c^\varepsilon w^\varepsilon = f^\varepsilon > 0 & \text{in } Q^k, \\ w^\varepsilon > 0 & \text{on } \partial_p Q^k. \end{cases}$$

Then $w^\varepsilon > 0$ in $\overline{Q}^k$ by Lemma 1.26, and as a result, $v = w \geqslant 0$ in $\overline{Q}^k$ by the fact that $\lim_{\varepsilon \to 0^+} w^\varepsilon = w$ in $C(\overline{Q}_T)$. By the arbitrariness of k we have $v \geqslant 0$ in Q_T, i.e., $u \geqslant -\sigma e^t$. It then follows that $u \geqslant 0$ by taking $\sigma \to 0^+$. □

Now we consider the second initial-boundary value problem

$$\begin{cases} \mathscr{L} u = f(x,t) \geqslant 0 & \text{in } Q_T, \\ \partial_{\boldsymbol{n}} u + bu = g(x,t) \geqslant 0 & \text{on } S_T, \\ u(x,0) = \varphi(x) \geqslant 0 & \text{in } \Omega, \end{cases} \tag{1.37}$$

where $b \in C^{1+\alpha, (1+\alpha)/2}(\overline{S}_T)$, $b \geqslant 0$ and $g \in W_p^{2,1}(Q_T)$ with $p > 1 + n/2$.

Theorem 1.42 *Let Ω be of class $C^{2+\alpha}$, the condition* **(A)** *hold and $b_i, c \in C(\overline{Q}_T)$. Assume that either $g > 0$ or $b > 0$ on $\overline{S}_T$, or $g \in C^{2+\alpha, 1+\alpha/2}(\overline{Q}_T)$. If $u \in W_p^{2,1}(Q_T)$ satisfies (1.37), then $u \geqslant 0$ in Q_T.*

Proof. We first deal with the case that either $g > 0$ or $b > 0$ on $\overline{S}_T$. Similar to the proof of Theorem 1.41 we have $f \in L^p(Q_T)$ and $u \in C^{\alpha, \alpha/2}(\overline{Q}_T)$, and we may assume $c \geqslant 0$. For any given $\sigma > 0$, the function $v = u + \sigma e^t$ satisfies

$$\begin{cases} \mathcal{L}v = (1+c)\sigma e^t + \mathcal{L}u = (1+c)\sigma e^t + f =: f_\sigma \geqslant \sigma & \text{in } Q_T, \\ \partial_{\boldsymbol{n}} v + bv = b\sigma e^t + g =: g_\sigma > 0 & \text{on } S_T, \\ v(x,0) = \sigma + \varphi(x) =: \varphi_\sigma \geqslant \sigma & \text{in } \Omega. \end{cases}$$

Since v is continuous in $\overline{Q}_T$, there exists $0 < \tau_0 \ll 1$ such that, for all $0 < \tau \leqslant \tau_0$,

$$v \geqslant \sigma/2 \quad \text{in } \overline{\Omega} \times [0, 3\tau].$$

For such fixed $\tau > 0$, take a truncated function $\eta \in C^\infty([0,T])$ so that $0 \leqslant \eta \leqslant 1$, $\eta \equiv 1$ in $[2\tau, T]$ and $\eta \equiv 0$ in $[0, \tau]$. Let $w = v\eta$. Then w satisfies

$$\begin{cases} \mathcal{L}w = v\eta_t + \eta f_\sigma =: f^* & \text{in } Q_T, \\ \partial_{\boldsymbol{n}} w + bw = \eta g_\sigma =: g^* & \text{on } S_T, \\ w(x,0) = \eta \varphi_\sigma = 0 & \text{in } \Omega. \end{cases}$$

Obviously, $f^* = f_\sigma$ in $\Omega \times [2\tau, T]$, $f^* = 0$ in $\Omega \times [0, \tau]$ and $g^* = g_\sigma$ on $\partial\Omega \times [2\tau, T]$, $g^* = 0$ on $\partial\Omega \times [0, \tau]$.

Take $a_{ij}^\varepsilon, b_i^\varepsilon, c^\varepsilon, f^\varepsilon \in C^{\alpha, \alpha/2}(\overline{Q}_T)$ and $g^\varepsilon \in C^{2+\alpha, 1+\alpha/2}(\overline{Q}_T)$ such that

(a) $f^\varepsilon = 0$ in $\Omega \times [0, \tau/2]$, $f^\varepsilon \geqslant \sigma/2$ in $\Omega \times [2\tau, T]$, $g^\varepsilon = 0$ on $\partial\Omega \times [0, \tau/2]$, $g^\varepsilon \geqslant \frac{1}{2} \min_{\overline{S}_T} g_\sigma > 0$ on $\partial\Omega \times [2\tau, T]$;

(b) $a_{ij}^\varepsilon \to a_{ij}, b_i^\varepsilon \to b_i, c^\varepsilon \to c$ in $C(\overline{Q}_T)$, $f^\varepsilon \to f^*$ in $L^p(Q_T)$ and $g^\varepsilon \to g^*$ in $W_p^{2,1}(Q_T)$ as $\varepsilon \to 0^+$.

By Theorem 1.17, the problem

$$\begin{cases} w_t^\varepsilon - a_{ij}^\varepsilon D_{ij} w^\varepsilon + b_i^\varepsilon D_i w^\varepsilon + c^\varepsilon w^\varepsilon = f^\varepsilon & \text{in } Q_T, \\ \partial_{\boldsymbol{n}} w^\varepsilon + bw^\varepsilon = g^\varepsilon & \text{on } S_T, \\ w^\varepsilon(x,0) = 0 & \text{in } \Omega \end{cases}$$

has a unique solution $w^\varepsilon \in C^{2+\alpha, 1+\alpha/2}(\overline{Q}_T)$. Similar to the proof of Theorem 1.41 we have $\lim_{\varepsilon \to 0^+} w^\varepsilon = w$ in $W_p^{2,1}(Q_T)$, and hence $\lim_{\varepsilon \to 0^+} w^\varepsilon = w$ in $C(\overline{Q}_T)$. Moreover, $w \geqslant \sigma/2$ in $\overline{\Omega} \times [2\tau, 3\tau]$. We may assume that $w^\varepsilon \geqslant \sigma/3$ in $\overline{\Omega} \times [2\tau, 3\tau]$. Consequently, w^ε satisfies

$$\begin{cases} w_t^\varepsilon - a_{ij}^\varepsilon D_{ij} w^\varepsilon + b_i^\varepsilon D_i w^\varepsilon + c^\varepsilon w^\varepsilon = f^\varepsilon > 0 & \text{in } \Omega \times [2\tau, T], \\ \partial_{\boldsymbol{n}} w^\varepsilon + bw^\varepsilon = g^\varepsilon > 0 & \text{on } \partial\Omega \times [2\tau, T], \\ w^\varepsilon(x, 2\tau) \geqslant \sigma/3 & \text{in } \overline{\Omega}. \end{cases}$$

The By Lemma 1.26, $w^\varepsilon > 0$ in $\overline{\Omega} \times [2\tau, T]$. Same as the proof of Theorem 1.41 we can conclude $u \geqslant 0$.

For the case $g \in C^{2+\alpha,\,1+\alpha/2}(\overline{Q}_T)$, we don't use g^ε to approximate g^*, but directly use g^*. The fact $g_\sigma \geqslant 0$ implies $g^* \geqslant 0$ on S_T. We can still get $w^\varepsilon > 0$ in $\overline{\Omega} \times [2\tau, T]$ by Lemma 1.26. □

If the condition $g > 0$ or $b > 0$ on $\overline{S}_T$, or $g \in C^{2+\alpha,\,1+\alpha/2}(\overline{Q}_T)$ in Theorem 1.42 can not be guaranteed, we do not know how to deal with it. However, when $a_{ij}(x,t) = a_{ij}(x)$ and $b_i(x,t) = b_i(x)$ are independent of time t, the maximum principle is still valid.

Theorem 1.43 *Let Ω be of class $C^{2+\alpha}$, the condition* **(A)** *hold and $b_i, c \in C(\overline{Q}_T)$. Assume that $a_{ij}(x,t) = a_{ij}(x)$, $b_i(x,t) = b_i(x)$ which do not depend on time t, and $u \in W_p^{2,1}(Q_T)$ satisfies (1.37). Then $u \geqslant 0$ in Q_T.*

Proof. Without loss of generality, we assume $c \geqslant 1$. As Ω is of class C^2, the following boundary value problem

$$\begin{cases} -a_{ij}(x)D_{ij}h + b_i(x)D_ih = 0 & \text{in } \Omega, \\ \partial_{\boldsymbol{n}}h = 1 & \text{on } \partial\Omega \end{cases}$$

has a positive solution $h \in W_p^2(\Omega)$. Take $0 < \varepsilon \ll 1$ and let $v = u + \varepsilon h$. Then

$$\begin{cases} \mathscr{L}v = \mathscr{L}u + \mathscr{L}(\varepsilon h) = f + c\varepsilon h \geqslant \varepsilon h & \text{in } Q_T, \\ \partial_{\boldsymbol{n}}v + bv = g + \varepsilon + b\varepsilon h \geqslant \varepsilon & \text{on } S_T, \\ v(x,0) = \varphi + \varepsilon h \geqslant \varepsilon h & \text{in } \Omega. \end{cases}$$

Similar to the proof of Theorem 1.42, we can show that $v \geqslant 0$, and hence $u \geqslant 0$ by letting $\varepsilon \to 0$. □

1.5   WEAK SOLUTION

Consider the following *divergence equation*

$$u_t - D_j(a_{ij}D_iu + c_ju) + b_iD_iu + cu = f + D_if_i \quad \text{in } Q_T.$$

Coefficients satisfy the following assumptions:

(i) there are positive constants λ and Λ such that

$$\lambda|y|^2 \leqslant a_{ij}(x,t)y_iy_j \leqslant \Lambda|y|^2, \quad \forall\, (x,t) \in \overline{Q}_T, \ y \in \mathbb{R}^n;$$

(ii) there exists $q > 1 + n/2$ such that $\|c_i^2 + b_i^2 + |c|\|_{q,Q_T} \leqslant \Lambda$;

(iii) $\|f_i\|_{2,Q_T} + \|f\|_{\frac{2(n+2)}{n+4},Q_T} < \infty$.

Define

$$Lu := -D_j(a_{ij}D_iu + c_ju) + b_iD_iu + cu,$$

$$\overset{\circ}{W}{}_2^{1,1}(Q_T) := \{u \in W_2^{1,1}(Q_T) : u = 0 \ \text{on } S_T\}.$$

For $\phi \in \overset{\circ}{W}_2^{1,1}(Q_T)$, we denote

$$\langle Lu, \phi \rangle = \int_\Omega [(a_{ij}D_i u + c_j u)D_j \phi + (b_i D_i u + cu)\phi]\mathrm{d}x,$$

and for $u, v \in L^2(Q_T)$, we denote

$$(u, v) = \int_\Omega u(x, t)v(x, t)\mathrm{d}x.$$

Set

$$V_2(Q_T) = L^\infty((0, T), L^2(\Omega)) \cap L^2((0, T), H^1(\Omega)),$$

$$V_2^{1,0}(Q_T) = C([0, T], L^2(\Omega)) \cap L^2((0, T), H^1(\Omega)),$$

$$\overset{\circ}{V}_2^{1,0}(Q_T) = \{u \in V_2^{1,0}(Q_T) : u(\cdot, t)|_{\partial\Omega} = 0 \ a.e. \ t \in (0, T)\}.$$

Definition 1.44 *Assume* $g \in L^2((0, T), H^1(\Omega))$, $u_0 \in L^2(\Omega)$. *Function* $u \in V_2(Q_T)$ *is called a* weak lower solution (upper solution) *of the problem*

$$\begin{cases} u_t + Lu = f + D_i f_i & in \ Q_T, \\ u = g & on \ S_T, \\ u(x, 0) = u_0(x) & in \ \Omega, \end{cases} \tag{1.38}$$

if $(u - g)_+ \in L^2((0, T), H_0^1(\Omega))$ $((g - u)_+ \in L^2((0, T), H_0^1(\Omega)))$, *and*

$$(u, \phi) + \int_0^t [\langle Lu, \phi \rangle - (u, \phi_t)]\mathrm{d}s \leqslant (\geqslant) (u_0, \phi(x, 0)) + \int_0^t [(f, \phi) - (f_i, D_i\phi)]\mathrm{d}s$$

for all $\phi \in \overset{\circ}{W}_2^{1,1}(Q_T)$ *with* $\phi \geqslant 0$, *and for all* $0 < t \leqslant T$.

If $u \in V_2(Q_T)$ *is both a weak lower solution and a weak upper solution of* (1.38), *then it is called the* weak solution *of* (1.38).

Theorem 1.45 (Existence and uniqueness of weak solutions [16, Chapter 3, Theorem 2.2, Theorem 3.1, Theorem 3.2]) *Let* **(i)**–**(iii)** *hold and* $g = 0$, $u_0 \in L^2(\Omega)$. *Then problem* (1.38) *admits a unique weak solution* $u \in \overset{\circ}{V}_2^{1,0}(Q_T)$, *and*

$$\|u\|_{V_2(Q_T)} \leqslant C(\|f_i\|_{2,Q_T} + \|f\|_{\frac{2(n+2)}{n+4}, Q_T} + \|u_0\|_{2,\Omega}).$$

Theorem 1.46 (Maximum principle of weak solutions [72, Theorem 6.25]) *Let the condition* **(i)** *hold. Assume* $f, f_i \in L^2(Q_T)$, $c, b_i, c_i \in L^\infty(Q_T)$, $g \in L^2((0, T), H^1(\Omega))$, $u_0 \in H^1(\Omega)$, *and*

$$\int_0^T [(c, \phi) + (c_j, D_j\phi)]\mathrm{d}t \geqslant 0, \quad \forall \phi \in C^1(\overline{Q}_T), \ \phi|_{S_T} = 0, \ \phi \geqslant 0,$$

$$\int_0^T [(f, \phi) - (f_i, D_i\phi)]\mathrm{d}t \leqslant 0, \quad \forall \phi \in C^1(\overline{Q}_T), \ \phi|_{S_T} = 0, \ \phi \geqslant 0,$$

$$g \leqslant 0 \ on \ S_T, \quad u_0 \leqslant 0 \ in \ \Omega.$$

If $u \in V_2(Q_T)$ *is a lower solution of* (1.38), *then* $u \leqslant 0$ *in* Q_T.

Now let us discuss a special case of (1.38):

$$u_t - D_j(a_{ij}(x,t)D_i u) + b_i(x,t)D_i u + cu = f(x,t) \quad \text{in} \quad Q_T.$$

It is assumed that $a_{ij}, b_i, c \in L^\infty(Q_T)$, $a_{ij} = a_{ji}$ in Q_T for all $1 \leqslant i, j \leqslant n$, and there exists a constant $\lambda > 0$ so that

$$a_{ij}(x,t)y_i y_j \geqslant \lambda |y|^2 \quad \text{for all} \quad (x,t) \in Q_T, \ y \in \mathbb{R}^n.$$

Define

$$\overset{\bullet}{C}{}^\infty(\overline{Q}_T) := \{u \in C^\infty(\overline{Q}_T) : u|_{\partial_p Q_T} = 0\},$$

and let $\overset{\bullet}{W}{}^{k,l}_p(Q_T)$ be closure of $\overset{\bullet}{C}{}^\infty(\overline{Q}_T)$ in $W^{k,l}_p(Q_T)$.

Theorem 1.47 (Maximum principle of weak solutions) *Let*

$$u \in C([0,T], L^2(\Omega)) \cap L^2((0,T), H^1(\Omega)) \cap L^\infty(\Omega \times (0,T)).$$

Assume $u|_{\partial_p Q_T} \leqslant 0 \, (\geqslant 0)$ and

$$\int_\Omega u(x,t)\phi(x,t)\mathrm{d}x + \int_{Q_t} [a_{ij}D_i u D_j \phi + b_i D_i u\phi + cu\phi - u\phi_\tau]\mathrm{d}x\mathrm{d}\tau \leqslant (\geqslant) 0$$

for all $0 < t \leqslant T$ and each test function $\phi \in \overset{\bullet}{W}{}^{1,1}_2(Q_T)$ with $\phi \geqslant 0$ in Q_T. Then $u \leqslant 0 \, (u \geqslant 0)$ in Q_T.

This theorem is a direct consequence of Theorem 2.7.

For further discussions on the existence, uniqueness, regularity, estimation, and maximum principle of weak solutions, readers can refer to monographs [16, 72].

EXERCISES

1.1 Show that if $(a_{ij})_{n \times n}$ is a non-negative definite matrix and $(b_{ij})_{n \times n}$ is a non-positive definite matrix, then the number $\sum_{i,j=1}^n a_{ij}b_{ij}$ is non-positive.

1.2 Let $1 \leqslant p < \infty$. Prove $W^{2,1}_p(\Omega \times [0,T]) \hookrightarrow C([0,T], L^p(\Omega))$.

1.3 Prove Theorem 1.3 for the case $\varepsilon > 0$.

1.4 Prove Theorem 1.19. Hint: first, take $p > n+2$ and Q_1 to satisfy $Q \Subset Q_1 \Subset Q_T$. Then (1.10) holds for $Q = Q_1$, and

$$\|u\|_{\infty,\overline{Q}_1} \leqslant C\|u\|_{W^{2,1}_p(Q_1)} \leqslant C\left(\|f\|_{p,Q_T} + \|u\|_{2,Q_T}\right) \leqslant C\left(|f|_{\alpha,\overline{Q}_T} + \|u\|_{2,Q_T}\right).$$

Second, prove $|u|_{2+\alpha,\overline{Q}} \leqslant C\left(|f|_{\alpha,\overline{Q}_T} + \|u\|_{\infty,\overline{Q}_1}\right)$.

1.5 Give an example to show that the condition $c \geqslant 0$ is necessary in Theorem 1.24.

1.6 Let $u \in C(\overline{Q}_T) \cap C^{2,1}(Q_T)$ satisfy

$$\begin{cases} u_t - \Delta u = -u^2 + cu & \text{in} \quad Q_T, \\ u = 0 & \text{on} \quad S_T, \\ u|_{t=0} = u_0(x) & \text{in} \quad \overline{\Omega}, \end{cases}$$

where $c \in C(\overline{Q}_T)$, $u_0 \in C(\overline{\Omega})$ and $u_0 \geqslant 0$. Prove that

$$0 \leqslant u \leqslant \max \left\{ \max_{\overline{Q}_T} |c|, \max_{\overline{\Omega}} u_0 \right\}.$$

1.7 Let u be a classical solution of initial value problem

$$\begin{cases} u_t - \Delta u = 0 & \text{in} \quad \mathbb{R}^n \times (0, T], \\ u(x, 0) = 0 & \text{in} \quad \mathbb{R}^n \end{cases}$$

and $0 < T < \infty$. Set $k_l = \sup_{|x| \leqslant l, 0 \leqslant t \leqslant T} |u(x, t)|$. Prove that if $\lim\limits_{l \to \infty} \dfrac{k_l}{l^2} = 0$, then $u \equiv 0$ in $\mathbb{R}^n \times [0, T]$.

1.8 Prove Theorem 1.32 and Theorem 1.35.

1.9 Please give an example to illustrate that the condition $c_2 \geqslant 0$ is necessary in both Theorem 1.34 and Theorem 1.35.

1.10 Give the details of the proof of Theorem 1.40.

CHAPTER 2

Comparison Principle, Regularity and Uniform Estimates

In this chapter we mainly study comparison principle, regularity and uniform estimates, and uniform bounds from bounded L^p norms. These techniques help us gain an important handle on solutions of nonlinear parabolic equations.

Comparison principle based on the maximum principle plays an important role in estimating upper and/or lower bounds of solutions and judging the sign of solution, as well as establishing the upper and lower solutions method.

Regularities and uniform estimates based on the L^p theory and Schauder theory are fundamental problems in the study of parabolic equations. The regularity theory is used to investigate smoothness of solutions (improving regularities of solutions). We will see that for parabolic equations with better coefficients, even if initial value or boundary value is not good enough, their solution will have better interior smoothness. The uniform estimate in time provides a basis for studying the convergence of a solution as time goes to infinity, and its limit is usually a solution of the corresponding equilibrium problem.

In order to prove global existence of classical or mild solutions of parabolic equations, some priori bounds in the uniform norm are needed. But usually it is easier to derive the estimate for some L^p-norm. Some interesting examples are given to show that one can use the structure of reaction terms to obtain uniform bounds from bounded L^p norms.

2.1 COMPARISON PRINCIPLE

The *comparison principle* or *comparison method* is an important tool in the study of partial differential equations, especially parabolic equations and elliptic equations. It is based on the maximum principle, and it is the basis of the upper and lower solutions method which will be introduced later.

2.1.1 Classical and strong solutions

In this subsection,

$$\mathscr{L} := \partial_t - a_{ij}(x,t)D_{ij} + b_i(x,t)D_i.$$

Theorem 2.1 *Let coefficients of $\mathscr{L}$ and Ω, a, b satisfy the conditions of Theorem 1.40 (2). Assume $\bar{u}, \underline{u} \in C(\overline{Q}_T) \cap C^{1,0}(Q_T) \cap W^{2,1}_{n+1,\,\text{loc}}(Q_T)$, and when $a \not\equiv 0$ we further assume $\bar{u}, \underline{u} \in C^{1,0}(\overline{\Omega} \times (0,T])$. Suppose*

$$\begin{cases} \mathscr{L}\bar{u} - f(x,t,\bar{u}) \geqslant \mathscr{L}\underline{u} - f(x,t,\underline{u}) & \text{in } Q_T, \\ a\partial_{\boldsymbol{n}}\bar{u} + b\bar{u} \geqslant a\partial_{\boldsymbol{n}}\underline{u} + b\underline{u} & \text{on } S_T, \\ \bar{u}(x,0) \geqslant \underline{u}(x,0) & \text{in } \Omega. \end{cases}$$

Denote $\underline{c} = \min_{\overline{Q}_T} \min\{\bar{u}, \underline{u}\}$ and $\bar{c} = \max_{\overline{Q}_T} \max\{\bar{u}, \underline{u}\}$. If f satisfies the Lipschitz condition in $u \in (\underline{c}, \bar{c})$, i.e., there exists a positive constant M such that

$$|f(x,t,u) - f(x,t,v)| \leqslant M|u-v| \quad \text{for all } (x,t) \in Q_T, \ u, v \in (\underline{c}, \bar{c}), \qquad (2.1)$$

then $\bar{u} \geqslant \underline{u}$ in Q_T. Moreover, $\bar{u} > \underline{u}$ in Q_T if $\bar{u}(x,0) \not\equiv \underline{u}(x,0)$.

Proof. For $(x,t) \in Q_T$, we take $\hat{c}(x,t) = M$ when $\bar{u}(x,t) \geqslant \underline{u}(x,t)$, and $\hat{c}(x,t) = -M$ when $\bar{u}(x,t) < \underline{u}(x,t)$, where M is given by (2.1). Let $u = \bar{u} - \underline{u}$. Then u satisfies

$$\begin{cases} \mathscr{L}u + \hat{c}(x,t)u \geqslant 0 & \text{in } Q_T, \\ \mathscr{B}u \geqslant 0 & \text{on } S_T, \\ u(x,0) \geqslant 0 & \text{in } \Omega. \end{cases}$$

By Theorem 1.40 (2) we have $u \geqslant 0$ in Q_T, and $u > 0$ in Q_T if $u(x,0) \not\equiv 0$. □

Theorem 2.2 *Let Ω be of class C^2, $\bar{u}, \underline{u} \in C(\overline{Q}_T) \cap W^{2,1}_p(Q_T)$ satisfy*

$$\begin{cases} \mathscr{L}\bar{u} - f(x,t,\bar{u}) \geqslant \mathscr{L}\underline{u} - f(x,t,\underline{u}) & \text{in } Q_T, \\ \bar{u} \geqslant \underline{u} & \text{on } S_T, \\ \bar{u}(x,0) \geqslant \underline{u}(x,0) & \text{in } \Omega. \end{cases}$$

If $p > 1 + n/2$, the condition **(A)** *holds, $b_i, c \in C(\overline{Q}_T)$ and $f_u(\cdot, u(\cdot)) \in C(\overline{Q}_T)$ for each $u \in C(\overline{Q}_T)$, then $\bar{u} \geqslant \underline{u}$ in Q_T.*

Theorem 2.3 *Let Ω be of class $C^{2+\alpha}$, the condition* **(A)** *hold, $b_i, c \in C(\overline{Q}_T)$, $b \in C^{1+\alpha,(1+\alpha)/2}(\overline{S}_T)$ and $b \geqslant 0$. Let $p > 1 + n/2$, $\bar{u}, \underline{u} \in C(\overline{Q}_T) \cap W^{2,1}_p(Q_T)$ satisfy*

$$\begin{cases} \mathscr{L}\bar{u} - f(x,t,\bar{u}) \geqslant \mathscr{L}\underline{u} - f(x,t,\underline{u}) & \text{in } Q_T, \\ \partial_{\boldsymbol{n}}\bar{u} + b\bar{u} \geqslant \partial_{\boldsymbol{n}}\underline{u} + b\underline{u} & \text{on } S_T, \\ \bar{u}(x,0) \geqslant \underline{u}(x,0) & \text{in } \Omega. \end{cases}$$

Assume $f_u(\cdot, u(\cdot)) \in C(\overline{Q}_T)$ for each $u \in C(\overline{Q}_T)$, and either $b > 0$ on $\overline{S}_T$, or $a_{ij} = a_{ij}(x)$ and $b_i = b_i(x)$. Then $\bar{u} \geqslant \underline{u}$ in Q_T.

2.1.2 Weak solutions

This subsection is concerned with weak solutions of the *divergence equation*

$$u_t - D_j(a_{ij}(x,t)D_iu) + b_i(x,t)D_iu = f(x,t,u) \quad \text{in} \quad Q_T, \tag{2.2}$$

where $a_{ij}, b_i \in L^\infty(Q_T)$, $a_{ij} = a_{ji}$ in Q_T for all $1 \leqslant i, j \leqslant n$, and there exists a constant $\lambda > 0$ so that

$$a_{ij}(x,t)y_iy_j \geqslant \lambda|y|^2 \quad \text{for all} \ (x,t) \in Q_T, \ y \in \mathbb{R}^n.$$

We define $\overset{\bullet}{W}{}_p^{k,l}(Q_T)$ as in §1.5.

Definition 2.4 *A* weak lower solution (*weak upper solution*) *of equation* (2.2) *is a measurable function u which satisfies*

$$u \in L^2((\sigma, T), H^1(\Omega')) \cap L^\infty(\Omega' \times (\sigma, T)), \quad \text{and} \ \ u_t, f(\cdot, u(\cdot)) \in L^2(\Omega' \times (\sigma, T))$$

for any $0 < \sigma < T$ *and any compact subset* Ω' *of* Ω, *and also satisfies*

$$\int_\Omega u(x,t)\phi(x,t) + \int_{Q_t}(a_{ij}D_iuD_j\phi + b_iD_iu\phi) \leqslant (\geqslant) \int_{Q_t}(f(x,\tau,u)\phi + u\phi_\tau)$$

for each $0 < t \leqslant T$ *and each test function* $\phi \in \overset{\bullet}{W}{}_2^{1,1}(Q_T)$ *with* $\phi \geqslant 0$ *in* Q_T.

A function u that is both a weak lower solution and a weak upper solution is called a weak solution *of* (2.2).

We should mention that in this definition, the lower (upper) solution may not be defined on $\partial_p Q_T$.

Theorem 2.5 (Comparison principle for weak solutions) *Let* $\bar{u}$ *and* $\underline{u}$ *be a weak upper solution and a weak lower solution of* (2.2), *respectively. Denote* $\underline{c} = \inf_{Q_T} \min\{\bar{u}, \underline{u}\}$ *and* $\bar{c} = \sup_{Q_T} \max\{\bar{u}, \underline{u}\}$. *Assume that* $f(x,t,u)$ *satisfies one side Lipschitz condition in* $u \in (\underline{c}, \bar{c})$, *i.e., there exists a constant* $k > 0$ *such that*

$$f(x,t,u) - f(x,t,v) \leqslant k(u-v) \quad \text{for all} \ (x,t) \in Q_T, \ u,v \in (\underline{c}, \bar{c}), \ u \geqslant v.$$

If

$$\limsup_{(x,t) \to \partial_p Q_T} (\underline{u} - \bar{u}) \leqslant 0, \tag{2.3}$$

then $\bar{u} \geqslant \underline{u}$ *in* Q_T.

Proof. Without loss of generality we assume $k = 0$. Let $u = \underline{u} - \bar{u}$, and $\phi \in \overset{\bullet}{W}{}_2^{1,1}(Q_T)$ be non-negative. Taking ϕ as a test function it yields that, for all $0 < t \leqslant T$,

$$\int_\Omega (u\phi)(x,t) + \int_{Q_t}[a_{ij}D_iuD_j\phi + b_iD_iu\phi] \leqslant \int_{Q_t}[f(x,\tau,\underline{u}) - f(x,\tau,\bar{u})]\phi + \int_{Q_t}u\phi_\tau. \tag{2.4}$$

For any given $0 < \varepsilon < 1$, let $v = (u - \varepsilon)_+$. It follows from the condition (2.3) that

$$\limsup_{(x,t)\to\partial_p Q_T} (\underline{u} - \bar{u} - \varepsilon) = \limsup_{(x,t)\to\partial_p Q_T} (\underline{u} - \bar{u}) - \varepsilon \leqslant -\varepsilon.$$

We can choose $t_\varepsilon \in (0, T)$ and $\Omega_\varepsilon \Subset \Omega$ with $t_\varepsilon \to 0^+$ and $\Omega_\varepsilon \to \Omega$ as $\varepsilon \to 0^+$, such that $v = 0$ in $Q_T \setminus \Omega_\varepsilon \times (2t_\varepsilon, T)$. Then $v \in \overset{\bullet}{W}{}_2^{1,1}(Q_T)$ is non-negative. As a result, (2.4) is valid with ϕ replaced by v. That is,

$$\int_\Omega uv + \int_{Q_t} [a_{ij}D_i u D_j v + b_i D_i uv] \leqslant \int_{Q_t} [f(x, \tau, \underline{u}) - f(x, \tau, \bar{u})]v + \int_{Q_t} uv_\tau$$

$$\leqslant \int_{Q_t} uv_\tau. \tag{2.5}$$

Notice that at the point where $v \neq 0$, we have $u - \varepsilon > 0$ and hence $v = u - \varepsilon$. As a consequence

$$\int_{Q_t} uv_\tau = \int_{Q_t} (u - \varepsilon)v_\tau + \varepsilon \int_{Q_t} v_\tau$$

$$= \int_{Q_t} vv_\tau + \varepsilon \int_{Q_t} v_\tau$$

$$= \frac{1}{2}\int_\Omega v^2 + \varepsilon \int_\Omega v.$$

Substituting this into (2.5) firstly, and letting $\varepsilon \to 0^+$ secondly, it yields

$$\int_\Omega u_+^2 + \int_{Q_t} [a_{ij}D_i u_+ D_j u_+ + b_i D_i u_+ u_+] \leqslant \frac{1}{2}\int_\Omega u_+^2. \tag{2.6}$$

Thanks to

$$\int_{Q_t} a_{ij}D_i u_+ D_j u_+ \geqslant \lambda \int_{Q_t} |Du_+|^2,$$

$$\left|\int_{Q_t} b_i D_i u_+ u_+\right| \leqslant \lambda \int_{Q_t} |Du_+|^2 + C\int_{Q_t} u_+^2,$$

it follows from (2.6) that

$$\frac{1}{2}\int_\Omega u_+^2 \leqslant C\int_0^t \int_\Omega u_+^2.$$

This implies $u_+ = 0$, i.e., $\underline{u} \leqslant \bar{u}$ almost everywhere in Q_T. □

Now we consider the first initial-boundary value problem of *divergence equation*

$$\begin{cases} u_t - D_j(a_{ij}(x,t)D_i u) + b_i(x,t)D_i u = f(x,t,u) & \text{in } Q_T, \\ u = g & \text{on } S_T, \\ u(x,0) = \varphi(x) & \text{in } \Omega, \end{cases} \tag{2.7}$$

where a_{ij} and b_i are the same as above, and $\varphi \in L^2(\Omega)$.

Definition 2.6 *A weak lower solution (weak upper solution) of problem* (2.7) *is a measurable function u such that*

$$u \in C([0,T], L^2(\Omega)) \cap L^2(0,T; H^1(\Omega)) \cap L^\infty(\Omega \times (0,T)),$$
$$f(\cdot, u(\cdot)) \in L^2(\Omega \times (0,T)),$$
$$u(x,0) \leqslant (\geqslant) \varphi(x) \ \text{in} \ \Omega, \quad u \leqslant (\geqslant) g \ \text{on} \ S_T,$$

and for all $0 < t \leqslant T$,

$$\int_\Omega u\phi|_{\tau=t} + \int_{Q_t} (a_{ij} D_i u D_j \phi + b_i D_i u \phi) \leqslant (\geqslant) \int_{Q_t} (f(x,\tau,u)\phi + u\phi_\tau)$$

holds for each test function $\phi \in \overset{\bullet}{W}{}^{1,1}_2(Q_T)$ *with* $\phi \geqslant 0$ *in* Q_T.

A function u that is both a weak lower solution and a weak upper solution is called a weak solution of (2.7).

Theorem 2.7 (Comparison principle for weak solutions) *Let* $\bar{u}, \underline{u}$ *be weak upper and lower solutions of* (2.7), *respectively. Denote* $\underline{c} = \inf_{Q_T} \min\{\bar{u}, \underline{u}\}$ *and* $\bar{c} = \sup_{Q_T} \max\{\bar{u}, \underline{u}\}$. *Assume that* $f(x,t,u)$ *satisfies one side Lipschitz condition in* $u \in (\underline{c}, \bar{c})$, *i.e., there exists a constant* $k > 0$ *such that*

$$f(x,t,u) - f(x,t,v) \leqslant k(u-v) \ \text{for all} \ (x,t) \in Q_T, \ u,v \in (\underline{c}, \bar{c}) \ \text{with} \ u \geqslant v.$$

Then $\bar{u} \geqslant \underline{u}$ *in* Q_T.

To prove this result, let us do some preparatory work first. For a given function g defined in Q_T, we make zero extension of g to $t < 0$ and $t > T$. We define the Steklov average of g as follows:

$$g^h(\cdot, t) = \frac{1}{h} \int_t^{t+h} g(\cdot, s) \mathrm{d}s, \quad h \neq 0.$$

It is easy to show that $(g^{-h})_t = (g_s)^{-h}$. Moreover, for the given $0 < h < s$, if $g(x,t) = 0$ for $t \leqslant 0$ and $t \geqslant s - h$, then

$$\int_{Q_s} u g^{-h} \mathrm{d}x\mathrm{d}t = \int_{Q_s} u^h g \mathrm{d}x\mathrm{d}t. \tag{2.8}$$

Lemma 2.8 *Let* $1 \leqslant p < \infty$ *and* $0 < h < \delta < T$. *Then we have*

(1) $u^h \in L^p(Q_{T-\delta})$ *and* $\lim_{h \to 0} u^h = u$ *in* $L^p(Q_{T-\delta})$ *if* $u \in L^p(Q_T)$;

(2) $Du^h \in L^p(Q_{T-\delta})$ *and* $\lim_{h \to 0} Du^h = Du$ *in* $L^p(Q_{T-\delta})$ *if* $Du \in L^p(Q_T)$.

Proof of Theorem 2.7. Without loss of generality, we assume $k = 0$. Let $u = \underline{u} - \bar{u}$, and $\phi \in \overset{\bullet}{W}{}^{1,1}_2(Q_T)$ be a non-negative function. Taking ϕ as a test function, we have that, for all $0 < t \leqslant T$, (2.4) holds.

Let $0 < h \ll 1$. For any $\zeta \in \overset{\bullet}{W}_2^{1,1}(Q_T)$ with $\zeta \geqslant 0$, and $\zeta = 0$ for $t \leqslant 0$ and $t \geqslant T - h$, taking $t = T$ and $\phi = \zeta^{-h}$ in (2.4) and applying (2.8), we have that, due to $\zeta^{-h}(x, T) = 0$,

$$\int_{Q_T} [(a_{ij} D_i u)^h D_j \zeta + (b_i D_i u)^h \zeta] \leqslant \int_{Q_T} (f(\cdot, \underline{u}) - f(\cdot, \bar{u}))^h \zeta + \int_{Q_T} u(\zeta^{-h})_\tau. \quad (2.9)$$

Note that $\zeta = 0$ for $t \leqslant 0$ and $t \geqslant T - h$. Using $(\zeta^{-h})_\tau = (\zeta_t)^{-h}$ and (2.8) firstly, and integrating by parts secondly, we obtain

$$\int_{Q_T} u(\zeta^{-h})_\tau = \int_{Q_T} u^h \zeta_\tau = -\int_{Q_T} (u^h)_\tau \zeta.$$

Therefore, in accordance with (2.9),

$$\int_{Q_T} [(a_{ij} D_i u)^h D_j \zeta + (b_i D_i u)^h \zeta] + \int_{Q_T} (u^h)_\tau \zeta \leqslant \int_{Q_T} (f(\cdot, \underline{u}) - f(\cdot, \bar{u}))^h \zeta. \quad (2.10)$$

Clearly, $(u^h)_+ = 0$ on S_T. For any given $0 < t < T$, we will show that (2.10) remains valid if ζ is replaced by $(u^h)_+ \xi(\tau)$, where $\xi(\tau) = 1$ for $\tau < t$ and $\xi(\tau) = 0$ for $\tau > t$. For this purpose, we fix ℓ large enough and choose a continuous non-negative function z_ℓ: z_ℓ is linear in $(0, 1/\ell) \cup (t - 1/\ell, t)$, $z_\ell = 0$ in $(-\infty, 0] \cup [t, \infty)$ and $z_\ell = 1$ in $(1/\ell, t - 1/\ell)$. Then the function $\zeta_\ell := (u^h)_+ z_\ell \in \overset{\bullet}{W}_2^{1,1}(Q_T)$, $\zeta_\ell \geqslant 0$, and $\zeta_\ell = 0$ for $\tau \leqslant 0$ and $\tau \geqslant T - h$ provided $h < T - t$. Therefore (2.10) holds for ζ_ℓ. Then we have by letting $\ell \to \infty$,

$$\int_{Q_t} [(a_{ij} D_i u)^h D_j (u^h)_+ + (b_i D_i u)^h (u^h)_+] + \int_{Q_t} (u^h)_\tau (u^h)_+$$

$$\leqslant \int_{Q_t} (f(\cdot, \underline{u}) - f(\cdot, \bar{u}))^h (u^h)_+. \quad (2.11)$$

It is easy to see that

$$\int_{Q_t} (u^h)_\tau (u^h)_+ = \int_{Q_t} ((u^h)_+)_\tau (u^h)_+ = \frac{1}{2} \left(\int_\Omega (u^h)_+^2 (x, t) - \int_\Omega (u^h)_+^2 (x, 0) \right),$$

and

$$\lim_{h \to 0} \int_\Omega (u^h)_+^2 (x, t) = \int_\Omega u_+^2 (x, t), \quad \lim_{h \to 0} \int_\Omega (u^h)_+^2 (x, 0) = \int_\Omega u_+^2 (x, 0) = 0$$

because $u \in C([0, T], L^2(\Omega))$. Substituting these conclusions into (2.11), and letting $h \to 0$ and then using Lemma 2.8 we find

$$\int_{Q_t} [a_{ij} D_i u D_j u_+ + b_i u_+ D_i u] + \frac{1}{2} \int_\Omega u_+^2 (x, t) \leqslant \int_{Q_t} [f(\cdot, \underline{u}) - f(\cdot, \bar{u})] u_+ \leqslant 0. \quad (2.12)$$

Noticing

$$\int_{Q_t} a_{ij} D_i u D_j u_+ = \int_{Q_t} a_{ij} D_i u_+ D_j u_+ \geqslant \lambda \int_{Q_t} |Du_+|^2,$$

$$\left| \int_{Q_t} b_i u_+ D_i u \right| \leqslant \lambda \int_{Q_t} |Du_+|^2 + C \int_{Q_t} u_+^2,$$

it can be deduced from (2.12) that

$$\frac{1}{2}\int_\Omega u_+^2(x,t) \leqslant C \int_{Q_t} u_+^2(x,\tau) = C \int_0^t \int_\Omega u_+^2(x,\tau).$$

This implies $u_+ = 0$, i.e., $\underline{u} \leqslant \bar{u}$ almost everywhere in Q_T. □

2.1.3 Quasilinear equations with nonlinear boundary conditions

In this subsection we consider the following initial-boundary value problem of quasi-linear equation with *nonlinear boundary conditions*

$$\begin{cases} u_t = \nabla(a(u)\nabla u) + f(u), & x \in \Omega, \ t > 0, \\ \partial_{\boldsymbol{n}} u = b(u), & x \in \partial\Omega, \ t > 0, \\ u(x,0) = u_0(x), & x \in \Omega. \end{cases} \tag{2.13}$$

Existence and uniqueness of solutions of (2.13) has been studied by Amann in [7].

Theorem 2.9 ([105]) *Assume $b \in C(\overline{\mathbb{R}}_+)$, $a, f \in C^1(\overline{\mathbb{R}}_+)$ and $a > 0$ in $\overline{\mathbb{R}}_+$. Let $u, v \in C^{2,1}(\overline{\Omega} \times (0,T)) \cap C(\overline{\Omega} \times [0,T])$, $u, v > 0$ in $\overline{\Omega} \times [0,T)$ and satisfy*

$$\begin{cases} u_t - \nabla(a(u)\nabla u) - f(u) \geqslant v_t - \nabla(a(v)\nabla v) - f(v) & in \ \ Q_T, \\ \partial_{\boldsymbol{n}} u - b(u) \geqslant \partial_{\boldsymbol{n}} v - b(v) & on \ \ S_T, \end{cases}$$

and $u(x,0) > v(x,0)$ for $x \in \overline{\Omega}$. If either $a'(s)$ is increasing in $s > 0$ or $a''(s)$ exists and is continuous, then $u > v$ in $\overline{\Omega} \times [0,T)$.

Proof. We first assume that $a'(s)$ is increasing in $s > 0$. Let $w = u - v$, then

$$w_t - a(u)\Delta w + [a(v) - a(u)]\Delta v - a'(u)(\nabla v + \nabla u) \cdot \nabla w$$

$$+ f(v) - f(u) \geqslant [a'(u) - a'(v)]|\nabla v|^2.$$

Write

$$g(v) - g(u) = (v - u)\int_0^1 g'(u + s(v-u))\mathrm{d}s,$$

and denote

$$a(x,t) = a(u(x,t)), \quad b_i(x,t) = -a'(u(x,t))(v_{x_i}(x,t) + u_{x_i}(x,t)),$$

$$c(x,t) = -\Delta v \int_0^1 a'(u + s(v-u))\mathrm{d}s - \int_0^1 f'(u + s(v-u))\mathrm{d}s.$$

Then c is bounded in $\Omega \times [\varepsilon, T - \varepsilon]$ for each fixed $0 < \varepsilon \ll 1$, and w satisfies

$$w_t - a\Delta w + b_i w_{x_i} + cw \geqslant [a'(u) - a'(v)]|\nabla v|^2.$$

If the statement was not valid, then there would exist $0 < t_0 < T$ and $x_0 \in \overline{\Omega}$ so that

$$u(x_0, t_0) = v(x_0, t_0) \ \text{ and } \ u - v > 0 \ \text{ in } \overline{\Omega} \times [0, t_0).$$

We claim that $x_0 \notin \Omega$. Otherwise, $[a'(u) - a'(v)]|\nabla v|^2 \geqslant 0$ in $\overline{\Omega} \times [0, t_0)$ since $u > v$ in this domain. Take $0 < \varepsilon < t_0$ and k so large that $k + c > 0$ in $\Omega \times [\varepsilon, t_0]$. Then the function $z = e^{-kt}w$ satisfies

$$\begin{cases} z_t - a\Delta z + b_i z_{x_i} + (k + c)z \geqslant 0 & \text{in } \overline{\Omega} \times [\varepsilon, t_0], \\ z(x_0, t_0) = 0, \quad z > 0 & \text{in } \overline{\Omega} \times [\varepsilon, t_0). \end{cases}$$

Corollary 1.25 avers $z \equiv 0$ in $\overline{\Omega} \times [\varepsilon, t_0)$. This is impossible, and hence $x_0 \notin \Omega$. Therefore, z satisfies

$$\begin{cases} z_t - a\Delta z + b_i z_{x_i} + (k + c)z \geqslant 0 & \text{in } \Omega \times [\varepsilon, t_0], \\ z = e^{-kt}w > 0 & \text{in } \Omega \times [\varepsilon, t_0), \\ z(x_0, t_0) = 0. \end{cases}$$

The Hopf boundary lemma asserts $\partial_{\boldsymbol{n}} z(x_0, t_0) < 0$, i.e., $\partial_{\boldsymbol{n}} w(x_0, t_0) < 0$. However,

$$\partial_{\boldsymbol{n}} w(x_0, t_0) = \partial_{\boldsymbol{n}} u(x_0, t_0) - \partial_{\boldsymbol{n}} v(x_0, t_0) \geqslant [b(u) - b(v)]\big|_{(x_0, t_0)} = 0.$$

This contradiction shows that $u > v$ in $\overline{\Omega} \times [0, T)$.

When $a''(s)$ exists and is continuous, we can write

$$[a'(u) - a'(v)]|\nabla v|^2 = \left(|\nabla v|^2 \int_0^1 a''(v + s(u - v)) \mathrm{d}s\right) w,$$

and then put it into $c(x, t)$. The remaining proof is the same as above. $\qquad\square$

2.2   REGULARITY AND UNIFORM ESTIMATES

Let Ω be of class $C^{2+\alpha}$. We first investigate the following initial-boundary value problem

$$\begin{cases} \mathscr{L}u = f(x, t) & \text{in } Q_T, \\ \mathscr{B}u = 0 & \text{on } S_T, \\ u(x, 0) = \varphi(x) & \text{in } \Omega, \end{cases} \tag{2.14}$$

where $\mathscr{B}u = u$, or $\mathscr{B}u = \partial_{\boldsymbol{n}}u + bu$ with $b \geqslant 0$ on S_T and $b \in C^{1+\alpha, (1+\alpha)/2}(\overline{S}_T)$, $\varphi \in W_p^2(\Omega)$ with $p > 1$ and

(i) $\varphi(x) = 0$ on $\partial\Omega$ when $\mathscr{B}u = u$;

(ii) $\partial_{\boldsymbol{n}}\varphi + b(x, 0)\varphi = 0$ on $\partial\Omega$ when $\mathscr{B}u = \partial_{\boldsymbol{n}}u + bu$.

Theorem 2.10 (Interior regularity and estimate) *Under above assumptions, we further assume that the function f and coefficients of $\mathscr{L}$ satisfy the requirement* **(C)**, *which is given in Chapter 1. Then* (2.14) *has a unique solution $u \in W_p^{2,1}(Q_T) \cap C^{2+\alpha, 1+\alpha/2}(\overline{\Omega} \times (0, T])$. Moreover, for any given $0 < \sigma < T$, there is a constant $C(\sigma, T)$ such that*

$$|u|_{2+\alpha, \overline{\Omega} \times [\sigma, T]} \leqslant C(\sigma, T) \left(|f|_{\alpha, \overline{Q}_T} + \|\varphi\|_{W_p^2(\Omega)}\right). \tag{2.15}$$

Proof. We only deal with the case $\mathscr{B}u = u$. Firstly, by Theorem 1.1, problem (2.14) has a unique solution $u \in W_p^{2,1}(Q_T)$, and

$$\|u\|_{W_p^{2,1}(Q_T)} \leqslant C \left(\|f\|_{p, Q_T} + \|\varphi\|_{W_p^2(\Omega)} \right) \leqslant C \left(|f|_{\alpha, \overline{Q}_T} + \|\varphi\|_{W_p^2(\Omega)} \right).$$

If $p > 1 + n/2$, then the embedding theorem (Theorem A.7 (1)) implies $u \in C^{\alpha, \alpha/2}(\overline{Q}_T)$ with $\alpha = 2 - (n+2)/p$, and

$$|u|_{\alpha, \overline{Q}_T} \leqslant C \|u\|_{W_p^{2,1}(Q_T)} \leqslant C \left(|f|_{\alpha, \overline{Q}_T} + \|\varphi\|_{W_p^2(\Omega)} \right).$$

If $p = 1 + n/2$, then the embedding theorem (Theorem A.6 (3)) implies $u \in L^q(Q_T)$ for any $1 \leqslant q < \infty$, and

$$\|u\|_{q, Q_T} \leqslant C \|u\|_{W_p^{2,1}(Q_T)} \leqslant C \left(|f|_{\alpha, \overline{Q}_T} + \|\varphi\|_{W_p^2(\Omega)} \right).$$

If $p < 1 + n/2$, then the embedding theorem (Theorem A.6 (2)) implies $u \in L^{p_1}(Q_T)$ with $p_1 = p(n+2)/(n+2-2p)$, and

$$\|u\|_{p_1, Q_T} \leqslant C \|u\|_{W_p^{2,1}(Q_T)} \leqslant C \left(|f|_{\alpha, \overline{Q}_T} + \|\varphi\|_{W_p^2(\Omega)} \right).$$

We only consider the case $p < 1 + n/2$. The following arguments are the same as those of Theorem 1.8. Take $0 < \varepsilon < \sigma/4$ and a truncated function $\eta \in C^\infty([0, T])$ such that $0 \leqslant \eta \leqslant 1$, $\eta \equiv 1$ in $[\varepsilon, T]$ and $\eta \equiv 0$ in $[0, \varepsilon/2]$. Then there exists $C' > 0$ such that $|\eta'(t)| \leqslant C'/\varepsilon$. As a result, the function $w = u\eta$ satisfies

$$\begin{cases} \mathscr{L}w = f(x,t)\eta(t) + u\eta'(t) & \text{in } Q_T, \\ w = 0 & \text{on } S_T, \\ w(x,0) = 0 & \text{in } \Omega. \end{cases} \tag{2.16}$$

This problem has a unique solution $w \in W_{p_1}^{2,1}(Q_T)$, and

$$\begin{aligned} \|w\|_{W_{p_1}^{2,1}(Q_T)} &\leqslant C \|f\eta + u\eta'\|_{p_1, Q_T} \\ &\leqslant C \left(|f|_{\alpha, \overline{Q}_T} + \|u\|_{p_1, Q_T} \right) \\ &\leqslant C \left(|f|_{\alpha, \overline{Q}_T} + \|\varphi\|_{W_p^2(\Omega)} \right) \end{aligned}$$

according to Theorem 1.1. As a consequence, $u \in W_{p_1}^{2,1}(\Omega \times (\varepsilon, T)) \hookrightarrow L^{p_2}(\Omega \times (\varepsilon, T))$ with $p_2 = p_1(n+2)/(n+2-2p_1)$. Repeating this process, we can show that $u \in W_q^{2,1}(\Omega \times (\sigma/4, T))$ for some $q > 1 + n/2$, and

$$\|u\|_{W_q^{2,1}(\Omega \times (\sigma/4, T))} \leqslant C \left(|f|_{\alpha, \overline{Q}_T} + \|\varphi\|_{W_p^2(\Omega)} \right).$$

Hence $u \in C^{\alpha, \alpha/2}(\overline{\Omega} \times [\sigma/4, T])$ for some $0 < \alpha < 1$, and

$$|u|_{\alpha, \overline{\Omega} \times [\sigma/4, T]} \leqslant C \|u\|_{W_q^{2,1}(\Omega \times (\sigma/4, T))} \leqslant C \left(|f|_{\alpha, \overline{Q}_T} + \|\varphi\|_{W_p^2(\Omega)} \right).$$

Take $\eta \in C^\infty([0,T])$ such that $0 \leqslant \eta \leqslant 1$, $\eta \equiv 1$ in $[\sigma, T]$ and $\eta \equiv 0$ in $[0, \sigma/2]$. Then there exists $C' > 0$ such that $|\eta'(t)| \leqslant C'/\sigma$ and $|\eta'|_{C^\alpha([0,T])} \leqslant C'/\sigma^{1+\alpha}$. The function $w = u\eta$ satisfies (2.16). According to Theorem 1.16, this problem has a unique solution $w \in C^{2+\alpha, 1+\alpha/2}(\overline{Q}_T)$, and

$$|w|_{2+\alpha, \overline{Q}_T} \leqslant C|f\eta + u\eta'|_{\alpha, \overline{Q}_T} \leqslant C\left(|f|_{\alpha, \overline{Q}_T} + \|\varphi\|_{W_p^2(\Omega)}\right).$$

This implies $u \in C^{2+\alpha, 1+\alpha/2}(\overline{\Omega} \times [\sigma, T])$ and (2.15) holds. $\qquad\square$

In the following, we attempt to give uniform estimates in time t of solution u and its derivatives for problem

$$\begin{cases} u_t + L[u] = f(x,t,u), & x \in \Omega, \quad t > 0, \\ B[u] = 0, & x \in \partial\Omega, \ t \geqslant 0, \\ u(x,0) = \varphi(x), & x \in \Omega, \end{cases} \tag{2.17}$$

where

$$L[u] = -a_{ij}(x)D_{ij}u + b_i(x)D_iu,$$

and $B[u] = u$, or $B[u] = \partial_{\boldsymbol{n}}u + b(x)u$ with $b \in C^{1+\alpha}(\partial\Omega)$ and $b \geqslant 0$.

Assume that $p > 1+n/2$, $\varphi \in W_p^2(\Omega)$ satisfies the compatibility conditions: $\varphi = 0$ on $\partial\Omega$ when $B[u] = u$, and $\partial_{\boldsymbol{n}}\varphi + b(x)\varphi = 0$ on $\partial\Omega$ when $B[u] = \partial_{\boldsymbol{n}}u + b(x)u$.

Function u is call a *global solution* of (2.17) if it is defined for all $(x,t) \in Q_\infty$ and satisfies (2.17) almost everywhere.

Theorem 2.11 (Regularity and uniform estimate) *Let Ω be of class $C^{2+\alpha}$, $a_{ij}, b_i \in C^\alpha(\overline{\Omega})$ and u be a global solution of (2.17) with $\underline{c} \leqslant u \leqslant \bar{c}$ for some $\underline{c}, \bar{c} \in \mathbb{R}$. Assume that $f \in L^\infty(Q_\infty \times [\underline{c}, \bar{c}])$ satisfies (2.1) with Q_T replaced by Q_∞. We further suppose $f(\cdot, u) \in C^{\alpha, \alpha/2}(\overline{\Omega} \times [h, h+3])$ uniformly in $u \in [\underline{c}, \bar{c}]$ and $h \geqslant 0$, i.e., there exists a constant C so that*

$$|f(x,t,u) - f(y,s,u)| \leqslant C(|x-y|^\alpha + |t-s|^{\alpha/2})$$

for all $(x,t), (y,s) \in \overline{\Omega} \times [h, h+3]$, $u \in [\underline{c}, \bar{c}]$ and all $h \geqslant 0$. Then, for any given $\tau > 0$, there is a constant $C(\tau)$ such that

$$\|u\|_{C^{2+\alpha, 1+\alpha/2}(\overline{\Omega} \times [\tau, \infty))} \leqslant C(\tau). \tag{2.18}$$

Proof. We only deal with the case $B[u] = \partial_{\boldsymbol{n}}u + b(x)u$. For any given $0 < T < \infty$, u solves the problem

$$\begin{cases} u_t + L[u] = f(x,t,u) & \text{in } Q_T, \\ B[u] = 0 & \text{on } S_T, \\ u(x,0) = \varphi(x) & \text{in } \Omega. \end{cases}$$

Thanks to $\underline{c} \leqslant u \leqslant \bar{c}$ and $f \in L^\infty(Q_\infty \times [\underline{c}, \bar{c}])$, we see that function $F(x,t) := f(x,t,u(x,t)) \in L^\infty(Q_\infty)$, and then $u \in W_p^{2,1}(Q_T)$ by the L^p theory (Theorem 1.2).

Therefore, $u \in C^{\alpha,\alpha/2}(\overline{Q}_T)$ due to $p > 1 + n/2$, and thereupon $f(x,t,u(x,t)) \in C^{\alpha,\alpha/2}(\overline{Q}_T)$. Thus, $u \in C^{2+\alpha,1+\alpha/2}(\overline{\Omega} \times (0,T])$ by Theorem 2.10.

Without loss of generality, we assume $\tau < 1$. For integer $k \geqslant 0$, we set

$$u^k(x,t) = u(x,k+t), \quad f^k(x,t) = f(x,k+t,u^k(x,t)) \quad \text{and} \quad \varphi^k(x) = u(x,k).$$

Then $\underline{c} \leqslant u^k, \varphi^k \leqslant \bar{c}$, f^k is bounded uniformly in k, and u^k satisfies

$$\begin{cases} u_t^k + L[u^k] = f^k(x,t), & x \in \Omega, \quad 0 < t \leqslant 3, \\ B[u^k] = 0, & x \in \partial\Omega, \ 0 < t \leqslant 3, \\ u^k(x,0) = \varphi^k(x), & x \in \Omega. \end{cases}$$

Clearly, $f^k \in C^{\alpha,\alpha/2}(\overline{\Omega} \times (0,3])$, $u^k \in C^{2+\alpha,1+\alpha/2}(\overline{\Omega} \times (0,3])$ for all integer $k \geqslant 0$. Take $\delta = \tau/2$ and $T = 3$ in Theorem 1.8. Then there is a constant C such that

$$\|u^k\|_{W_p^{2,1}(\Omega \times (\tau/2,3))} \leqslant C \left(\|f^k\|_{p,\Omega \times (0,3)} + \|u^k\|_{p,\Omega \times (0,3)} \right) \leqslant C$$

for all $k \geqslant 0$. Since $p > 1 + n/2$, the embedding theorem declares $|u^k|_{\alpha,\overline{\Omega} \times [\tau/2,3]} \leqslant C_1$. Assumptions on f show that $|f^k|_{\alpha,\overline{\Omega} \times [\tau/2,3]} \leqslant C_2$. Taking $t_0 = \tau/2$, $t_1 = \tau$ and $T = 3$ in (1.19), we get

$$|u^k|_{2+\alpha,\overline{\Omega} \times [\tau,3]} \leqslant C \left(|f^k|_{\alpha,\overline{\Omega} \times [\tau/2,3]} + |\underline{c}| + |\bar{c}| \right) \leqslant C_3 \quad \text{for all } k \geqslant 0.$$

Thus we have, noticing $u^k(x,t) = u(x,k+t)$,

$$\|u\|_{C^{2+\alpha,1+\alpha/2}(\overline{\Omega} \times [k+\tau,k+3])} \leqslant C_3 \quad \text{for all } k \geqslant 0.$$

Estimate (2.18) is obtained immediately. □

Remark 2.12 *An analogous theorem is valid for system version of (2.17).*

It can be seen from the proof of Theorem 2.11 that the following theorem holds.

Theorem 2.13 (Regularity and uniform estimate) *Let Ω be of class C^2, $a_{ij} \in C(\overline{\Omega})$, $b_i \in L^\infty(\Omega)$ and u be a global solution of (2.17) with $\underline{c} \leqslant u \leqslant \bar{c}$ for some $\underline{c}, \bar{c} \in \mathbb{R}$. Assume that $f \in L^\infty(Q_\infty \times [\underline{c},\bar{c}])$ satisfies (2.1) with Q_T replaced by Q_∞. Then, for any given $\tau > 0$, $p > 1$ and $0 < \alpha < 1$, there exist constants $C(\tau,p)$ and $C(\tau,\alpha)$ such that*

$$\|u\|_{W_p^{2,1}(\Omega \times [k+\tau,k+3])} \leqslant C(\tau,p) \quad \text{for all } k \geqslant 0,$$

and

$$\|u\|_{C^{1+\alpha,(1+\alpha)/2}(\overline{\Omega} \times [\tau,\infty))} \leqslant C(\tau,\alpha).$$

Theorem 2.11 and Theorem 2.13 play an important role in understanding longtime behaviours of solutions.

2.3 UNIFORM BOUNDS FROM BOUNDED L^p NORMS

In order to get the global existence of solutions of parabolic equations, some priori bounds of the L^∞ norm are needed. But for some interesting examples, it is easy to get L^p-bounds for some $1 \leqslant p < \infty$. Some special structure of the reaction term can be used to obtain uniform bounds. Here we shall state a result of Rothe ([95]).

Suppose that Ω is of class C^2 and $0 < T \leqslant \infty$. Let u satisfy

$$\begin{cases} u_t - d\Delta u = f(x,t,u), & x \in \Omega, \ 0 < t < T, \\ u(x,0) = u_0(x), & x \in \Omega, \end{cases}$$

and either

$$\partial_n u = 0, \quad x \in \partial\Omega, \ 0 \leqslant t < T,$$

or

$$u = 0, \quad x \in \partial\Omega, \ 0 \leqslant t < T.$$

We further assume that either

$$u \geqslant 0 \quad \text{and} \quad f(x,t,u) \leqslant c(x,t)(1+u)^m,$$

or

$$|f(x,t,u)| \leqslant c(x,t)(1+|u|)^m.$$

Define $U_0 = \|u_0\|_{\infty,\Omega}$,

$$C_q(T) = \sup_{0 \leqslant t < T} \|c(\cdot,t)\|_{q,\Omega} \quad \text{and} \quad U_p(T) = \sup_{0 \leqslant t < T} \|u(\cdot,t)\|_{p,\Omega}.$$

Theorem 2.14 (Uniform bound from bounded L^p norm [95, Proposition 1]) *Assume that $m \geqslant 1$, $q \geqslant 1$ and $p \geqslant 1/2$ satisfy*

$$m < 1 + p(2/n - 1/q).$$

Set

$$\eta = \exp\left(\frac{(m-1)(1+2/n)}{1+p(2/n-1/q)-m}\right),$$

$$\theta = \frac{1+2/n}{1+p(2/n-1/q)-m} \exp\left(\frac{(m-1)(1+2/n)}{1+2p(2/n-1/q)-m}\right).$$

Then there exists a constant $K = K(n,m,q,p,\Omega)$ so that

$$\sup_{0 \leqslant t < T} \|u(\cdot,t)\|_{\infty,\Omega} \leqslant K \left(\max\{1, C_q(T)\}\right)^\theta \left(\max\{1, U_0, U_p(T)\}\right)^\eta.$$

In applications, the most common situation is $q = \infty$. In such a case, if $m < 1 + 2p/n$ then uniform bound of $\|u(\cdot,t)\|_{\infty,\Omega}$ with respect to time t can be established once we get that of $\|u(\cdot,t)\|_{p,\Omega}$.

In the following three examples we always assume that Ω is of class C^2, initial data $u_0, v_0 \in C^2(\overline{\Omega})$ and $u_0, v_0 \geqslant 0$, $\not\equiv 0$ in Ω.

Example 2.1   *A typical example is a prey-predator model with homogeneous Neumann boundary conditions*

$$\begin{cases} u_t - d_1 \Delta u = u(a - u - hv), & x \in \Omega, \quad t > 0, \\ v_t - d_2 \Delta v = v(ku - b), & x \in \Omega, \quad t > 0, \\ \partial_{\boldsymbol{n}} u = \partial_{\boldsymbol{n}} v = 0, & x \in \partial\Omega, \ t > 0, \\ u(x,0) = u_0(x), \ \ v(x,0) = v_0(x), & x \in \Omega, \end{cases}$$

where d_1, d_2, a, h, k and b are positive constants, and $\partial_{\boldsymbol{n}} u_0 = \partial_{\boldsymbol{n}} v_0 = 0$ on $\partial\Omega$.

Let $T_{\max}$ denote the *maximal existence time*. Applying the strong maximum principle and comparison arguments, we assert

$$0 < u \leqslant \max\left\{a, \max_{\overline{\Omega}} u_0(x)\right\} =: B, \ \ v > 0 \ \ \text{in} \ \overline{\Omega} \times (0, T_{\max}).$$

Therefore

$$\begin{aligned} \frac{\mathrm{d}}{\mathrm{d}t} \int_\Omega (ku + hv) &= -b \int_\Omega (ku + hv) + \int_\Omega ku(a + b - u) \\ &\leqslant -b \int_\Omega (ku + hv) + k(a + b)B|\Omega|, \end{aligned}$$

which implies

$$\int_\Omega (ku + hv) \leqslant \int_\Omega (ku_0 + hv_0) + \frac{k(a + b)B|\Omega|}{b} =: C$$

for all $0 \leqslant t < T_{\max}$. Certainly

$$\sup_{0 \leqslant t < T_{\max}} \|v(\cdot, t)\|_{1,\Omega} \leqslant C/h.$$

Set $c(x,t) = ku(x,t) - b$. Then $\sup_{0 \leqslant t < T_{\max}} \|c(\cdot,t)\|_{\infty,\Omega} \leqslant kB + b$. We apply Theorem 2.14 to the equation for v with $q = \infty$ and $p = m = 1$, and then infer $\sup_{0 \leqslant t < T_{\max}} \|v(\cdot,t)\|_{\infty,\Omega} \leqslant C' < \infty$. As a consequence, $T_{\max} = \infty$, i.e., solution (u, v) exists globally in time t and is bounded uniformly.

Example 2.2   *The second example is a generalized autocatalytic chemical reaction model with homogeneous Dirichlet boundary conditions*

$$\begin{cases} u_t - d_1 \Delta u = -ku^r v^{m+\sigma}, & x \in \Omega, \quad t > 0, \\ v_t - d_2 \Delta v = hu^{r+\rho} v^m, & x \in \Omega, \quad t > 0, \\ u = v = 0, & x \in \partial\Omega, \ t > 0, \\ u(x,0) = u_0(x), \ \ v(x,0) = v_0(x), & x \in \Omega, \end{cases}$$

where $k, h > 0$, $r, m \geqslant 1$ and $\sigma, \rho \geqslant 0$ are constants, and $u_0 = v_0 = 0$ on $\partial\Omega$.

In the same way as above it can be shown that $u, v > 0$ and u is bounded in $\Omega \times (0, T_{\max})$. Here we only deal with the case that $\sigma, \rho > 0$. Multiplying two differential equations by u^ρ and v^σ, respectively, and then integrating the result over Ω by parts, we derive

$$\frac{1}{\rho + 1} \frac{\mathrm{d}}{\mathrm{d}t} \int_\Omega u^{\rho+1} = -d_1 \rho \int_\Omega |\nabla u|^2 u^{\rho-1} - k \int_\Omega u^{r+\rho} v^{m+\sigma},$$

$$\frac{1}{\sigma + 1} \frac{\mathrm{d}}{\mathrm{d}t} \int_\Omega v^{\sigma+1} = -d_2 \sigma \int_\Omega |\nabla v|^2 v^{\sigma-1} + h \int_\Omega u^{r+\rho} v^{m+\sigma}.$$

And then

$$\frac{\mathrm{d}}{\mathrm{d}t} \int_\Omega \left(\frac{h}{\rho + 1} u^{\rho+1} + \frac{k}{\sigma + 1} v^{\sigma+1} \right) \leqslant 0.$$

Therefore

$$\int_\Omega \left(\frac{h}{\rho + 1} u^{\rho+1} + \frac{k}{\sigma + 1} v^{\sigma+1} \right) \leqslant \int_\Omega \left(\frac{h}{\rho + 1} u_0^{\rho+1} + \frac{k}{\sigma + 1} v_0^{\sigma+1} \right),$$

which implies $\sup_{0 \leqslant t < T_{\max}} \|v(\cdot, t)\|_{\sigma+1, \Omega} \leqslant C < \infty$. When $m < 1 + 2(\sigma + 1)/n$, by applying Theorem 2.14 to the equation of v with $q = \infty$ and $p = \sigma + 1$ we ascertain $\sup_{0 \leqslant t < T_{\max}} \|v(\cdot, t)\|_{\infty, \Omega} \leqslant C$. As a result, $T_{\max} = \infty$ and $\sup_{0 \leqslant t < \infty} \|v(\cdot, t)\|_{\infty, \Omega} \leqslant C$.

Example 2.3 *The third example is a Brusselator system with homogeneous Neumann boundary conditions*

$$\begin{cases} u_t - d_1 \Delta u = a - (b+1)u + u^2 v, & x \in \Omega, \ t > 0, \\ v_t - d_2 \Delta v = bu - u^2 v, & x \in \Omega, \ t > 0, \\ \partial_{\boldsymbol{n}} u = \partial_{\boldsymbol{n}} v = 0, & x \in \Omega, \ t > 0, \\ u(x, 0) = u_0(x), \quad v(x, 0) = v_0(x), & x \in \Omega, \end{cases} \tag{2.19}$$

where coefficients are positive constants, $u_0, v_0 > 0$ in $\overline{\Omega}$ and $\partial_{\boldsymbol{n}} u_0 = \partial_{\boldsymbol{n}} v_0 = 0$ on $\partial\Omega$.

Theorem 2.15 ([95]) *Let (u, v) be a solution of (2.19). If $1 \leqslant n \leqslant 3$, then there exists a constant C depending on a, b and initial data such that*

$$0 < u(x, t) \leqslant C \quad \text{and} \quad 0 < v(x, t) \leqslant C \quad \text{for all } x \in \overline{\Omega}, \ t \geqslant 0.$$

Proof. In the following positive constants C_i depend only on a, b and initial data. By comparison arguments one can deduce $u \geqslant \delta > 0$ and $v > 0$ in $\overline{\Omega} \times [0, T_{\max})$. For any given $0 < \tau < T_{\max}$, it is not hard to show that

$$\max_{\overline{Q}_\tau} v(x, t) \leqslant \max\{\max_{\overline{\Omega}} v_0(x), \ b/\delta\} =: C_1, \tag{2.20}$$

and subsequently, $v \leqslant C_1$ in $\overline{\Omega} \times [0, T_{\max})$. It yields

$$a - (b+1)u + u^2 v \leqslant C_2 (1+u)^2.$$

Adding up the first two equations of (2.19) and integrating the result, we have

$$\frac{d}{dt}\int_\Omega (u+v) = a|\Omega| - \int_\Omega u,$$

and hence

$$\int_\Omega u \leqslant C_3 \quad \text{for all } t \in [0, T_{\max}).$$

Multiplying the equation $(u+v)_t - \Delta(d_1 u + d_2 v) = a - u$ by $u + v$ and integrating the result over Ω, we conclude

$$\frac{d}{dt}\int_\Omega \frac{(u+v)^2}{2} + \int_\Omega (d_1|\nabla u|^2 + (d_1+d_2)\nabla u \cdot \nabla v + d_2|\nabla v|^2) = \int_\Omega (a-u)(u+v).$$

The quadratic form in ∇u and ∇v is indefinite, however it can be estimated by

$$d_1|\nabla u|^2 + (d_1+d_2)\nabla u \cdot \nabla v + d_2|\nabla v|^2 \geqslant -\frac{(d_1-d_2)^2}{4d_1}|\nabla v|^2.$$

Consequently

$$\frac{d}{dt}\int_\Omega \frac{(u+v)^2}{2} + \int_\Omega (u+v)^2 - \frac{(d_1-d_2)^2}{4d_1}\int_\Omega |\nabla v|^2 \leqslant \int_\Omega (a+v)(u+v) \leqslant C_4. \quad (2.21)$$

On the other hand, multiplying the second equation of (2.19) by v and integrating the result, one has

$$\frac{d}{dt}\int_\Omega \frac{v^2}{2} + d_2 \int_\Omega |\nabla v|^2 = \int_\Omega uv(b-uv) \leqslant C_5.$$

This combined with (2.21) allows us to deduce

$$\left(1 + \frac{d}{dt}\right)\int_\Omega \left(2(u+v)^2 + \frac{(d_1-d_2)^2}{2d_1 d_2}v^2\right) \leqslant C_6,$$

which follows

$$\int_\Omega u^2 \leqslant C_7 \quad \text{for all } t \in [0, T_{\max}).$$

When $n = 1, 2, 3$, Theorem 2.14 can be applied to the equation for u with $q = \infty$ and $p = m = 2$, and then estimate $\sup_{0\leqslant t < T_{\max}} \|u(\cdot, t)\|_{\infty, \Omega} \leqslant C$ is obtained. Therefore $T_{\max} = \infty$ and $\sup_{0\leqslant t < \infty} \|u(\cdot, t)\|_{\infty, \Omega} \leqslant C$. The proof is complete. ☐

Another related works for obtaining L^∞-bounds from L^p-bounds were given in [4, 5] and [93, Theorem 16.4, Proposition 51.34]. Particularly, for a more general problem

$$\begin{cases} u_t - \Delta u = f(x, t, u, \nabla u), & x \in \Omega, \quad t > 0, \\ u = 0, & x \in \partial\Omega, \ t > 0, \\ u(x, 0) = u_0(x) \geqslant 0, & x \in \Omega, \end{cases} \quad (2.22)$$

where $f \in C^1$ satisfies growth condition $|f| \leqslant C(1+|u|^m+|\nabla u|^r)$ with $m < 1+2p/n$ and $r < 1+p/(n+p)$, the L^p-bound for solution of (2.22) guarantees the L^∞-bound. The similar conclusion is still valid when we consider (2.22) with nonlinear Neumann boundary condition $\partial_{\boldsymbol{n}} u = g(x, t, u)$ instead of homogeneous Dirichlet condition, provided that $g \in C^1$ satisfies growth condition $|g(\cdot, u)| \leqslant C(1+|u|^s)$ with $s < 1+p/n$ (see [8, 92, 93]).

EXERCISES

2.1 Prove Theorems 2.2 and 2.3.

2.2 Assume that $g \geqslant 0$ in $\Omega \times S_T$. Let $u, v \in C(\overline{Q}_T) \cap C^{2,1}(Q_T)$, $u \geqslant 0$ and u, v satisfy

$$
\begin{cases}
u_t - \Delta u - u^2 \displaystyle\int_\Omega u \mathrm{d}y \geqslant v_t - \Delta v - v^2 \int_\Omega v \mathrm{d}y & \text{in } Q_T, \\[2mm]
u - \displaystyle\int_\Omega g(y,x,t)u(y,t)\mathrm{d}y \geqslant v - \int_\Omega g(y,x,t)v(y,t)\mathrm{d}y & \text{on } S_T, \\[2mm]
u(x,0) \geqslant v(x,0) & \text{in } \Omega.
\end{cases}
$$

Prove $u \geqslant v$ in Q_T.

2.3 Prove (2.8).

2.4 Prove Lemma 2.8.

2.5 Assume that $u \in C([0,T], L^2(\Omega))$. Prove

$$
\lim_{h \to 0} \int_\Omega (u^h)_+^2(x,t) = \int_\Omega u_+^2(x,t).
$$

2.6 Prove Theorem 2.10 when $\mathscr{B}u = \partial_{\boldsymbol{n}} u + bu$.

2.7 Use the boundary condition $u = 0$ instead of $\partial_{\boldsymbol{n}} u + b(x)u = 0$ and assume $f(x,t,0) = 0$. Prove Theorem 2.11.

2.8 For system version of (2.17), state and prove a theorem same as Theorem 2.11.

2.9 Use $u_t - a_{ij}(x,t)D_{ij}u + b_i(x,t)D_i u$ instead of $u_t + L[u]$ in (2.17) and assume that $a_{ij}, b_i \in L^\infty(Q_\infty)$, and a_{ij} is uniformly continuous in Q_∞. Prove Theorem 2.13.

2.10 Let Γ_1 and Γ_2 be given in (1.3). Assume that, in (2.17), $B[u] = \partial_{\boldsymbol{n}} u + b(x)u$ on $\Gamma_1 \times \mathbb{R}_+$ with $b \geqslant 0$, $b \in C^{1+\alpha}(\Gamma_1)$, and $B[u] = u$ on $\Gamma_2 \times \mathbb{R}_+$. State and prove the conclusions similar to Theorem 2.11 and Theorem 2.13.

2.11 Let (u,v) be a local strong solution of

$$
\begin{cases}
u_t - d_1 \Delta u = -kuv^m, & x \in \Omega, \quad t > 0, \\
v_t - d_2 \Delta v = kuv^m, & x \in \Omega, \quad t > 0, \\
\partial_{\boldsymbol{n}} u = \partial_{\boldsymbol{n}} v = 0, & x \in \partial\Omega, \ t > 0, \\
u(x,0) = u_0(x), \quad v(x,0) = v_0(x), & x \in \Omega,
\end{cases}
$$

where $k > 0$ and $m \geqslant 1$. Prove that (u,v) exists globally and is bounded.

2.12 Prove estimate (2.20).

CHAPTER 3

Semilinear Parabolic Equations

This chapter is devoted to studying initial-boundary value problems of semilinear parabolic equations. Main contents are the upper and lower solutions method, monotone properties and convergence. Some examples will be given as applications.

It is shown by applying the fixed point theorem that if a problem admits the upper and lower solutions, then it must have a unique solution located between the upper and lower solutions. Then, by means of the upper and lower solutions, the monotone iterative sequences are constructed to get the unique solution between the upper and lower solutions. The monotone properties and convergence of solutions can also be studied via the upper and lower solutions method. This chapter further introduces the weak upper and lower solutions method, and it is proved that the minimum function of two upper solutions is a weak upper solution and the maximum function of two lower solutions is a weak lower solution.

Let us consider the initial-boundary value problem

$$\begin{cases} \mathscr{L}u = f(x,t,u) & \text{in } Q_T, \\ \mathscr{B}u = g & \text{on } S_T, \\ u(x,0) = \varphi(x) & \text{in } \Omega, \end{cases} \tag{3.1}$$

where $\mathscr{L}u = u_t - a_{ij}D_{ij}u + b_i D_i u$. As a consequence of Theorems 2.1–2.3, we have the following uniqueness result.

Theorem 3.1 (Uniqueness) *Under conditions of Theorem 2.1, problem* (3.1) *has at most one solution in the class of* $C(\overline{Q}_T) \cap C^{1,0}(Q_T) \cap W^{2,1}_{n+1,\,\mathrm{loc}}(Q_T)$ *when* $\mathscr{B}u = u$, *and in the class of* $C^{1,0}(\overline{Q}_T) \cap W^{2,1}_{n+1,\,\mathrm{loc}}(Q_T)$ *when* $\mathscr{B}u = \partial_{\boldsymbol{n}}u + bu$ *with* $b \geqslant 0$. *Under conditions of Theorem 2.2 when* $\mathscr{B}u = u$, *and conditions of Theorem 2.3 when* $\mathscr{B}u = \partial_{\boldsymbol{n}}u + bu$ *with* $b \geqslant 0$, *problem* (3.1) *has at most one solution in the class of* $C(\overline{Q}_T) \cap W^{2,1}_p(Q_T)$.

3.1 THE UPPER AND LOWER SOLUTIONS METHOD

The *upper and lower solutions method* provides a unified and powerful tool for the study of differential equations including ordinary differential equations, partial differential equations, functional differential equations, and combinations thereof. Many monographs and textbooks have an introduction to the upper and lower solutions method ([59, 88, 99, 124]), and there are also many papers devoted to this topic. The notion of upper and lower solutions was used by Keller and Cohen [57] in special cases, and general results were due to [6, 97].

The so-called *upper and lower solutions method* is, by defining and constructing proper upper and lower solutions, to ensure the existence of solutions. The upper and lower solutions method provides us with estimates of solutions as well.

Definition 3.2 *Let $u \in C(\overline{Q}_T) \cap C^{2,1}(Q_T)$ when $\mathscr{B}u = u$, and $u \in C^{1,0}(\overline{Q}_T) \cap C^{2,1}(Q_T)$ when $\mathscr{B}u = \partial_{\boldsymbol{n}} u + bu$. Such a function u is referred to as an* upper solution *(a* lower solution*) of* (3.1) *if*

$$
\begin{cases}
\mathscr{L}u \geqslant (\leqslant) f(x,t,u) & in \quad Q_T, \\
\mathscr{B}u \geqslant (\leqslant) g & on \quad S_T, \\
u(x,0) \geqslant (\leqslant) \varphi(x) & in \quad \Omega.
\end{cases}
$$

We have the following theorem as a direct consequence of Theorem 2.1.

Theorem 3.3 (Ordering of upper and lower solutions) *Let $a_{ij}, b_i \in L^\infty(Q_T)$, $\mathscr{L}$ be parabolic in Q_T, and Ω be of class C^2 when $\mathscr{B}u = \partial_{\boldsymbol{n}} u + bu$. Let $\bar{u}$ and $\underline{u}$ be an upper solution and a lower solution of* (3.1)*, respectively. If f satisfies* (2.1) *with $\underline{c} = \min_{\overline{Q}_T} \min\{\bar{u}, \underline{u}\}$ and $\bar{c} = \max_{\overline{Q}_T} \max\{\bar{u}, \underline{u}\}$, then $\bar{u} \geqslant \underline{u}$ in Q_T.*

Remark 3.4 *The upper and lower solutions defined in Definition 3.2 are called the classical upper and lower solutions. We can also define the strong upper and lower solutions (e.g., the space $C^{2,1}(Q_T)$ is replaced by $W_p^{2,1}(Q_T)$ or $W_{p,\mathrm{loc}}^{2,1}(Q_T)$) and Theorem 3.3 still holds provided that the comparison principle is valid. For example, under conditions of Theorem 2.1, the upper and lower solutions $\bar{u}, \underline{u} \in C^{1,0}(\overline{Q}_T) \cap C^{2,1}(Q_T)$ can be reduced to $\bar{u}, \underline{u} \in C^{1,0}(\overline{Q}_T) \cap W_{n+1,\mathrm{loc}}^{2,1}(Q_T)$ when $\mathscr{B}u = \partial_{\boldsymbol{n}} u + bu$, and Theorem 3.3 holds.*

Moreover, the upper and lower solutions method established later for classical upper and lower solutions is still valid for such strong upper and lower solutions.

We state two requirements.

(G1) The condition **(A)** holds, $\varphi \in W_p^2(\Omega)$ with $p > 1 + n/2$, $b_i \in L^\infty(Q_T)$ and Ω is of class C^2. Moreover,

    (i) $g \in W_p^{2,1}(Q_T)$ and $\varphi(x) = g(x,0)$ on $\partial\Omega$ when $\mathscr{B}u = u$;

    (ii) $g \in W_q^{2,1}(Q_T)$ with $q > n+2$ and $q \geqslant p$, $b \in C^{1,1/2}(\overline{S}_T)$ and $\partial_{\boldsymbol{n}}\varphi + b(x,0)\varphi = g(x,0)$ on $\partial\Omega$ when $\mathscr{B}u = \partial_{\boldsymbol{n}} u + bu$.

(G2) Conditions **(C)** and **(E)** hold, Ω is of class $C^{2+\alpha}$, $b \in C^{1+\alpha,\,(1+\alpha)/2}(\overline{S}_T)$, and (i) and (ii) of **(G1)** hold.

We first use the Schauder fixed point theorem to prove the existence of solutions.

Theorem 3.5 (The upper and lower solutions method) *Let the condition* **(G1)** *hold, and $\bar{u}$ and $\underline{u}$ be an upper solution and a lower solution of (3.1), respectively. Define $\underline{c}$ and $\bar{c}$ as in Theorem 3.3. If f satisfies (2.1) and $f \in L^\infty(Q_T \times \langle \underline{c}, \bar{c} \rangle)$, then, in the interval $\langle \underline{u}, \bar{u} \rangle$, problem (3.1) has a unique solution u and $u \in W_p^{2,1}(Q_T)$, where*

$$\langle \underline{u}, \bar{u} \rangle = \left\{ u \in C(\overline{Q}_T) : \underline{u} \leqslant u \leqslant \bar{u} \ \text{ in } Q_T \right\}.$$

Proof. By Theorem 3.3, $\bar{u} \geqslant \underline{u}$. We only deal with the case $\mathscr{B}u = u$.

Step 1. For any given $v \in \langle \underline{u}, \bar{u} \rangle$, the linear problem

$$\begin{cases} \mathscr{L}u + Mu = f(x,t,v(x,t)) + Mv(x,t) & \text{in } Q_T, \\ u = g & \text{on } S_T, \\ u(x,0) = \varphi & \text{in } \Omega \end{cases}$$

has a unique solution $u \in W_p^{2,1}(Q_T)$, and

$$\|u\|_{W_p^{2,1}(Q_T)} \leqslant C \left(\|f(\cdot, v(\cdot))\|_{p,Q_T} + M\|v\|_{p,Q_T} + \|\varphi\|_{W_p^2(\Omega)} + \|g\|_{W_p^{2,1}(Q_T)} \right)$$

$$\leqslant C_1 \left(\|f\|_{\infty,Q_T \times \langle \underline{c}, \bar{c} \rangle} + |\underline{c}| + |\bar{c}| + \|\varphi\|_{W_p^2(\Omega)} + \|g\|_{W_p^{2,1}(Q_T)} \right)$$

$$=: C_2 \ \text{ independent of } v \in \langle \underline{u}, \bar{u} \rangle$$

by the L^p theory (Theorem 1.1). Note that $p > 1 + n/2$. Using the above estimate and the embedding theorem, we have in addition that $|u|_{\alpha,\overline{Q}_T} \leqslant C_3$ for all $v \in \langle \underline{u}, \bar{u} \rangle$. Moreover, similar to Theorem 2.10 we have $u \in C^{1+\alpha,\,(1+\alpha)/2}(Q_T) \cap W_{n+1,\,\text{loc}}^{2,1}(Q_T)$.

We define an operator $\mathscr{F}$ from $\langle \underline{u}, \bar{u} \rangle$ to $C^{\alpha,\alpha/2}(\overline{Q}_T)$ by $\mathscr{F}(v) = u$. Then $|\mathscr{F}(v)|_{\alpha,\overline{Q}_T} \leqslant C_3$ for all $v \in \langle \underline{u}, \bar{u} \rangle$.

Step 2. We claim that $\mathscr{F} : \langle \underline{u}, \bar{u} \rangle \to C(\overline{Q}_T)$ is a compact operator. Recalling the estimate $|\mathscr{F}(v)|_{\alpha,\overline{Q}_T} \leqslant C_3$ for all $v \in \langle \underline{u}, \bar{u} \rangle$, it suffices to prove that $\mathscr{F} : \langle \underline{u}, \bar{u} \rangle \to C(\overline{Q}_T)$ is continuous. For $v_i \in \langle \underline{u}, \bar{u} \rangle$, $i = 1, 2$, let $u_i = \mathscr{F}(v_i)$ and $w = u_1 - u_2$. Then w satisfies

$$\begin{aligned} \mathscr{L}w + Mw = f(x,t,v_1) + Mv_1 - f(x,t,v_2) - Mv_2 & \quad \text{in } Q_T, \\ w = 0 & \quad \text{on } S_T, \\ w(x,0) = 0 & \quad \text{in } \Omega. \end{aligned}$$

The assumption on f implies

$$|f(x,t,v_1) + Mv_1 - f(x,t,v_2) - Mv_2| \leqslant 2M|v_1 - v_2|.$$

Then we have

$$\|w\|_{W_p^{2,1}(Q_T)} \leqslant 2MC\|v_1 - v_2\|_{p,Q_T} \leqslant C_1\|v_1 - v_2\|_\infty$$

by the L^p theory (Theorem 1.1). As $p > 1 + n/2$, the embedding theorem avers

$$|w|_{\alpha,\overline{Q}_T} \leqslant C_1\|v_1 - v_2\|_\infty \implies \|w\|_{C(\overline{Q}_T)} \leqslant C_1\|v_1 - v_2\|_\infty,$$

where C_1 is independent of v_1 and v_2. Thus $\mathscr{F} : \langle \underline{u}, \bar{u} \rangle \to C(\overline{Q}_T)$ is continuous.

Step 3. Now we prove $\mathscr{F} : \langle \underline{u}, \bar{u} \rangle \to \langle \underline{u}, \bar{u} \rangle$. Suppose $v \in \langle \underline{u}, \bar{u} \rangle$ and $u = \mathscr{F}(v)$. Let $w = \bar{u} - u$. Then w satisfies

$$\mathscr{L}w + Mw \geqslant f(x,t,\bar{u}) - f(x,t,v) + M(\bar{u} - v) \geqslant 0 \quad \text{in } Q_T,$$
$$w \geqslant 0 \qquad\qquad\qquad\qquad\qquad \text{on } S_T,$$
$$w(x,0) \geqslant 0 \qquad\qquad\qquad\qquad\qquad \text{in } \Omega.$$

The maximum principle gives $w \geqslant 0$, i.e., $u \leqslant \bar{u}$. Likewise, $\underline{u} \leqslant u$. Consequently $\mathscr{F} : \langle \underline{u}, \bar{u} \rangle \to \langle \underline{u}, \bar{u} \rangle$. Note that $\langle \underline{u}, \bar{u} \rangle$ is a bounded and closed convex set of $C(\overline{Q}_T)$. It follows that $\mathscr{F}$ has at least one fixed point $u \in \langle \underline{u}, \bar{u} \rangle$ by the Schauder fixed point theorem (Corollary A.3). Clearly, such a u solves problem (3.1), and $u \in W_p^{2,1}(Q_T)$ because $u = \mathscr{F}(u) \in W_p^{2,1}(Q_T)$.

Step 4. The uniqueness of solutions follows from Theorem 3.1. $\qquad\square$

Now we use the monotone iterative method to prove existence of solutions. This approach is convenient for numerical calculations.

Theorem 3.6 (The upper and lower solutions method) *Let the condition* **(G2)** *hold, and $\bar{u}$ and $\underline{u}$ be an upper solution and a lower solution of (3.1), respectively. Define $\underline{c}$ and $\bar{c}$ as in Theorem 3.3. If f satisfies (2.1) and $f(\cdot, u) \in C^{\alpha,\alpha/2}(\overline{Q}_T)$ is uniformly with respect to $u \in \langle \underline{u}, \bar{u} \rangle$, then there exist two sequences $\{\bar{u}_i\}$ and $\{\underline{u}_i\}$ such that*

$$\underline{u} \leqslant \underline{u}_1 \leqslant \ldots \leqslant \underline{u}_i \leqslant \bar{u}_i \leqslant \ldots \leqslant \bar{u}_1 \leqslant \bar{u} \quad \text{for all } i \geqslant 2,$$

and

$$\lim_{i\to\infty} \underline{u}_i = \lim_{i\to\infty} \bar{u}_i =: u$$

in $C^{1+\alpha,(1+\alpha)/2}(\overline{Q}_T) \cap C^{2+\beta,1+\beta/2}(Q)$ for any $Q \Subset Q_T$ and any $0 < \beta < \alpha$, here u is the unique solution of (3.1) in the interval $\langle \underline{u}, \bar{u} \rangle$.

Furthermore, $u \in C^{2+\alpha,1+\alpha/2}(\overline{Q}_T)$ if either $\mathscr{B}u = \partial_{\boldsymbol{n}}u + bu$, or $\mathscr{B}u = u$ and the following compatibility condition holds:

$$g_t - a_{ij}D_{ij}\varphi + b_i D_i\varphi = f(x,t,\varphi(x)) \quad \text{on } \partial\Omega \times \{0\}. \tag{3.2}$$

Proof. By Theorem 3.3, it holds $\bar{u} \geqslant \underline{u}$.

Step 1. Note that **(G2)** implies **(G1)** for each $p > 1$. In the proof of Theorem 3.5 we have obtained an operator $\mathscr{F} : \langle \underline{u}, \bar{u} \rangle \to W_p^{2,1}(Q_T)$ satisfying $\mathscr{F}(\underline{u}) \geqslant \underline{u}$ and $\mathscr{F}(\bar{u}) \leqslant \bar{u}$, and $\mathscr{F}(v)$ is in a bounded set of $W_p^{2,1}(Q_T)$ when v varies in $\langle \underline{u}, \bar{u} \rangle$. Using the same argument as in Step 3 in proving Theorem 3.5, we can prove that $\mathscr{F}$ is monotonically increasing, i.e., $v, w \in \langle \underline{u}, \overline{u} \rangle$ with $v \geqslant w$ implies $\mathscr{F}(v) \geqslant \mathscr{F}(w)$.

Step 2. Define two sequences $\{\overline{u}_i\}$ and $\{\underline{u}_i\}$ by

$$\overline{u}_1 = \mathscr{F}(\overline{u}), \quad \overline{u}_{i+1} = \mathscr{F}(\overline{u}_i), \quad \underline{u}_1 = \mathscr{F}(\underline{u}), \quad \underline{u}_{i+1} = \mathscr{F}(\underline{u}_i).$$

Since $\mathscr{F}$ is monotonically increasing, it follows that, by the inductive process,

$$\underline{u} \leqslant \underline{u}_1 \leqslant \ldots \leqslant \underline{u}_i \leqslant \ldots \overline{u}_i \leqslant \ldots \overline{u}_1 \leqslant \overline{u}.$$

Then we have

$$\lim_{i \to \infty} \overline{u}_i = \hat{u} \quad \text{and} \quad \lim_{i \to \infty} \underline{u}_i = \tilde{u} \quad \text{pointwisely in } \overline{Q}_T.$$

We now show that $\lim_{i \to \infty} \overline{u}_i = \hat{u}$ and $\lim_{i \to \infty} \underline{u}_i = \tilde{u}$ in $C^{1+\alpha, (1+\alpha)/2}(\overline{Q}_T) \cap C^{2+\beta, 1+\beta/2}(Q)$ for any $Q \Subset Q_T$ and $0 < \beta < \alpha$. Once this is done, it is not difficult to see that both $\hat{u}$ and $\tilde{u}$ are solutions of (3.1) by letting $i \to \infty$. We only handle $\hat{u}$ since $\tilde{u}$ can be handled in a similar way. Owing to $0 < \alpha < 1$, we can take a large p such that $0 < \alpha < 1 - (n+2)/p$.

Using the uniform estimate of $\|\overline{u}_i\|_{W_p^{2,1}(Q_T)}$ and embedding theorem (Theorem A.7 (3)) we can show that $\overline{u}_i \to \hat{u}$ in $C^{1+\alpha, (1+\alpha)/2}(\overline{Q}_T)$ for all $0 < \alpha < 1 - (n+2)/p$.

Set $f_i(x, t) = f(x, t, \overline{u}_i(x, t))$. Recall that f satisfies (2.1) and $f(\cdot, u) \in C^{\alpha, \alpha/2}(\overline{Q}_T)$ is uniformly with respect to $u \in \langle \underline{u}, \overline{u} \rangle$. We see that $f_i \in C^{\alpha, \alpha/2}(\overline{Q}_T)$ and $|f_i|_{\alpha, \overline{Q}_T} \leqslant C$ which is independent of i. Hence, $\overline{u}_i \in C^{2+\alpha, 1+\alpha/2}(Q_T)$ and for any $Q \Subset Q_T$,

$$|\overline{u}_i|_{2+\alpha, \overline{Q}} \leqslant C \left(|f_i|_{\alpha, \overline{Q}_T} + \|\overline{u}_i\|_{p, Q_T} \right) \leqslant C(Q)$$

by Theorem 1.19. Therefore $\overline{u}_i \to \hat{u}$ in $C^{2+\beta, 1+\beta/2}(Q)$ for any $0 < \beta < \alpha$.

Step 3. The uniqueness of solutions is guaranteed by Theorem 3.1.

Step 4. If $\mathscr{B}u = \partial_{\boldsymbol{n}} u + bu$, then $u \in C^{2+\alpha, 1+\alpha/2}(\overline{Q}_T)$ by Theorem 1.17. If $\mathscr{B}u = u$ and (3.2) holds, then $u \in C^{2+\alpha, 1+\alpha/2}(\overline{Q}_T)$ by Theorem 1.16. □

3.2 SOME EXAMPLES

We first give a basic fact which will be often used later.

Lemma 3.7 *Let Ω be of class C^1, $u, v \in C^1(\overline{\Omega})$ satisfy $u = v = 0$ on $\partial\Omega$ and $v > 0$ in Ω. If $\partial_{\boldsymbol{n}} v < 0$ on $\partial\Omega$, then there exists a constant $K > 0$ such that $u \leqslant Kv$ in $\overline{\Omega}$.*

Proof. Owing to $u, v \in C^1(\overline{\Omega})$ and $\partial_{\boldsymbol{n}} v < 0$ on $\partial\Omega$, there exists $K_1 > 0$ for which $\partial_{\boldsymbol{n}} u - K_1 \partial_{\boldsymbol{n}} v > 0$ on $\partial\Omega$. Noting $u - K_1 v = 0$ on $\partial\Omega$, we can find a Ω-neighbourhood V of $\partial\Omega$ such that $u - K_1 v \leqslant 0$ in V. Because of $u, v \in C^1(\Omega \setminus V)$ and $v > 0$ in $\Omega \setminus V$, there is a positive constant K_2 such that $u \leqslant K_2 v$ in $\Omega \setminus V$. Take $K = K_1 + K_2$. Then the desired conclusion holds. See Fig. 3.1. □

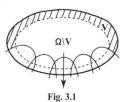

Fig. 3.1

Example 3.1 *Let $\underline{u}$ be a lower solution of* (3.1). *Assume that* **(G1)** *holds, and f satisfies* (2.1) *for $\underline{c} = \inf_{Q_T} \underline{u}(x,t)$ and any $\bar{c} > \underline{c}$. If $f \in L^\infty(Q_T \times (\underline{c}, \infty))$, then problem* (3.1) *has a unique solution u, and*

$$u \geqslant \underline{u} \quad in \ \overline{Q}_T. \tag{3.3}$$

In fact, let $C = \sup_{\overline{Q}_T \times (\underline{c}, \infty)} f(x, t, u)$. Then the linear problem

$$\begin{cases} \mathscr{L}u = C & in \ Q_\infty, \\ \mathscr{B}u = g & on \ S_\infty, \\ u(x, 0) = \varphi(x) & in \ \Omega \end{cases}$$

admits a unique solution denoted by $\bar{u}$. Obviously, $\bar{u}$ is an upper solution of (3.1). Hence (3.1) has a unique solution u and (3.3) holds by the upper and lower solutions method.

In the sequel of this section, we always assume that Ω is of class C^2, the initial datum $u_0 \in W_p^2(\Omega)$ with $p > n + 2$ and satisfies $\partial_{\boldsymbol{n}} u_0 + b u_0 = 0$ on $\partial\Omega$.

Example 3.2 *Let $a > 0$ be a constant and $u_0 \geqslant 0, \not\equiv 0$. Then the problem*

$$\begin{cases} u_t - \Delta u = u(a - u), & x \in \Omega, \quad t > 0, \\ \partial_{\boldsymbol{n}} u = 0, & x \in \partial\Omega, \ t > 0, \\ u(x, 0) = u_0(x), & x \in \Omega \end{cases} \tag{3.4}$$

has a unique positive global solution u, and

$$\lim_{t \to \infty} u(x, t) = a \quad uniformly \ in \ \overline{\Omega}.$$

In fact, for any given $z_0 > 0$, the following initial value problem

$$z'(t) = z(a - z), \quad z(0) = z_0$$

has a unique positive solution $z = z(t; z_0)$ which satisfies $\lim_{t \to \infty} z(t; z_0) = a$. Take $k = \max_{\overline{\Omega}} u_0(x)$. Then $k > 0$ and $\bar{u} = z(t; k)$ is an upper solution of (3.4). Obviously, $\underline{u} = 0$ is a lower solution of (3.4). So, (3.4) has a unique solution u, and

$$0 \leqslant u(x, t) \leqslant z(t; k) \quad for \ all \ x \in \overline{\Omega}, \ t \geqslant 0$$

by the upper and lower solutions method. Moreover, the maximum principle confirms

$$u(x, t) > 0, \quad x \in \overline{\Omega}, \ t > 0.$$

Take $\delta > 0$. Then $u(x, \delta) > 0$ in $\overline{\Omega}$ and the function $w(x, t) = u(x, t + \delta)$ satisfies

$$\begin{cases} w_t - \Delta w = w(a - w), & x \in \Omega, \quad t > 0, \\ \partial_{\boldsymbol{n}} w = 0, & x \in \partial\Omega, \ t > 0, \\ w(x, 0) = u(x, \delta), & x \in \Omega. \end{cases}$$

Let us define
$$m = \min_{\overline{\Omega}} u(x, \delta), \quad M = \max_{\overline{\Omega}} u(x, \delta).$$

Then $m, M > 0$, and comparison argument leads to
$$z(t; m) \leqslant w(x, t) = u(x, t + \delta) \leqslant z(t; M) \quad \text{for all } t \geqslant 0.$$

Notice $\lim_{t \to \infty} z(t; m) = \lim_{t \to \infty} z(t; M) = a$. It follows that $\lim_{t \to \infty} u(x, t) = a$ uniformly in $\overline{\Omega}$.

Example 3.3 *Let us deal with the following problem*
$$\begin{cases} u_t - \Delta u = c(e^{au} - 1), & x \in \Omega, \quad t > 0, \\ \partial_n u + bu = 0, & x \in \partial\Omega, \ t > 0, \\ u(x, 0) = u_0(x), & x \in \Omega, \end{cases} \tag{3.5}$$

where $u_0(x) > 0$, a, c and b are constants, meanwhile, $a, c > 0$ and $b \geqslant 0$.

We first discuss the local solution of (3.5). Take $B = \max_{\overline{\Omega}} u_0(x)$, and $A > 0$ is a constant determined later. Set $\overline{u} = At + B$. Then $\overline{u}$ is an upper solution of (3.5) in Q_T as long as
$$A \geqslant c(e^{a(At+B)} - 1) \quad \text{for } 0 < t \leqslant T.$$
Fix $A > c(e^{a(1+B)} - 1)$. Then the above inequality holds for $T = 1/A$. Obviously, $\underline{u} = 0$ is a lower solution of (3.5). Hence, problem (3.5) has a unique solution u defined in $\Omega \times (0, 1/A]$ by the upper and lower solutions method.

Let λ_1 be the *first eigenvalue* of
$$\begin{cases} -\Delta\phi = \lambda\phi & \text{in } \Omega, \\ \partial_n\phi + b\phi = 0 & \text{on } \partial\Omega, \end{cases} \tag{3.6}$$

and ϕ_1 be the *positive eigenfunction* corresponding to λ_1. Then $\phi_1(x) > 0$ in $\overline{\Omega}$. We call (λ_1, ϕ_1) the *first eigen-pair* of (3.6).

Theorem 3.8 *Assume $ac < \lambda_1$. When the initial datum u_0 is suitably small, the unique solution u of (3.5) satisfies $\lim_{t \to \infty} u(x, t) = 0$ uniformly in $\overline{\Omega}$.*

Proof. To this aim, we will seek an upper solution of (3.5) with the form: $\overline{u} = \rho e^{-kt}\phi_1(x)$, where ρ and k are positive constants to be determined. It suffices to verify that
$$\begin{cases} \overline{u}_t - \Delta\overline{u} \geqslant c(e^{a\overline{u}} - 1), & x \in \Omega, \quad t > 0, \\ \partial_n\overline{u} + b\overline{u} \geqslant 0, & x \in \partial\Omega, \ t > 0, \\ \overline{u}(x, 0) \geqslant u_0(x), & x \in \Omega. \end{cases} \tag{3.7}$$

A straightforward computation shows that (3.7) is satisfied as long as
$$\begin{cases} -k\rho e^{-kt}\phi_1 + \lambda_1\rho e^{-kt}\phi_1 \geqslant c(\exp\{a\rho e^{-kt}\phi_1\} - 1), & x \in \Omega, \ t > 0, \\ \rho\phi_1(x) \geqslant u_0(x), & x \in \Omega. \end{cases} \tag{3.8}$$

A simple analysis indicates that, for any small $\varepsilon > 0$, there exists $\rho_0(\varepsilon) > 0$ such that, when $0 < \rho \leqslant \rho_0(\varepsilon)$,

$$\exp\{a\rho e^{-kt}\phi_1(x)\} - 1 \leqslant (1 + \varepsilon)a\rho e^{-kt}\phi_1(x) \quad \text{for all } x \in \overline{\Omega}, \ t \geqslant 0.$$

Therefore the first inequality of (3.8) holds provided

$$-k\rho e^{-kt}\phi_1(x) + \lambda_1\rho e^{-kt}\phi_1(x) \geqslant c(1 + \varepsilon)a\rho e^{-kt}\phi_1(x) \quad \text{for all } x \in \overline{\Omega}, \ t \geqslant 0,$$

which is equivalent to

$$\lambda_1 - k \geqslant c(1 + \varepsilon)a.$$

As $ac < \lambda_1$, we can select $0 < k, \varepsilon \ll 1$ so that the above inequality holds. For such determined constants k and ε, we take $\rho_0(\varepsilon) > 0$ as above and $\bar{u} = \rho e^{-kt}\phi_1(x)$ with $0 < \rho \leqslant \rho_0(\varepsilon)$. Then $\bar{u}$ is an upper solution of (3.5) if $u_0 \leqslant \rho\phi_1$ in $\overline{\Omega}$, which in turn implies $0 \leqslant u(x,t) \leqslant \rho e^{-kt}\phi_1(x) \to 0$ as $t \to \infty$. □

Example 3.4 *Now we study a problem as follows*

$$\begin{cases} u_t - \Delta u = u(1 - u)(u - a), & x \in \Omega, \quad t > 0, \\ \partial_n u = 0, & x \in \partial\Omega, \ t > 0, \\ u(x, 0) = u_0(x) \geqslant 0, & x \in \Omega, \end{cases} \quad (3.9)$$

where $0 < a < 1$ is a constant.

Theorem 3.9 *Problem (3.9) has a unique non-negative global solution u, and*

(1) *if $\max_{\overline{\Omega}} u_0(x) < a$, then $\lim_{t\to\infty} u(x,t) = 0$ uniformly in $\overline{\Omega}$;*

(2) *if $\min_{\overline{\Omega}} u_0(x) > a$, then $\lim_{t\to\infty} u(x,t) = 1$ uniformly in $\overline{\Omega}$.*

Proof. Take $K > 1 + \max_{\overline{\Omega}} u_0(x)$. Then any solution u of (3.9) satisfies $0 \leqslant u \leqslant K$ by the maximum principle. Obviously, $\bar{u} = K$ and $\underline{u} = 0$ are an upper solution and a lower solution of (3.9), respectively. Therefore (3.9) has a unique solution u which satisfies $0 \leqslant u \leqslant K$ by the upper and lower solutions method.

Let $z_0 > 0$ and $z(t; z_0)$ be the unique solution of

$$z' = z(1 - z)(z - a), \quad t > 0; \quad z(0) = z_0.$$

Then $\lim_{t\to\infty} z(t; z_0) = 0$ when $z_0 < a$, while $\lim_{t\to\infty} z(t; z_0) = 1$ when $z_0 > a$. Take $M = \max_{\overline{\Omega}} u_0(x)$ and $m = \min_{\overline{\Omega}} u_0(x)$. Then $z(t; M)$ and $z(t; m)$ are an upper solution and a lower solution of (3.9), respectively. So the unique solution u of (3.9) satisfies $z(t; m) \leqslant u(x, t) \leqslant z(t; M)$ by the upper and lower solutions method.

If $M < a$, then $m < a$ and $\lim_{t\to\infty} z(t; M) = \lim_{t\to\infty} z(t; m) = 0$. Accordingly, $\lim_{t\to\infty} u(x, t) = 0$ uniformly in $\overline{\Omega}$. If $m > a$, then $M > a$ and $\lim_{t\to\infty} z(t; m) = \lim_{t\to\infty} z(t; M) = 1$. Accordingly, $\lim_{t\to\infty} u(x, t) = 1$ uniformly in $\overline{\Omega}$. □

3.3 MONOTONICITY AND CONVERGENCE

In this section, we discuss monotonicity and convergence of solutions to the following problem

$$
\begin{cases}
u_t + L[u] = f(x, u), & x \in \Omega, \quad t > 0, \\
B[u] = g(x), & x \in \partial\Omega, \ t > 0, \\
u(x, 0) = \varphi(x), & x \in \Omega,
\end{cases}
\tag{3.10}
$$

where

$$
L[u] = -a_{ij}(x)D_{ij}u + b_i(x)D_i u,
$$

and $B[u] = u$, or $B[u] = \partial_{\boldsymbol{n}}u + b(x)u$ with $b \in C^{1+\alpha}(\partial\Omega)$ and $b \geqslant 0$, a_{ij} satisfies the condition (**A**), $b_i \in C(\overline{\Omega})$.

The equilibrium problem of (3.10) is

$$
\begin{cases}
L[u] = f(x, u) & \text{in } \Omega, \\
B[u] = g(x) & \text{on } \partial\Omega.
\end{cases}
\tag{3.11}
$$

Let us define an upper (a lower) solution of (3.11) as follows.

Definition 3.10 *Let $u \in C(\overline{\Omega}) \cap C^2(\Omega)$ when $B[u] = u$, and $u \in C^1(\overline{\Omega}) \cap C^2(\Omega)$ when $B[u] = \partial_{\boldsymbol{n}}u + b(x)u$. Such a function u is called an* upper solution (*a* lower solution) *of (3.11) if*

$$
\begin{cases}
L[u] \geqslant (\leqslant) f(x, u) & \text{in } \Omega, \\
B[u] \geqslant (\leqslant) g(x) & \text{on } \partial\Omega.
\end{cases}
$$

When u is both upper and lower solutions, we say that u is a solution of (3.11).

An upper solution (a lower solution) is called a strict upper solution (strict lower solution) *if it is not a solution.*

We first study the monotonicity of solutions in time t.

Theorem 3.11 (Monotonicity in time t) *Assume that Ω is of class C^2, and f satisfies (2.1) for any $\underline{c} < \bar{c}$. If the initial datum φ of (3.10) is a lower (an upper) solution of (3.11), and problem (3.10) has a solution $u \in C(\overline{Q_T}) \cap C^{2,1}(Q_T)$ for some $T > 0$. Then u is monotonically increasing (decreasing) in time t, i.e., $0 \leqslant s < t < T$ implies $u(x, s) \leqslant (\geqslant) u(x, t)$ for all $x \in \Omega$.*

Furthermore, if the initial datum φ of (3.10) is a strict lower (upper) solution of (3.11), then u is strictly monotonically increasing (decreasing) in time t, i.e., $0 \leqslant s < t < T$ implies $u(x, s) < (>) u(x, t)$ for all $x \in \Omega$.

Proof. Step 1. Obviously, φ is a lower solution of (3.10). Because the function f

is locally Lipschitz continuous in u, we have that $u(x, t) \geqslant \varphi(x)$ in $\overline{\Omega} \times [0, T)$ by Theorem 3.3. Fix $0 < \delta < T$ and consider the problem

$$\begin{cases} w_t + L[w] = f(x, w), & x \in \Omega, \quad 0 < t < T - \delta, \\ B[w] = g(x), & x \in \partial\Omega, \ 0 < t < T - \delta, \\ w(x, 0) = u(x, \delta), & x \in \Omega. \end{cases} \tag{3.12}$$

Obviously, $w(x, t) := u(x, \delta + t)$ is a solution of (3.12). Note $u(x, \delta) \geqslant \varphi(x)$. We have $w \geqslant u$ in $\Omega \times [0, T - \delta)$ by applying the comparison principle to (3.10) and (3.12). That is, $u(x, \delta + t) \geqslant u(x, t)$ for all $x \in \Omega$, $0 \leqslant t < T - \delta$.

Step 2. In addition, if φ is a strict lower solution of (3.11), then there is $x_0 \in \Omega$ such that $L[\varphi](x_0) < f(x_0, \varphi(x_0))$, or there is $x_0 \in \partial\Omega$ such that $B[\varphi](x_0) < g(x_0)$. Let $v(x, t) = u(x, t) - \varphi(x)$. Then

$$\begin{cases} v_t + L[v] + Mv \geqslant 0, & x \in \Omega, \quad 0 < t < T, \\ B[v] \geqslant 0, & x \in \partial\Omega, \ 0 < t < T, \\ v(x, 0) = 0, & x \in \Omega, \end{cases}$$

and for any $0 < t < T$, we have $v_t + L[v] + Mv > 0$ at (x_0, t) when $x_0 \in \Omega$, and $B[v] > 0$ at (x_0, t) when $x_0 \in \partial\Omega$, where M is the Lipschitz constant of f given by (2.1). The strong maximum principle gives $v > 0$, i.e., $u(x, t) > \varphi(x)$ for $x \in \Omega$ and $0 < t < T$. Similar to the step 1, it can be shown that u is strictly monotonically increasing in time t. □

Now, we study the convergence of solutions as $t \to \infty$. Here, we only consider the case $\partial_n u + b(x)u = 0$.

Theorem 3.12 (Convergence of solution) *Let conditions of Theorem 2.11 be fulfilled with $f = f(x, u)$. We further assume that the unique solution u of (3.10) is monotonically increasing (decreasing) in time t. Then (3.11) has at least one solution denoted by $u_s(x)$, and*

$$\lim_{t \to \infty} \|u(\cdot, t) - u_s(\cdot)\|_{C^{2+\beta}(\overline{\Omega})} = 0 \text{ for any } 0 < \beta < \alpha. \tag{3.13}$$

Proof. Clearly, the point-wise limit $\lim_{t \to \infty} u(x, t) = u_s(x)$ exists since u is bounded and monotone in time t.

On the other hand, the set $\{u(\cdot, t)\}_{t \geqslant 1}$ is compact in $C^{2+\beta}(\overline{\Omega})$ owing to the estimate (2.18). So, $\lim_{t \to \infty} u(\cdot, t) = u_s(\cdot)$ in $C^{2+\beta}(\overline{\Omega})$, which implies that $u_s \in C^{2+\beta}(\overline{\Omega})$ and (3.13) holds. Recalling (2.18), we also have $\lim_{t \to \infty} u_t(x, t) = 0$ for any $x \in \Omega$. Thus $u_s(x)$ solves (3.11) by letting $t \to \infty$ in the first two equations of (3.10). □

Theorem 3.13 (Convergence of solution) *Let conditions of Theorem 2.11 be fulfilled with $f = f(x, u)$. Suppose that (3.11) admits a unique solution u_s, and φ and ψ are a lower solution and an upper solution of (3.11), respectively, which satisfy $\varphi \leqslant u_s \leqslant \psi$ in Ω. Let u_φ and u_ψ be solutions of (3.10) respectively with initial data φ and ψ. Then, for any $0 < \beta < \alpha$,*

$$u_\varphi(\cdot, t) \nearrow u_s(\cdot) \text{ and } u_\psi(\cdot, t) \searrow u_s(\cdot) \text{ in } C^{2+\beta}(\overline{\Omega}) \text{ as } t \to \infty.$$

Proof. Based on Theorem 3.11, u_φ and u_ψ are, respectively, monotonically increasing and decreasing in time t. The comparison principle confirms $u_\varphi \leqslant u_\psi$ as well. Therefore, $\varphi \leqslant u_\varphi \leqslant u_\psi \leqslant \psi$. The desired conclusions can be followed immediately from Theorem 3.12. $\qquad\square$

Theorem 3.14 *Under conditions of Theorem 2.11 with $f = f(x, u)$, we further assume that (3.11) admits a unique solution $u_s(x) \in C^{2+\alpha}(\overline{\Omega})$. Let u be the unique solution of (3.10). Then, for any $0 < \beta < \alpha$, the following three kinds of convergence are equivalent:*

$$\lim_{t\to\infty} u(x, t) = u_s(x) \quad \text{a.e. in } \Omega, \tag{3.14}$$

$$\lim_{t\to\infty} \|u(x, t) - u_s(x)\|_{C^{2+\beta}(\overline{\Omega})} = 0, \tag{3.15}$$

$$\lim_{t\to\infty} \|u(x, t) - u_s(x)\|_{H^1(\Omega)} = 0.$$

Proof. It suffices to show that (3.14) implies (3.15). Assume that (3.14) holds. Owing to $C^{2+\alpha}(\overline{\Omega}) \overset{C}{\hookrightarrow} C^{2+\beta}(\overline{\Omega})$ for $0 < \beta < \alpha$, the limit (3.15) can be deduced from the estimate (2.18). $\qquad\square$

Same as the proof of Theorem 3.14, we have the following conclusion.

Theorem 3.15 *Under requirements of Theorem 2.13 with $f = f(x, u)$, we further assume that problem (3.11) admits a unique solution $u_s(x) \in C^{1+\alpha}(\overline{\Omega})$. Let u be the unique solution of (3.10). Then, for any $0 < \beta < \alpha$, the following three kinds of convergence are equivalent:*

$$\lim_{t\to\infty} u(x, t) = u_s(x) \quad \text{a.e. in } \Omega,$$

$$\lim_{t\to\infty} \|u(x, t) - u_s(x)\|_{C^{1+\beta}(\overline{\Omega})} = 0,$$

$$\lim_{t\to\infty} \|u(x, t) - u_s(x)\|_{H^1(\Omega)} = 0.$$

Let us study the following problem as an example

$$\begin{cases} u_t - \Delta u = u(a - u), & x \in \Omega, \quad t > 0, \\ u = 0, & x \in \partial\Omega, \ t > 0, \\ u(x, 0) = u_0(x) \geqslant 0, \not\equiv 0, & x \in \Omega, \end{cases} \tag{3.16}$$

where Ω is of class $C^{2+\alpha}$, $u_0 \in C^2(\overline{\Omega})$ and $u_0 = 0$ on $\partial\Omega$. Its equilibrium problem is

$$\begin{cases} -\Delta u = u(a - u), & x \in \Omega, \\ u = 0, & x \in \partial\Omega. \end{cases} \tag{3.17}$$

In accordance with the upper and lower solutions method, and the regularity theory (Theorem 2.10), it can be shown that (3.16) has a unique non-negative global solution $u \in C^{2+\alpha, 1+\alpha/2}(\overline{\Omega} \times \mathbb{R}_+)$. Moreover, $u \leqslant \max\{a, \max_{\overline{\Omega}} u_0(x)\}$ by the maximum principle.

Let (λ, ϕ) be the first eigen-pair of (1.33). It is well known that, when $a > \lambda$, (3.17) has a unique positive solution $u_s \in C^{2+\alpha}(\overline{\Omega})$, and $\partial_n u_s < 0$ on $\partial\Omega$.

Theorem 3.16 *Let u be the unique non-negative global solution of (3.16). If $a > \lambda$, then $\lim_{t \to \infty} u(x,t) = u_s(x)$ in $C^{2+\beta}(\overline{\Omega})$ for any $0 < \beta < \alpha$.*

Proof. In accordance with the maximum principle and the Hopf boundary lemma, $u > 0$ in $\Omega \times \mathbb{R}_+$ and $\partial_n u < 0$ on $\partial\Omega \times \mathbb{R}_+$. Without loss of generality, we may assume that $u_0 > 0$ in Ω and $\partial_n u_0 < 0$ on $\partial\Omega$.

Note that $\phi > 0$ in Ω and $\partial_n \phi < 0$ on $\partial\Omega$. We can find two constants $0 < \varepsilon \ll 1$ and $k > 1$ for which $\varepsilon\phi(x) \leqslant u_0(x) \leqslant k u_s(x)$ in $\overline{\Omega}$.

It is easy to verify that $\varepsilon\phi$ and ku_s are respectively a lower solution and an upper solution of (3.17) as long as $0 < \varepsilon \ll 1$. Let u_1 and u_2 be solutions of (3.16) with, respectively, initial data $\varepsilon\phi$ and ku_s. Then $u_1 \leqslant u \leqslant u_2$ by the comparison principle. Theorem 3.13 asserts that $u_i(x,t) \to u_s(x)$ in $C^{2+\beta}(\overline{\Omega})$ as $t \to \infty$. $\qquad\square$

Theorem 3.17 *Let u be the unique non-negative global solution of (3.16). If $a \leqslant \lambda$, then $\lim_{t \to \infty} u(x,t) = 0$ in $C^{2+\beta}(\overline{\Omega})$ for any $0 < \beta < \alpha$.*

Proof. Normalize ϕ by $\int_\Omega \phi(x) = 1$. Multiplying the first equation of (3.16) by ϕ and integrating the result over Ω, we conclude

$$\int_\Omega u_t \phi + \lambda \int_\Omega u\phi = a \int_\Omega u\phi - \int_\Omega u^2 \phi \leqslant a \int_\Omega u\phi - \left(\int_\Omega u\phi \right)^2.$$

Set $f(t) = \int_\Omega u(x,t)\phi(x)$. Then

$$f'(t) \leqslant -f^2(t) \text{ and } f(t) > 0 \quad \text{for } t > 0; \quad f(0) > 0,$$

which implies $f(t) \to 0$, i.e., $\int_\Omega u(x,t)\phi(x) \to 0$ as $t \to \infty$. Notice that u is bounded, in view of Theorem 2.11 we can find t_i with $t_i \to \infty$, and $\tilde{u} \in C^{2+\beta}(\overline{\Omega})$ such that $u(x,t_i) \to \tilde{u}(x)$ in $C^{2+\beta}(\overline{\Omega})$. Combining this with $\int_\Omega u(x,t_i)\phi(x) \to 0$, we derive $\tilde{u} \equiv 0$. That is, $u(x,t_i) \to 0$ in $C^{2+\beta}(\overline{\Omega})$ as $i \to \infty$. This also implies $u(x,t) \to 0$ in $C^{2+\beta}(\overline{\Omega})$ as $t \to \infty$. $\qquad\square$

3.4 THE WEAK UPPER AND LOWER SOLUTIONS METHOD

For the sake of simplicity, we only focus on (2.7), and rewrite it here:

$$\begin{cases} u_t - D_j(a_{ij}D_i u) + b_i D_i u = f(x,t,u) & \text{in } Q_T, \\ u = g & \text{on } S_T, \\ u(x,0) = \varphi(x) & \text{in } \Omega, \end{cases} \quad (3.18)$$

where $a_{ij}, b_i \in L^\infty(Q_T)$, $a_{ij} = a_{ji}$ in Q_T for all $1 \leqslant i,j \leqslant n$, and there exists a constant $\lambda > 0$ so that

$$a_{ij}(x,t)y_i y_j \geqslant \lambda |y|^2 \text{ for all } (x,t) \in Q_T, \, y \in \mathbb{R}^n.$$

Also, *weak lower solution, weak upper solution* and *weak solution* of (3.18) are described in Definition 2.6.

We first focus on the weak upper and lower solutions method. For convenience and to facilitate understanding, we give the conclusion and proof under relatively strong conditions.

Theorem 3.18 (Weak upper and lower solutions method) *Assume that the condition* **(G1)** *holds with $p \geqslant 2$, and $f \in L^\infty(Q_T \times (\underline{c}, \bar{c}))$ satisfies (2.1) for any constants $\underline{c} < \bar{c}$. Let $\bar{u}$ and $\underline{u}$ be respectively the bounded weak upper solution and weak lower solution of (3.18). Then, in the interval $\langle \underline{u}, \bar{u} \rangle$, problem (3.18) has a unique solution $u \in W_p^{2,1}(Q_T)$, where*

$$\langle \underline{u}, \bar{u} \rangle = \{u \in L^\infty(Q_T) : \underline{u} \leqslant u \leqslant \bar{u} \ \text{ in } Q_T\}.$$

Proof. Firstly, $\underline{u} \leqslant \bar{u}$ by Theorem 2.7. Let us define

$$\mathscr{L}u = u_t - D_j(a_{ij}D_i u) + b_i D_i u.$$

For any given $v \in \langle \underline{u}, \bar{u} \rangle$, the linear problem

$$\begin{cases} \mathscr{L}u + Mu = f(x, t, v(x, t)) + Mv(x, t) & \text{in } Q_T, \\ u = g & \text{on } S_T, \\ u(x, 0) = \varphi(x) & \text{in } \Omega \end{cases} \tag{3.19}$$

has a unique solution $u \in W_p^{2,1}(Q_T)$ by the L^p theory (Theorem 1.1). As in the proof of Theorem 3.5 we conclude that

$$\|u\|_{W_p^{2,1}(Q_T)} \leqslant C \ \text{ and } \ |u|_{\alpha, \overline{Q}_T} \leqslant C \ \text{ for all } v \in \langle \underline{u}, \bar{u} \rangle,$$

and the operator $\mathscr{F} : \langle \underline{u}, \bar{u} \rangle \to C(\overline{Q}_T)$ defined by $\mathscr{F}(v) = u$ is compact.

Next we prove $\mathscr{F} : \langle \underline{u}, \bar{u} \rangle \to \langle \underline{u}, \bar{u} \rangle$. Suppose $v \in \langle \underline{u}, \bar{u} \rangle$ and $u = \mathscr{F}(v)$. Clearly, u is a weak solution of (3.19). Note that $\bar{u}$ and $\underline{u}$ are respectively the weak upper solution and weak lower solution of (3.18), and the function $f(\cdot, s) + Ms$ is increasing in s and $\underline{u} \leqslant v \leqslant \bar{u}$. It follows that $\bar{u}$ and $\underline{u}$ are respectively the weak upper solution and weak lower solution of (3.19), and hence $\underline{u} \leqslant u \leqslant \bar{u}$ in Q_T by Theorem 2.7. Consequently $\mathscr{F} : \langle \underline{u}, \bar{u} \rangle \to \langle \underline{u}, \bar{u} \rangle$.

Because $\langle \underline{u}, \bar{u} \rangle$ is a bounded and closed convex set of $L^\infty(Q_T)$, the operator $\mathscr{F}$ has at least one fixed point $u \in \langle \underline{u}, \bar{u} \rangle$ by the Schauder fixed point theorem (Corollary A.3), which in turn solves problem (3.18). Clearly $u \in W_p^{2,1}(Q_T)$ due to $u = \mathscr{F}(u) \in W_p^{2,1}(Q_T)$.

The uniqueness follows from Theorem 2.7. □

In applications, we usually encounter such a situation that two better upper solutions (lower solutions) u_1 and u_2 have been constructed, however they do not meet our requirements, and $\min\{u_1, u_2\}$ ($\max\{u_1, u_2\}$) might be suitable for our problems. Naturally, we want to know whether $\min\{u_1, u_2\}$ ($\max\{u_1, u_2\}$) is a weak upper (lower) solution. We will answer this question in the following.

Theorem 3.19 *Assume that $p \geqslant 2$ and $f \in L^\infty(Q_T \times (\underline{c}, \bar{c}))$ for any constants $\underline{c} < \bar{c}$. If $\bar{u}_1, \bar{u}_2 \in W_p^{2,1}(Q_T) \cap C(\overline{Q}_T)$ are two upper solutions of (3.18), then $\bar{u} = \min\{\bar{u}_1, \bar{u}_2\}$ is a weak upper solution of (3.18). If $\underline{u}_1, \underline{u}_2 \in W_p^{2,1}(Q_T) \cap C(\overline{Q}_T)$ are two lower solutions of (3.18), then $\underline{u} = \max\{\underline{u}_1, \underline{u}_2\}$ is a weak lower solution of (3.18).*

Proof. We only prove the conclusion for weak lower solutions, and the assertion for weak upper solutions is analogous.

Step 1. Let $\underline{u}_1, \underline{u}_2 \in W_p^{2,1}(Q_T) \cap C(\overline{Q}_T)$ be two lower solutions of (3.18). Then they are weak lower solutions of (3.18). To simplify notations, we denote $\underline{u}_i$ by u_i ($i = 1, 2$), and $\underline{u}$ by u in the sequel. It will be illustrated that u is a weak lower solution of (3.18).

Write $u = u_1 + (u_2 - u_1)_+$. We find that, for any $0 < t \leqslant T$,

$$Du = \begin{cases} Du_1 & \text{in } A_1^t := \{(x,s) \in Q_t : u_1(x,s) > u_2(x,s)\}, \\ Du_1 = Du_2 & \text{in } A_0^t := \{(x,s) \in Q_t : u_1(x,s) = u_2(x,s)\}, \\ Du_2 & \text{in } A_2^t := \{(x,s) \in Q_t : u_2(x,s) > u_1(x,s)\}. \end{cases}$$

As $p \geqslant 2$, it is easy to see that

$$u \in C([0,T], L^2(\Omega)) \cap L^2((0,T), H^1(\Omega)) \cap L^\infty(\Omega \times (0,T)),$$
$$f(\cdot, u(\cdot)) \in L^2(\Omega \times (0,T)),$$
$$u(x,0) \leqslant \varphi(x) \text{ in } \Omega, \quad u \leqslant g \text{ on } S_T.$$

It remains to show that, for all $0 < t \leqslant T$,

$$\int_{Q_t} (a_{ij} D_i u D_j \phi + b_i D_i u \phi) \leqslant \int_{Q_t} (f(x,s,u)\phi + u\phi_s) - \int_\Omega u\phi|_{s=t} \qquad (3.20)$$

for each test function $\phi \in \overset{\bullet}{W}_2^{1,1}(Q_T)$ with $\phi \geqslant 0$ in Q_T.

Step 2. Let $v = u_2 - u_1 \in W_2^{2,1}(Q_T)$. Thanks to classical density results for $W_2^{2,1}(Q_T)$, there exists a sequence $\{v_l\} \subset C^2(\overline{Q}_T)$ such that $v_l \to v$ in $W_2^{2,1}(Q_T)$ as $l \to \infty$. By passing to a sub-sequence, we may assume that

$$\|v_l - v\|_{W_2^{2,1}(Q_T)} \leqslant l^{-3} \text{ for all } l \geqslant 1. \qquad (3.21)$$

Select a function $\theta : \mathbb{R} \to \mathbb{R}$ satisfying

    (a) $\theta \in C^\infty(\mathbb{R})$;

    (b) θ is nondecreasing in $\mathbb{R}$;

    (c) $0 \leqslant \theta(\tau) \leqslant 1$;

    (d) $\theta(\tau) = 0$ for $\tau \leqslant 0$, $\theta(\tau) = 1$ for $\tau \geqslant 1$.

Set $\theta_l(\tau) = \theta(l\tau)$. Obviously, θ_l meets with (a)–(c) above, and

$$\theta_l(\tau) = 0 \text{ for } \tau \leqslant 0, \quad \theta_l(\tau) = 1 \text{ for } \tau \geqslant 1/l.$$

Let $M = \max_{[0,1]}\{\theta' + |\theta''|\}$. Then we have

$$0 \leqslant \theta_l'(\tau) = l\theta'(l\tau) \leqslant Ml, \quad |\theta_l''(\tau)| \leqslant Ml^2. \qquad (3.22)$$

Now for any $\phi \in \overset{\bullet}{C}{}^{\infty}(\overline{Q}_T)$ with $\phi \geq 0$, we define

$$\phi_1 = \phi_1^l = (1 - \theta_l(v_l))\phi, \quad \phi_2 = \phi_2^l = \theta_l(v_l)\phi.$$

It is easy to see that $\phi_1, \phi_2 \in \overset{\bullet}{W}{}_2^{1,1}(Q_T)$ are non-negative. Since $u_k \ (k = 1, 2)$ are weak lower solutions of (3.18), by the definition we see that for $k = 1, 2$,

$$\int_{Q_t} (a_{ij}D_iu_kD_j\phi_k + b_iD_iu_k\phi_k) \leq \int_{Q_t} [f(\cdot, u_k)\phi_k + u_k\phi_{k,s}] - \int_{\Omega} u_k\phi_k|_{s=t}.$$

That is,

$$\int_{Q_t} \{a_{ij}D_iu_1[(1 - \theta_l(v_l))D_j\phi - \phi\theta_l'(v_l)D_jv_l] + b_iD_iu_1(1 - \theta_l(v_l))\phi\}$$
$$\leq \int_{Q_t} [f(\cdot, u_1)\phi_1 + u_1\phi_{1,s}] - \int_{\Omega} u_1\phi_1|_{s=t} \tag{3.23}$$

and

$$\int_{Q_t} \{a_{ij}D_iu_2[\theta_l(v_l)D_j\phi + \phi\theta_l'(v_l)D_jv_l] + b_iD_iu_2\theta_l(v_l)\phi\}$$
$$\leq \int_{Q_t} [f(\cdot, u_2)\phi_2 + u_2\phi_{2,s}] - \int_{\Omega} u_2\phi_2|_{s=t}. \tag{3.24}$$

It then follows from (3.23) and (3.24) that

$$\int_{Q_t} a_{ij}D_iu_1D_j\phi + \int_{Q_t} a_{ij}D_iv[\theta_l(v_l)D_j\phi + \phi\theta_l'(v_l)D_jv_l]$$
$$+ \int_{Q_t} [b_iD_iu_1\phi + b_iD_iv\theta_l(v_l)\phi] - \int_{Q_t} [u_1\phi_s + v\theta_l(v_l)\phi_s + v\theta_l'(v_l)v_{l,s}\phi]$$
$$\leq \int_{Q_t} f(\cdot, u_1)\phi + \int_{Q_t} [f(\cdot, u_2) - f(\cdot, u_1)]\theta_l(v_l)\phi - \int_{\Omega} (u_1\phi + v\theta_l(v_l)\phi)|_{s=t}. \tag{3.25}$$

Step 3. Since $f(x, t, u_2) = f(x, t, u_1)$ for a.e. $(x, t) \in A_0^t$, we conclude

$$\int_{Q_t} [f(\cdot, u_2) - f(\cdot, u_1)]\theta_l(v_l)\phi = \int_{A_1^t \cup A_2^t} [f(\cdot, u_2) - f(\cdot, u_1)]\theta_l(v_l)\phi.$$

Notice $v > 0$ in A_2^t. It follows that, by passing to a sub-sequence, $\theta_l(v_l) \to 1$ a.e. in A_2^t as $l \to \infty$. Similarly, $\theta_l(v_l) \to 0$ a.e. in A_1^t as $l \to \infty$. Due to the dominated convergence theorem, we find that

$$\lim_{l \to \infty} \int_{A_2^t} [f(\cdot, u_2) - f(\cdot, u_1)]\theta_l(v_l)\phi = \int_{A_2^t} [f(\cdot, u_2) - f(\cdot, u_1)]\phi$$

and

$$\lim_{l \to \infty} \int_{A_1^t} [f(\cdot, u_2) - f(\cdot, u_1)]\theta_l(v_l)\phi = 0.$$

Therefore

$$\lim_{l\to\infty}\int_{Q_t}[f(\cdot,u_2)-f(\cdot,u_1)]\theta_l(v_l)\phi=\int_{A_2^t}[f(\cdot,u_2)-f(\cdot,u_1)]\phi. \tag{3.26}$$

Analogously, we have

$$\begin{cases} \displaystyle\lim_{l\to\infty}\int_{Q_t}a_{ij}D_iv\theta_l(v_l)D_j\phi=\int_{A_2^t}a_{ij}D_ivD_j\phi,\\[4mm] \displaystyle\lim_{l\to\infty}\int_{Q_t}b_iD_iv\theta_l(v_l)\phi=\int_{A_2^t}b_iD_iv\phi,\\[4mm] \displaystyle\lim_{l\to\infty}\int_{Q_t}v\theta_l(v_l)\phi_s=\int_{A_2^t}v\phi_s,\\[4mm] \displaystyle\lim_{l\to\infty}\int_\Omega v\theta_l(v_l)\phi|_{s=t}=\int_{A_2^t\cap\{s=t\}}v\phi. \end{cases} \tag{3.27}$$

Next, we use $\theta_l'(v_l)\geqslant 0$ and $\phi\geqslant 0$ to obtain

$$\int_{Q_t}a_{ij}D_iv\phi\theta_l'(v_l)D_jv_l=\int_{Q_t}[a_{ij}D_iv\theta_l'(v_l)D_jv\phi+a_{ij}D_iv\theta_l'(v_l)D_j(v_l-v)\phi]$$

$$\geqslant\int_{Q_t}a_{ij}D_iv\theta_l'(v_l)D_j(v_l-v)\phi. \tag{3.28}$$

By (3.21) and (3.22) it yields

$$\lim_{l\to\infty}\|D_j(v_l-v)\theta_l'(v_l)\phi\|_{2,Q_t}=0,$$

and hence

$$\lim_{l\to\infty}\int_{Q_t}a_{ij}D_iv\theta_l'(v_l)D_j(v_l-v)\phi=0.$$

Consequently, from (3.28) we derive

$$\liminf_{l\to\infty}\int_{Q_t}a_{ij}D_iv\phi\theta_l'(v_l)D_jv_l\geqslant 0. \tag{3.29}$$

Analogously, we have

$$\lim_{l\to\infty}\int_{Q_t}v\theta_l'(v_l)v_{l,s}\phi=\lim_{l\to\infty}\int_{A_2^t}\theta_l'(v_l)vv_s\phi. \tag{3.30}$$

We claim

$$\lim_{l\to\infty}\int_{A_2^t}\theta_l'(v_l)vv_s\phi=0. \tag{3.31}$$

In fact, write

$$\theta_l'(v_l)=\theta_l'(v_l)-\theta_l'(v)+\theta_l'(v)=\theta_l''(v^*)(v_l-v)+\theta_l'(v).$$

Noticing (3.21), (3.22), and $v, \phi \in L^\infty(Q_T)$, we have

$$\lim_{l \to \infty} \int_{A_2^t} \theta_l''(v^*)(v_l - v)vv_s\phi = 0$$

as above. Owing to $|\theta_l'(v)v| \leqslant M$ in A_2^t and $\lim_{l \to \infty} \theta_l'(v(x,t))v(x,t) = 0$ for each $(x,t) \in A_2^t$, it can be obtained by the dominated convergence theorem that

$$\lim_{l \to \infty} \int_{A_2^t} \theta_l'(v)vv_s\phi = 0.$$

Therefore (3.31) holds.

Step 4. We now take $l \to \infty$ in (3.25) and make use of (3.26), (3.27), (3.29), (3.30) and (3.31) to infer that

$$\int_{Q_t} a_{ij}D_iu_1D_j\phi + \int_{A_2^t} a_{ij}D_ivD_j\phi + \int_{Q_t} b_iD_iu_1\phi + \int_{A_2^t} b_iD_iv\phi - \int_{Q_t} u_1\phi_s - \int_{A_2^t} v\phi_s$$

$$\leqslant \int_{Q_t} f(\cdot, u_1)\phi + \int_{A_2^t} [f(\cdot, u_2) - f(\cdot, u_1)]\phi - \int_\Omega u_1\phi|_{s=t} - \int_{A_2^t \cap \{s=t\}} v\phi, \qquad (3.32)$$

which implies (3.20) (cf. Exercise 3.12).

For the given $\phi \in \overset{\bullet}{W}_2^{1,1}(Q_T)$ with $\phi \geqslant 0$, we can choose $\phi_\ell \in \overset{\bullet}{C}^\infty(\overline{Q}_T)$ with $\phi_\ell \geqslant 0$, such that $\phi_\ell \to \phi$ in $W_2^{1,1}(Q_T)$. Then (3.20) holds for all such ϕ_ℓ. Letting $\ell \to \infty$ we conclude that (3.20) holds for such ϕ. □

Remark 3.20 *Assume that conditions of Theorems* 3.18 *and* 3.19 *hold.*

(1) *We have* $\max\{\underline{u}_1, \underline{u}_2\} \leqslant \min\{\bar{u}_1, \bar{u}_2\}$.

(2) *For* $\underline{u}_i, \bar{u}_i$, *using the upper and lower solutions method we know that problem* (3.18) *has a unique solution* $u \in W_p^{2,1}(Q_T)$ *satisfying* $\underline{u}_i \leqslant u \leqslant \bar{u}_i$, $i = 1, 2$. *Hence,* $\max\{\underline{u}_1, \underline{u}_2\} \leqslant u \leqslant \min\{\bar{u}_1, \bar{u}_2\}$.

EXERCISES

3.1 Prove that, for the upper and lower solutions defined in Remark 3.4, Theorem 3.3 still holds. Hint: apply Theorems 1.41, 1.42 and 1.43.

3.2 Let $u_0 \in C^2(\overline{\Omega})$ and $u_0 = 0$ on $\partial\Omega$. Using the upper and lower solutions method to prove that there is $T > 0$ such that problem

$$\begin{cases} u_t - \Delta u = u^2 & \text{in} \quad Q_T \\ u = 0 & \text{on} \quad S_T \\ u(x, 0) = u_0(x) & \text{in} \quad \Omega \end{cases}$$

has a unique solution.

3.3 Let (λ, ϕ) be the first eigen-pair of (1.33). Assume $a > \lambda$ and $0 < \varepsilon < a - \lambda$. Prove that the solution u of

$$\begin{cases} u_t - \Delta u = au - u^2, & x \in \Omega, \ t > 0, \\ u = 0, & x \in \partial\Omega, \ t > 0, \\ u(x,0) = \varepsilon\phi(x), & x \in \Omega \end{cases}$$

is strictly increasing in time t.

3.4 Consider an initial-boundary value problem

$$\begin{cases} u_t - \Delta u = u(a - u), & x \in \Omega, \ t > 0, \\ \mathscr{B}u = 0, & x \in \partial\Omega, \ t > 0, \\ u(x,0) = u_0(x) \geqslant 0, & x \in \Omega, \end{cases}$$

where $\mathscr{B}u = \partial_n u$ or $\mathscr{B}u = u$, a is a constant, $u_0 \in C^1(\overline{\Omega}) \cap W_p^2(\Omega)$ with $p > 1$ and $\mathscr{B}u_0 = 0$ on $\partial\Omega$.

(1) Prove that this problem has a unique global solution u and u satisfies $0 \leqslant u \leqslant \max\{a, \max_{\overline{\Omega}} u_0\}$. Especially, please explain requirements of the Hopf boundary lemma when we use it.

(2) Let λ be the first eigenvalue of

$$-\Delta\phi = \lambda\phi \ \text{in} \ \Omega, \quad \mathscr{B}\phi = 0 \ \text{on} \ \partial\Omega.$$

Prove that $a < \lambda$ implies

$$\lim_{t \to \infty} u(x,t) = 0 \ \text{uniformly in} \ \overline{\Omega}.$$

3.5 Consider problem

$$\begin{cases} u_t - \Delta u = au - u^k, & x \in \Omega, \ t > 0, \\ u = 0, & x \in \partial\Omega, \ t > 0, \\ u(x,0) = u_0(x) \geqslant 0, \not\equiv 0, & x \in \Omega, \end{cases} \quad \text{(P3.1)}$$

where constant $k > 1$, Ω is of class $C^{2+\alpha}$, $u_0 \in C(\overline{\Omega}) \cap W_p^2(\Omega)$ with $p > 1$ and $u_0 = 0$ on $\partial\Omega$.

(1) Use the upper and lower solutions method and regularity theory to show that problem (P3.1) has a unique global solution $u \in C^{2+\alpha,1+\alpha/2}(\overline{\Omega} \times \mathbb{R}_+)$.

(2) Let λ be the first eigenvalue of (1.33) and $a > \lambda$. Let u_s be the unique positive solution of

$$\begin{cases} -\Delta u = au - u^k \ \text{in} \ \Omega \\ u = 0 \qquad \text{on} \ \partial\Omega. \end{cases}$$

Prove that the unique solution u of (P3.1) satisfies $\lim_{t \to \infty} u(x,t) = u_s(x)$ in $C^2(\overline{\Omega})$.

3.6 Apply the regularity and uniform estimates given in §2.2 to prove that the unique solution of (3.4) satisfies $\lim_{t\to\infty} u(x,t) = a$ in $C^2(\overline{\Omega})$.

3.7 Apply the regularity and uniform estimates given in §2.2 to prove Theorem 3.9.

3.8 Consider the following linear problem

$$\begin{cases} u_t - \Delta u = au, & x \in \Omega, \quad t > 0, \\ u = 0, & x \in \partial\Omega, \ t > 0, \\ u(x,0) = u_0(x), & x \in \Omega, \end{cases}$$

where $a > 0$ is a constant and $u_0 \in W_p^2(\Omega)$ with $p > 1$ and $u_0 \geqslant 0, \not\equiv 0$. Let (λ, ϕ) be the first eigen-pair of (1.33). Prove the following conclusions.

(1) Solution u of this problem exists uniquely and globally in time t and $u \in C^{2+\alpha,1+\alpha/2}(\Omega \times \mathbb{R}_+)$. Moreover, $u > 0$ in $\Omega \times \mathbb{R}_+$ and $\partial_n u < 0$ on $\partial\Omega \times \mathbb{R}_+$.

(2) If $a < \lambda$, then $\lim_{t\to\infty} u(x,t) = 0$ uniformly in $\overline{\Omega}$.

(3) If $a = \lambda$, then for any $\tau > 0$, there exist positive constants $\varepsilon < K$ such that $\varepsilon\phi(x) \leqslant u(x,t) \leqslant K\phi(x)$ in $\overline{\Omega} \times [\tau, \infty)$.

(4) If $a > \lambda$, then $\lim_{t\to\infty} u(x,t) = \infty$ for $x \in \Omega$.

3.9 Let μ_1 be the principal eigenvalue of

$$\begin{cases} L[\varphi] = \mu\varphi & \text{in } \Omega, \\ B[\varphi] = 0 & \text{on } \partial\Omega \end{cases}$$

and φ_1 be the corresponding positive eigenfunction, where the operators L and B are as in (3.10). Let $u_s(x)$ be a solution of (3.11). Assume that there exist $\delta > 0$ and $0 < \theta < \mu_1$ such that

$$|f_u(x,u)| \leqslant \mu_1 - \theta \quad \text{for } x \in \overline{\Omega}, \ u_s(x) - \delta \leqslant u \leqslant u_s(x) + \delta.$$

Prove that there exists $\rho > 0$ such that when $|\varphi(x) - u_s(x)| \leqslant \rho\varphi_1(x)$, problem (3.10) has a unique solution u_φ, and

$$|u_\varphi(x,t) - u_s(x)| \leqslant \rho e^{-(\mu_1-\theta)t}\varphi_1(x) \quad \text{for all } (x,t) \in \overline{\Omega} \times \overline{\mathbb{R}}_+.$$

3.10 Let $q \in L^\infty(\Omega)$ and $\lambda_1(q)$ be the principal eigenvalue of the operator $-\Delta + q$ in Ω with the homogeneous Dirichlet boundary condition. Assume that $f \geqslant 0$, $g \geqslant 0$ and $u_0 \geqslant 0, \not\equiv 0$. Prove that if $\lambda_1(q) < 0$ then solution u of

$$\begin{cases} u_t - \Delta u + q(x)u = f(x,t) & \text{in } Q_\infty \\ u = g(x,t) & \text{on } S_\infty \\ u(x,0) = u_0(x) & \text{in } \Omega \end{cases}$$

satisfies $\lim_{t\to\infty} u(x,t) = \infty$ uniformly in any compact subset of Ω.

3.11 Let λ be the first eigenvalue of (1.33) and $f \in C^1(\mathbb{R})$ satisfy

$$\limsup_{|u| \to \infty} \frac{f(u)}{u} < \lambda.$$

Prove that the following problem

$$\begin{cases} u_t - \Delta u = f(u), & x \in \Omega, \quad t > 0, \\ u = 0, & x \in \partial\Omega, \quad t > 0, \\ u(x,0) = u_0(x), & x \in \Omega \end{cases}$$

has a unique global solution u and u is bounded.

3.12 Prove that (3.32) implies (3.20).

3.13 Let Γ_1 and Γ_2 be given in (1.3). Set

$$\mathscr{B}u = \partial_{\boldsymbol{n}} u + bu, \ b \geqslant 0 \ \text{ on } \Gamma_1 \times (0, T], \quad \text{and} \quad \mathscr{B}u = u \ \text{ on } \Gamma_2 \times (0, T],$$

and consider initial-boundary value problem

$$\begin{cases} \mathscr{L}u = f(x,t,u) & \text{in } Q_T, \\ \mathscr{B}u = g & \text{on } S_T, \\ u(x,0) = \varphi(x) & \text{in } \Omega. \end{cases}$$

Establish the upper and lower solutions method (definitions of upper and lower solutions, existence and uniqueness of solutions derived from the upper and lower solutions, monotonicity and convergence of solution).

CHAPTER 4

Weakly Coupled Parabolic Systems

As a continuation of the previous chapter, this chapter deals with the initial-boundary value problems of weakly coupled semilinear parabolic systems. First, the L^p theory, Schauder theory and Schauder fixed point theorem are used to investigate the local existence and uniqueness of solutions to initial-boundary value problems of weakly coupled nonlinear systems. Second, by applying the Schauder fixed point theorem it is proved that if a problem admits a pair of coupled upper and lower solutions, then it must have a unique solution located between upper and lower solutions. Third, by means of the coupled upper and lower solutions, we construct the monotone iterative sequences to obtain the unique solution. Examples are included to illustrate this method.

4.1   LOCAL EXISTENCE AND UNIQUENESS OF SOLUTIONS

In this section we intend to use the L^p theory, Schauder theory and Schauder fixed point theorem to study the local existence and uniqueness of solutions to initial-boundary value problem of *weakly coupled parabolic systems*. Let us consider the following initial-boundary value problem

$$\begin{cases} \mathscr{L}_k u_k = f_k\,(x,t,u) & \text{in } Q_T, \\ \mathscr{B}_k u_k = g_k & \text{on } S_T, \\ u_k(x,0) = \varphi_k(x) & \text{in } \Omega, \\ k = 1,\ldots,m, \end{cases} \tag{4.1}$$

where Ω is of class C^2, $u = (u_1,\ldots,u_m)$, and

$$\mathscr{L}_k = \partial_t - a_{ij}^k(x,t)D_{ij} + b_i^k(x,t)D_i, \quad \mathscr{B}_k = a^k \partial_{\boldsymbol{n}} + b^k(x,t), \quad 1 \leqslant k \leqslant m.$$

Assume that either (i) $a^k = 0$, $b^k = 1$; or (ii) $a^k = 1$, $b^k \geqslant 0$ and $b^k \in C^{1,1/2}(\overline{S}_{T_0})$ for some $T_0 > 0$. Define

$$f = (f_1,\ldots,f_m), \quad g = (g_1,\ldots,g_m), \varphi = (\varphi_1,\ldots,\varphi_m).$$

Let us discuss the strong solution (solution in $[W_p^{2,1}(Q_T)]^m$) first. We further assume that coefficients of $\mathscr{L}_k$ are bounded and satisfy the condition (**A**) which is given in §1.1, $g \in [W_p^{2,1}(Q_{T_0})]^m$ and $\varphi \in [W_p^2(\Omega)]^m$ with $p > 1$. Define

$$U_{T_0}(\varphi) = \left\{ u \in [L^p(Q_{T_0})]^m : \|u - \varphi\|_{[L^p(Q_{T_0})]^m} \leqslant 1 \right\}.$$

Theorem 4.1 (Local existence and uniqueness) *Assume that, for each $1 \leqslant k \leqslant m$, $f_k : Q_{T_0} \times U_{T_0}(\varphi) \to L^p(Q_{T_0})$ is continuous with respect to $u \in U_{T_0}(\varphi)$, i.e., $\|f_k(\cdot, u^i) - f_k(\cdot, u)\|_{p, Q_{T_0}} \to 0$ when $u^i, u \in U_{T_0}(\varphi)$ and $\|u^i - u\|_{[L^p(Q_{T_0})]^m} \to 0$. If compatibility conditions*

$$\mathscr{B}_k \varphi_k = g_k \ \ for \ \ x \in \partial\Omega, \ t = 0 \ \ and \ all \ 1 \leqslant k \leqslant m$$

hold, then there exists $0 < T \leqslant T_0$ such that problem (4.1) has at least one solution $u \in [W_p^{2,1}(Q_T)]^m$.

Furthermore, if f satisfies the Lipschitz condition *with respect to $u \in U_{T_0}(\varphi)$, i.e., there is a constant $M > 0$ such that, for each $1 \leqslant k \leqslant m$,*

$$\|f_k(\cdot, u) - f_k(\cdot, v)\|_{p, Q_{T_0}} \leqslant M \|u - v\|_{[L^p(Q_{T_0})]^m} \ \ for \ all \ \ u, v \in U_{T_0}(\varphi),$$

then the solution is unique.

Proof. By our assumption, there is a constant $C_1 = C_1(T_0) > 0$ such that

$$\|f_k(\cdot, u)\|_{p, Q_{T_0}} \leqslant C_1 \ \ for \ all \ \ u \in U_{T_0}(\varphi).$$

For the given $u \in U_T(\varphi)$, we set $u = \varphi$ in $\Omega \times [T, T_0]$. Then $u \in U_{T_0}(\varphi)$, and the linear problem

$$\begin{cases} \mathscr{L}_k v_k = f_k(x, t, u) & \text{in} \ Q_{T_0}, \\ \mathscr{B}_k v_k = g_k & \text{on} \ S_{T_0}, \\ v_k(x, 0) = \varphi_k(x) & \text{in} \ \Omega \end{cases}$$

has a unique solution $v_k \in W_p^{2,1}(Q_{T_0})$, and

$$\begin{aligned} \|v_k\|_{W_p^{2,1}(Q_{T_0})} &\leqslant C \left(\|f_k(\cdot, u)\|_{p, Q_{T_0}} + \|g_k\|_{W_p^{2,1}(Q_{T_0})} + \|\varphi_k\|_{W_p^2(\Omega)} \right) \\ &\leqslant C \left(C_1 + \|g_k\|_{W_p^{2,1}(Q_{T_0})} + \|\varphi_k\|_{W_p^2(\Omega)} \right) \end{aligned} \tag{4.2}$$

by Theorem 1.7. We set $v = (v_1, \ldots, v_m)$, and define an operator $\mathscr{F} : U_T(\varphi) \to [L^p(Q_{T_0})]^m \subset [L^p(Q_T)]^m$ by letting $\mathscr{F}(u) = v$.

Let $u^i, u \in U_T(\varphi)$ and $\|u^i - u\|_{[L^p(Q_T)]^m} \to 0$ as $i \to \infty$. Set $u^i = u = \varphi$ in $\Omega \times [T, T_0]$. Then $\|u^i - u\|_{[L^p(Q_{T_0})]^m} = \|u^i - u\|_{[L^p(Q_T)]^m} \to 0$ as $i \to \infty$, and $f_k(\cdot, u^i) = f_k(\cdot, u)$ in $\Omega \times [T, T_0]$. Thus

$$\|f_k(\cdot, u^i) - f_k(\cdot, u)\|_{p, Q_T} = \|f_k(\cdot, u^i) - f_k(\cdot, u)\|_{p, Q_{T_0}} \to 0 \tag{4.3}$$

as $i \to \infty$ by our assumption. Let $v^i = \mathscr{F}(u^i)$ and $v = \mathscr{F}(u)$. Then $v^i - v$ satisfies, for each $1 \leqslant k \leqslant m$,

$$
\begin{cases}
\mathscr{L}_k(v_k^i - v_k) = f_k(x, t, u^i) - f_k(x, t, u) & \text{in } Q_T, \\
\mathscr{B}_k(v_k^i - v_k) = 0 & \text{on } S_T, \\
v_k^i(x, 0) - v_k(x, 0) = 0 & \text{in } \Omega.
\end{cases}
$$

By using the L^p theory (Theorem 1.7) and (4.3) in turn, it can be obtained that

$$
\|v_k^i - v_k\|_{W_p^{2,1}(Q_T)} \leqslant C\|f_k(\cdot, u^i) - f_k(\cdot, u)\|_{p, Q_T} \to 0
$$

as $i \to \infty$, which implies

$$
\|\mathscr{F}(u^i) - \mathscr{F}(u)\|_{[W_p^{2,1}(Q_T)]^m} \to 0 \ \text{ as } i \to \infty.
$$

Therefore, $\mathscr{F}$ is continuous. This, together with the estimate (4.2), concludes that $\mathscr{F} : U_T(\varphi) \to [L^p(Q_T)]^m$ is compact since $[W_p^{2,1}(Q_T)]^m \overset{C}{\hookrightarrow} [L^p(Q_T)]^m$.

Now, we prove $\mathscr{F} : U_T(\varphi) \to U_T(\varphi)$ provided $0 < T \ll 1$. In fact, the estimate (4.2) implies that, for $v = \mathscr{F}(u)$ and each $1 \leqslant k \leqslant m$,

$$
\|v_k - \varphi_k\|_{W_p^{2,1}(Q_{T_0})} \leqslant C \ \text{ for all } u \in U_T(\varphi).
$$

We can find $q > p$ such that $W_p^{2,1}(Q_{T_0}) \hookrightarrow L^q(Q_{T_0})$. Then $\|v_k - \varphi_k\|_{q, Q_{T_0}} \leqslant C$, and

$$
\begin{aligned}
\|v_k - \varphi_k\|_{p, Q_T} &\leqslant |Q_T|^{(q-p)/(pq)} \|v_k - \varphi_k\|_{q, Q_T} \\
&\leqslant CT^{(q-p)/(pq)} \|v_k - \varphi_k\|_{q, Q_{T_0}} \ \text{ for all } u \in U_T(\varphi).
\end{aligned}
$$

This shows that $\mathscr{F}(u) \in U_T(\varphi)$ if $0 < T \ll 1$.

Therefore, $\mathscr{F}$ has at least one fixed point $u \in U_T(\varphi)$ by the Schauder fixed point theorem (Corollary A.3), and then problem (4.1) has at least one solution $u \in U_T(\varphi)$. Clearly, $u \in [W_p^{2,1}(Q_T)]^m$ on account that $u = \mathscr{F}(u)$.

In the following we prove the uniqueness under assumption that f satisfies the Lipschitz condition with respect to $u \in U_{T_0}(\varphi)$. Let $0 < T \ll 1$ and $u, w \in [W_p^{2,1}(Q_T)]^m \cap U_T(\varphi)$ be two solutions of (4.1). We extend u and w by defining

$$
u = w = \varphi \ \text{ in } \Omega \times [T, T_0].
$$

Obviously, $u, w \in U_{T_0}(\varphi)$. Thus, we can define

$$
\tilde{u} = \mathscr{F}(u) \in [L^p(Q_{T_0})]^m, \quad \tilde{w} = \mathscr{F}(w) \in [L^p(Q_{T_0})]^m.
$$

It is easy to see that $u, \tilde{u}$ are solutions of linear initial-boundary value problem

$$
\begin{cases}
\mathscr{L}_k U_k = f_k(x, t, u(x, t)) & \text{in } Q_T, \\
\mathscr{B}_k U_k = g_k & \text{on } S_T, \\
U_k(x, 0) = \varphi_k(x) & \text{in } \Omega, \\
k = 1, \dots, m,
\end{cases}
$$

and $w, \tilde{w}$ are solutions of linear initial-boundary value problem

$$\begin{cases} \mathscr{L}_k W_k = f_k(x, t, w(x, t)) & \text{in } Q_T, \\ \mathscr{B}_k W_k = g_k & \text{on } S_T, \\ W_k(x, 0) = \varphi_k(x) & \text{in } \Omega, \\ k = 1, \dots, m. \end{cases}$$

The uniqueness of solutions to linear initial-boundary value problem implies

$$\tilde{u} = u, \quad \tilde{w} = w \text{ in } \overline{Q}_T.$$

For each $1 \leqslant k \leqslant m$,

$$\begin{aligned} \|\tilde{u}_k - \tilde{w}_k\|_{W_p^{2,1}(Q_{T_0})} &\leqslant C \|f_k(\cdot, u) - f_k(\cdot, w)\|_{p, Q_{T_0}} \\ &\leqslant C \|u - w\|_{[L^p(Q_{T_0})]^m} \\ &= C \|u - w\|_{[L^p(Q_T)]^m}. \end{aligned}$$

Similar to the above, $\|\tilde{u}_k - \tilde{w}_k\|_{q, Q_{T_0}} \leqslant C \|\tilde{u}_k - \tilde{w}_k\|_{W_p^{2,1}(Q_{T_0})}$ for some $q > p$. Then

$$\begin{aligned} \|u_k - w_k\|_{p, Q_T} &= \|\tilde{u}_k - \tilde{w}_k\|_{p, Q_T} \\ &\leqslant C T^{(q-p)/(pq)} \|\tilde{u}_k - \tilde{w}_k\|_{q, Q_T} \\ &\leqslant C T^{(q-p)/(pq)} \|\tilde{u}_k - \tilde{w}_k\|_{q, Q_{T_0}} \\ &\leqslant C T^{(q-p)/(pq)} \|u - w\|_{[L^p(Q_T)]^m} \end{aligned}$$

for each $1 \leqslant k \leqslant m$. This implies $u = w$ if $0 < T \ll 1$. $\qquad\square$

Next, we deal with the classical solution (solution in $[C^{2+\alpha, 1+\alpha/2}(\overline{Q}_T)]^m$). Suppose that, for Q_{T_0}, coefficients of $\mathscr{L}_k$ satisfy the condition **(C)** which is given in §1.1, $g_k \in C^{2+\alpha, 1+\alpha/2}(\overline{Q}_{T_0})$ and $\varphi_k \in C^{2+\alpha}(\overline{\Omega})$. Define

$$V_{T_0}(\varphi) := \{u \in [C(\overline{Q}_{T_0})]^m : \|u - \varphi\|_{[L^\infty(Q_{T_0})]^m} \leqslant 1\}.$$

Theorem 4.2 *Let Ω be of class $C^{2+\alpha}$ and $b \in [C^{1+\alpha, (1+\alpha)/2}(\overline{S}_{T_0})]^m$. Assume that f is Hölder continuous with respect to $(x, t) \in \overline{Q}_{T_0}$ and $u \in V_{T_0}(\varphi)$. If compatibility conditions*

$$\partial_{\boldsymbol{n}} \varphi_k + b^k \varphi_k = g_k \ \text{for} \ x \in \partial\Omega, \ t = 0 \ \text{if} \ a^k = 1,$$

$$\varphi_k = g_k, \ g_{kt} - a_{ij}^k D_{ij} \varphi_k + b_i^k D_i \varphi_k = f_k(\cdot, \varphi_k) \ \text{for} \ x \in \partial\Omega, \ t = 0 \ \text{if} \ a^k = 0$$

hold, then there exists $0 < T \leqslant T_0$ such that problem (4.1) has at least one solution $u \in [C^{2+\alpha, 1+\alpha/2}(Q_T)]^m$.

Furthermore, if f satisfies the Lipschitz condition with respect to $u \in V_{T_0}(\varphi)$, i.e., there is a constant $M > 0$ such that, for each $1 \leqslant k \leqslant m$,

$$|f_k(x, t, u) - f_k(x, t, v)| \leqslant M |u - v| \ \text{for all} \ (x, t) \in \overline{Q}_{T_0}, \ u, v \in V_{T_0}(\varphi),$$

then the solution is unique.

The proof of Theorem 4.2 is left to readers as an exercise.

4.2   THE UPPER AND LOWER SOLUTIONS METHOD

In §3.1 we have mentioned that the upper and lower solutions method is a powerful tool for the study of differential equations. Moreover, the upper and lower solutions method not only provides existence but also gives estimates of solutions.

For the scalar case, we have seen that the upper and lower solutions method is completely based on comparison principle. Regarding systems, in §1.3.3 we specifically emphasized that the maximum principle holds only for certain special circumstances. For this reason, when investigating systems, we cannot use the same manner as in the scalar case to define upper and lower solutions, we need to define them according to the structure of systems.

Define $[u]_k = (u_1, \ldots, u_{k-1}, u_{k+1}, \ldots, u_m)$. Then (4.1) can be written as

$$\begin{cases} \mathscr{L}_k u_k = f_k (x, t, u_k, [u]_k) & \text{in } Q_T, \\ \mathscr{B}_k u_k = g_k & \text{on } S_T, \\ u_k(x, 0) = \varphi_k(x) & \text{in } \Omega, \\ 1 \leqslant k \leqslant m. \end{cases} \tag{4.4}$$

It is assumed that coefficients of $\mathscr{L}_k$ are bounded in Q_T and satisfy the condition **(A)** which is given in §1.1.

In this section we always assume

$$\bar{u}_k, \underline{u}_k \in C(\overline{Q}_T) \cap C^{2,1}(Q_T) \quad \text{if } a^k = 0,$$
$$\bar{u}_k, \underline{u}_k \in C^{1,0}(\overline{Q}_T) \cap C^{2,1}(Q_T) \quad \text{if } a^k = 1.$$

For two given functions w, z, satisfying $w \leqslant z$, i.e., $w_k \leqslant z_k$ for all $1 \leqslant k \leqslant m$, let us define an order interval

$$\langle w, z \rangle = \left\{ u \in [C(\overline{Q}_T)]^m : w_k \leqslant u_k \leqslant z_k, \ 1 \leqslant k \leqslant m \right\}.$$

Definition 4.3   *For $\bar{u} = (\bar{u}_1, \ldots, \bar{u}_m)$ and $\underline{u} = (\underline{u}_1, \ldots, \underline{u}_m)$, we define*

$$\underline{\psi}_k = \min\{\bar{u}_k, \underline{u}_k\}, \quad \bar{\psi}_k = \max\{\bar{u}_k, \underline{u}_k\}, \ 1 \leqslant k \leqslant m,$$
$$\underline{\psi} = (\underline{\psi}_1, \ldots, \underline{\psi}_m), \quad \bar{\psi} = (\bar{\psi}_1, \ldots, \bar{\psi}_m).$$

We say that $(\bar{u}, \underline{u})$ is a pair of coupled upper and lower solutions of (4.4) if the following holds:

$$\begin{cases} \mathscr{L}_k \bar{u}_k \geqslant f_k (x, t, \bar{u}_k, [u]_k) & \text{for all } [\underline{\psi}]_k \leqslant [u]_k \leqslant [\bar{\psi}]_k & \text{in } Q_T, \\ \mathscr{L}_k \underline{u}_k \leqslant f_k (x, t, \underline{u}_k, [u]_k) & \text{for all } [\underline{\psi}]_k \leqslant [u]_k \leqslant [\bar{\psi}]_k & \text{in } Q_T, \\ \mathscr{B}_k \bar{u}_k \geqslant g_k \geqslant \mathscr{B}_k \underline{u}_k & \text{on } S_T, \\ \bar{u}_k(x, 0) \geqslant \varphi_k(x) \geqslant \underline{u}_k(x, 0) & \text{in } \Omega, \\ 1 \leqslant k \leqslant m. \end{cases} \tag{4.5}$$

Sometimes we call that $\bar{u}$ and $\underline{u}$ are the coupled upper and lower solutions of (4.4).

Theorem 4.4 (Ordering of upper and lower solutions) *Let $(\bar{u}, \underline{u})$ be a pair of coupled upper and lower solutions of* (4.4). *We define* $\underline{c}_k = \min_{\overline{Q}_T} \min\{\bar{u}_k, \underline{u}_k\}$, $\bar{c}_k = \max_{\overline{Q}_T} \max\{\bar{u}_k, \underline{u}_k\}$ *for* $1 \leqslant k \leqslant m$, *and*

$$\underline{c} = (\underline{c}_1, \ldots, \underline{c}_m) \quad and \quad \bar{c} = (\bar{c}_1, \ldots, \bar{c}_m).$$

Suppose that there exists a constant $M > 0$ such that

$$|f_k(x, t, u) - f_k(x, t, v)| \leqslant M|u - v| \tag{4.6}$$

for all $(x, t) \in Q_T$, $\underline{c} \leqslant u, v \leqslant \bar{c}$ and $1 \leqslant k \leqslant m$. Then $\bar{u} \geqslant \underline{u}$ in Q_T.

Proof. Set $w_k = \bar{u}_k - \underline{u}_k$. Then $\mathscr{B}_k w_k \geqslant 0$ on S_T and $w_k(x, 0) \geqslant 0$ in Ω. Taking $[u]_k = [\bar{u}]_k$ in the first two inequalities of (4.5), and $u = \bar{u}$, $v = (\underline{u}_k, [\bar{u}]_k)$ in (4.6), we have

$$\mathscr{L}_k w_k \geqslant f_k(x, t, \bar{u}_k, [\bar{u}]_k) - f_k(x, t, \underline{u}_k, [\bar{u}]_k) \geqslant -M|w_k| \quad \text{in} \ \ Q_T.$$

By means of Lemma 1.26, it turns out that $w_k \geqslant 0$, i.e., $\bar{u}_k \geqslant \underline{u}_k$ in Q_T. □

4.2.1 Existence and uniqueness of solutions

Theorem 4.5 (The upper and lower solutions method) *Assume that, for each $1 \leqslant k \leqslant m$, functions g_k, φ_k, b^k and coefficients of $\mathscr{L}_k$ satisfy the condition* **(G1)** *given in §3.1. Let $(\bar{u}, \underline{u})$ be a pair of coupled upper and lower solutions of* (4.4), $\underline{c}$ *and* $\bar{c}$ *be defined in Theorem 4.4 and $f_k \in L^\infty(Q_T \times \langle \underline{c}, \bar{c} \rangle)$ satisfy* (4.6). *Then, in the order interval $\langle \underline{u}, \bar{u} \rangle$,* (4.4) *has a unique solution $u \in [W_p^{2,1}(Q_T)]^m$.*

If we further assume that, for each $1 \leqslant k \leqslant m$, functions g_k, φ_k, b^k and coefficients of $\mathscr{L}_k$ satisfy the condition **(G2)** *given in §3.1, and $f_k(\cdot, u) \in C^{\alpha, \alpha/2}(\overline{Q}_T)$ is uniform in $u \in \langle \underline{c}, \bar{c} \rangle$. Then such a solution $u \in [C^{1+\alpha,(1+\alpha)/2}(\overline{Q}_T) \cap C^{2+\alpha, 1+\alpha/2}(Q_T)]^m$. In addition, we have $u_k \in C^{2+\alpha, 1+\alpha/2}(\overline{Q}_T)$ if either $a^k = 1$, or $a^k = 0$ and*

$$g_{kt} - a_{ij}^k D_{ij}\varphi_k + b_i^k D_i\varphi_k = f_k(x, t, \varphi_k) \quad on \ \ \partial\Omega \times \{0\}.$$

Proof. By Theorem 4.4, $\underline{u} \leqslant \bar{u}$. The following proof will be divided into several steps.

Step 1. For any given $v \in \langle \underline{u}, \bar{u} \rangle$, function $f_k(\cdot, v) + Mv_k \in L^\infty(Q_T)$. Then, by the L^p theory, the linear problem

$$\begin{cases} \mathscr{L}_k u_k + M u_k = f_k(x, t, v(x, t)) + Mv_k(x, t) & \text{in} \ \ Q_T, \\ \mathscr{B}_k u_k = g_k & \text{on} \ \ S_T, \\ u_k(x, 0) = \varphi_k(x) & \text{in} \ \ \Omega \end{cases}$$

has a unique solution $u_k \in W_p^{2,1}(Q_T)$, and

$$\begin{aligned} \|u_k\|_{W_p^{2,1}(Q_T)} &\leqslant C \left(\|f_k(\cdot, v)\|_{p, Q_T} + M\|v_k\|_{p, Q_T} + \|g_k\|_{W_p^{2,1}(Q_T)} + \|\varphi_k\|_{W_p^2(\Omega)} \right) \\ &\leqslant C_1 \left(\|f_k\|_{\infty, Q_T \times \langle \underline{c}, \bar{c} \rangle} + |\underline{c}| + |\bar{c}| + \|g_k\|_{W_p^{2,1}(Q_T)} + \|\varphi_k\|_{W_p^2(\Omega)} \right) \\ &\leqslant C_2 \ \text{for all} \ v \in \langle \underline{u}, \bar{u} \rangle. \end{aligned}$$

As $p > 1 + n/2$, the embedding theorem causes

$$|u_k|_{\alpha, \overline{Q}_T} \leqslant C_3 \quad \text{for all} \quad v \in \langle \underline{u}, \bar{u} \rangle, \quad 1 \leqslant k \leqslant m. \tag{4.7}$$

Moreover, similar to Theorem 2.10 we have $u_k \in C^{1+\alpha, (1+\alpha)/2}(Q_T) \cap W^{2,1}_{n+1, \text{loc}}(Q_T)$ when $a^k = 0$, and $u_k \in C^{1+\alpha, (1+\alpha)/2}(\overline{\Omega} \times (0, T]) \cap W^{2,1}_{q, \text{loc}}(Q_T)$ when $a^k = 1$.

Step 2. Define an operator $\mathscr{F}$ by

$$\mathscr{F}(v) = (u_1, \ldots, u_m) =: u.$$

We shall show that $\mathscr{F} : \langle \underline{u}, \bar{u} \rangle \to [C(\overline{Q}_T)]^m$ is compact. Recall the estimate (4.7). It suffices to prove that $\mathscr{F} : \langle \underline{u}, \bar{u} \rangle \to [C(\overline{Q}_T)]^m$ is continuous. For $v^i \in \langle \underline{u}, \bar{u} \rangle$, we put $u^i = \mathscr{F}(v^i)$, $i = 1, 2$, and $w_k = u_k^1 - u_k^2$. Then w_k satisfies

$$\mathscr{L}_k w_k + M w_k = f_k(x, t, v^1) + M v_k^1 - f_k(x, t, v^2) - M v_k^2 \quad \text{in} \quad Q_T,$$

$$\mathscr{B}_k w_k = 0 \quad \text{on} \quad S_T \quad \text{and} \quad w_k(x, 0) = 0 \quad \text{in} \quad \overline{\Omega}.$$

The assumption on f_k demonstrates

$$|f_k(x, t, v^1) + M v_k^1 - f_k(x, t, v^2) - M v_k^2| \leqslant 2M |v^1 - v^2|.$$

Hence, by the L^p estimate and embedding theorem, we have

$$|w_k|_{\alpha, Q_T} \leqslant C \|w_k\|_{W^{2,1}_p(Q_T)} \leqslant 2MC \|v^1 - v^2\|_{p, Q_T} \leqslant C' \|v^1 - v^2\|_{\infty}.$$

This illustrates that $\mathscr{F} : \langle \underline{u}, \bar{u} \rangle \to [C(\overline{Q}_T)]^m$ is continuous.

Step 3. Now we prove $\mathscr{F} : \langle \underline{u}, \bar{u} \rangle \to \langle \underline{u}, \bar{u} \rangle$. Suppose $v \in \langle \underline{u}, \bar{u} \rangle$ and $\mathscr{F}(v) = u$. Thanks to $\underline{u} \leqslant v \leqslant \bar{u}$, we see that

$$\begin{cases} \mathscr{L}_k \bar{u}_k + M \bar{u}_k \geqslant f_k(x, t, \bar{u}_k, [v]_k) + M \bar{u}_k & \text{in} \quad Q_T, \\ \mathscr{B}_k \bar{u}_k \geqslant g_k & \text{on} \quad S_T, \\ \bar{u}_k(x, 0) \geqslant \varphi_k(x) & \text{in} \quad \Omega. \end{cases}$$

Therefore, function $w_k = \bar{u}_k - u_k$ satisfies

$$\mathscr{L}_k w_k + M w_k \geqslant f_k(x, t, \bar{u}_k, [v]_k) - f_k(x, t, v_k, [v]_k) + M(\bar{u}_k - v_k) \geqslant 0 \quad \text{in} \quad Q_T,$$

$\mathscr{B}_k w_k \geqslant 0$ on S_T, and $w_k(x, 0) \geqslant 0$ in $\overline{\Omega}$. Theorem 1.40 asserts $w_k \geqslant 0$, i.e., $u_k \leqslant \bar{u}_k$. Likewise, $\underline{u}_k \leqslant u_k$. This accomplishes that $\mathscr{F} : \langle \underline{u}, \bar{u} \rangle \to \langle \underline{u}, \bar{u} \rangle$.

Step 4. Note that $\langle \underline{u}, \bar{u} \rangle$ is a bounded and closed convex set of $[C(\overline{Q}_T)]^m$, by the Schauder fixed point theorem (Corollary A.3), $\mathscr{F}$ has at least one fixed point $u \in \langle \underline{u}, \bar{u} \rangle$, which in turn implies that u solves (4.4) and $u \in [W^{2,1}_p(Q_T)]^m$ owing to $u = \mathscr{F}(u) \in [W^{2,1}_p(Q_T)]^m$.

Step 5. The proof of uniqueness of solutions is analogous to that of Theorem 4.4.

Step 6. We have known the solution $u \in [W^{2,1}_p(Q_T)]^m \hookrightarrow [C^{\alpha, \alpha/2}(\overline{Q}_T)]^m$ due to $p > 1 + n/2$. Set $F_k(x, t) = f_k(x, t, u(x, t))$. Then, under our further assumptions, it can be seen that $F_k \in C^{\alpha, \alpha/2}(\overline{Q}_T)$. Hence, $u_k \in C^{2+\alpha, 1+\alpha/2}(Q_T)$ by Theorem 1.19. Furthermore, if $a^k = 1$ then $u_k \in C^{2+\alpha, 1+\alpha/2}(\overline{Q}_T)$ by Theorem 1.17, if $a^k = 0$ and the compatibility condition holds then $w_k \in C^{2+\alpha, 1+\alpha/2}(\overline{Q}_T)$ by Theorem 1.16. □

4.2.2 Monotone iterative

In Definition 4.3, we acknowledge that the requirements for upper and lower solutions are relatively strong. However, for some special cases, we can simplify requirements on upper and lower solutions and obtain better results.

Suppose that $[u]_k$ is an union of two disjoint subsets $[u]_{i_k}$ and $[u]_{d_k}$, which consist of i_k and d_k components, respectively, with $i_k + d_k = m - 1$. Under this decomposition, we can rewrite (4.4) in the form

$$\begin{cases} \mathscr{L}_k u_k = f_k\,(x, t, u_k, [u]_{i_k}, [u]_{d_k}) & \text{in } Q_T, \\ \mathscr{B}_k u_k = g_k & \text{on } S_T, \\ u_k(x, 0) = \varphi_k(x) & \text{in } \Omega, \\ 1 \leqslant k \leqslant m. \end{cases} \tag{4.8}$$

Definition 4.6 *Let $w \leqslant z$. The system $f = (f_1, \ldots, f_m)$ is said to be* mixed quasi-monotonous *in the order interval $\langle w, z \rangle$ if there exists a decomposition as above such that for each $1 \leqslant k \leqslant m$ and any fixed $(x, t) \in Q_T$, the function $f_k\,(x, t, u_k, [u]_{i_k}, [u]_{d_k})$ is monotonically increasing in $[u]_{i_k}$ and monotonically decreasing in $[u]_{d_k}$ for all $u \in \langle w, z \rangle$.*

If $i_k = 0$ for all k, we say that f is quasi-monotonically decreasing. *If $d_k = 0$ for all k, we say that f is* quasi-monotonically increasing.

Definition 4.7 *Give $\underline{u}$ and $\bar{u}$, we define $\underline{\psi}$ and $\bar{\psi}$ as in Definition 4.3. Assume that the system $f = (f_1, \ldots, f_m)$ is mixed quasi-monotonous in $\langle \underline{\psi}, \bar{\psi} \rangle$. We say that $(\bar{u}, \underline{u})$ is* a pair of coupled upper and lower solutions *of (4.8) if the following holds:*

$$\begin{cases} \mathscr{L}_k \bar{u}_k \geqslant f_k\,(x, t, \bar{u}_k, [\bar{u}]_{i_k}, [\underline{u}]_{d_k}) & \text{in } Q_T, \\ \mathscr{L}_k \underline{u}_k \leqslant f_k\,(x, t, \underline{u}_k, [\underline{u}]_{i_k}, [\bar{u}]_{d_k}) & \text{in } Q_T, \\ \mathscr{B}_k \bar{u}_k \geqslant g_k \geqslant \mathscr{B}_k \underline{u}_k & \text{on } S_T, \\ \bar{u}_k(x, 0) \geqslant \varphi_k(x) \geqslant \underline{u}_k(x, 0) & \text{in } \Omega, \\ 1 \leqslant k \leqslant m. \end{cases}$$

Sometimes we call that $\bar{u}$ and $\underline{u}$ are coupled upper and lower solutions *of (4.8).*

Let us emphasize that, in fact, the coupled upper and lower solutions are defined together with the mixed quasi-monotonicity of the system. Besides, it is easy to verify that Definition 4.3 and Definition 4.7 are equivalent if f is mixed quasi-monotonous in $\langle \underline{\psi}, \bar{\psi} \rangle$ and satisfies the Lipschitz condition (4.6).

Theorem 4.8 (Ordering of upper and lower solutions) *Under conditions of Definition 4.7, if f satisfies the Lipschitz condition (4.6), then $\bar{u} \geqslant \underline{u}$ in Q_T.*

Proof. First, derivatives $\dfrac{\partial f_k}{\partial u_j}$ exist almost everywhere and are bounded, and $\dfrac{\partial f_k}{\partial u_j} \geqslant 0$

for $u_j \in [u]_{i_k}$, $\dfrac{\partial f_k}{\partial u_j} \leqslant 0$ for $u_j \in [u]_{d_k}$ by assumptions. Set $w_k = \bar{u}_k - \underline{u}_k$. Then w_k satisfies

$$\mathscr{L}_k w_k \geqslant f_k \left(\cdot, \bar{u}_k, [\bar{u}]_{i_k}, [\underline{u}]_{d_k}\right) - f_k \left(\cdot, \underline{u}_k, [\underline{u}]_{i_k}, [\bar{u}]_{d_k}\right)$$

$$= \frac{\partial f_k(\cdot)}{\partial u_k} w_k + \sum_{u_j \in [u]_{i_k}} \frac{\partial f_k(\cdot)}{\partial u_j} w_j - \sum_{u_j \in [u]_{d_k}} \frac{\partial f_k(\cdot)}{\partial u_j} w_j.$$

Define

$$h_{kk} = -\frac{\partial f_k}{\partial u_k}, \quad h_{kj} = -\frac{\partial f_k}{\partial u_j} \text{ if } u_j \in [u]_{i_k}, \text{ and } h_{kj} = \frac{\partial f_k}{\partial u_j} \text{ if } u_j \in [u]_{d_k}.$$

Then $h_{kj} \in L^\infty(Q_T)$, $h_{kj} \leqslant 0$ when $k \neq j$, and

$$\mathscr{L}_k w_k + \sum_{j=1}^m h_{kj} w_j \geqslant 0 \quad \text{in } Q_T.$$

Obviously, $\mathscr{B}_k w_k \geqslant 0$ on S_T and $w_k(x,0) \geqslant 0$ in $\overline{\Omega}$. Theorem 1.31 declares $w_k \geqslant 0$, i.e., $\bar{u}_k \geqslant \underline{u}_k$ in Q_T. $\qquad \square$

Theorem 4.9 (The upper and lower solutions method) *Suppose that, for each $1 \leqslant k \leqslant m$, functions g_k, φ_k, b^k and coefficients of $\mathscr{L}_k$ satisfy the condition* **(G2)** *given in §3.1. Let $(\bar{u}, \underline{u})$ be a pair of coupled upper and lower solutions of (4.8), $\underline{c}$ and $\bar{c}$ be defined in Theorem 4.4 and $f_k(\cdot, u) \in C^{\alpha, \alpha/2}(\overline{Q}_T)$ be uniform in $u \in \langle \underline{c}, \bar{c} \rangle$. Assume that f has mixed quasi-monotonicity property in $\langle \underline{u}, \bar{u} \rangle$, and f satisfies (4.6). Then there exist two monotone sequences $\{\underline{u}^i\}$ and $\{\bar{u}^i\}$ satisfying*

$$\underline{u} \leqslant \underline{u}^i \leqslant \underline{u}^{i+1} \leqslant \bar{u}^{i+1} \leqslant \bar{u}^i \leqslant \bar{u} \quad \text{for all } i \geqslant 1.$$

In addition,

$$\lim_{i \to \infty} \underline{u}^i = \lim_{i \to \infty} \bar{u}^i = u \in [W_p^{2,1}(Q_T) \cap C^{2+\alpha, 1+\alpha/2}(Q_T)]^m,$$

and u is the unique solution of (4.8) in the order interval $\langle \underline{u}, \bar{u} \rangle$.

Furthermore, $u_k \in C^{2+\alpha, 1+\alpha/2}(\overline{Q}_T)$ if either $a^k = 1$, or $a^k = 0$ and

$$g_{kt} - a_{ij}^k D_{ij} \varphi_k + b_i^k D_i \varphi_k = f_k(x, t, \varphi_k(x)), \quad x \in \partial\Omega, \ t = 0.$$

Proof. First, $\underline{u} \leqslant \bar{u}$ by Theorem 4.8. The proof will be divided into five steps.

Step 1. Constructions of $\underline{u}^i$ and $\bar{u}^i$. For any given $v, w \in \langle \underline{u}, \bar{u} \rangle$ and each $1 \leqslant k \leqslant m$, the linear initial-boundary value problem

$$\begin{cases} \mathscr{L}_k u_k + M u_k = f_k \left(x, t, v_k, [v]_{i_k}, [w]_{d_k}\right) + M v_k & \text{in } Q_T, \\ \mathscr{B}_k u_k = g_k & \text{on } S_T, \\ u_k(x, 0) = \varphi_k(x) & \text{in } \Omega \end{cases}$$

admits a unique solution $u_k \in W_p^{2,1}(Q_T)$ by the L^p theory. Define an operator $\mathscr{F}$ by $\mathscr{F}(v,w) = u$, and set

$$\underline{u}^1 = \mathscr{F}(\underline{u}, \bar{u}), \quad \bar{u}^1 = \mathscr{F}(\bar{u}, \underline{u}), \quad \underline{u}^{i+1} = \mathscr{F}(\underline{u}^i, \bar{u}^i), \quad \text{and} \quad \bar{u}^{i+1} = \mathscr{F}(\bar{u}^i, \underline{u}^i).$$

We claim that sequences $\{\underline{u}^i\}$ and $\{\bar{u}^i\}$ are monotonic and convergent. In fact, since $\bar{u}$ satisfies

$$\begin{cases} \mathscr{L}_k \bar{u}_k + M\bar{u}_k \geqslant f_k\left(x, t, \bar{u}_k, [\bar{u}]_{i_k}, [\underline{u}]_{d_k}\right) + M\bar{u}_k & \text{in } Q_T, \\ \mathscr{B}_k \bar{u}_k \geqslant g_k & \text{on } S_T, \\ u_k(x,0) \geqslant \varphi_k(x) & \text{in } \Omega, \\ 1 \leqslant k \leqslant m, \end{cases}$$

it is easy to see that function $v_k = \bar{u}_k - \bar{u}_k^1$ satisfies

$$\begin{cases} \mathscr{L}_k v_k + Mv_k \geqslant 0 & \text{in } Q_T, \\ \mathscr{B}_k v_k \geqslant 0 & \text{on } S_T, \\ v_k(x,0) \geqslant 0 & \text{in } \Omega. \end{cases}$$

The maximum principle gives $v_k \geqslant 0$, i.e., $\bar{u}_k \geqslant \bar{u}_k^1$. In the same way, $\underline{u}_k \leqslant \underline{u}_k^1$.

Let $w_k = \bar{u}_k^1 - \underline{u}_k^1$. Noting $\underline{u} \leqslant \bar{u}$, the monotonicity of f_k shows that w_k satisfies

$$\begin{aligned} \mathscr{L}_k w_k + Mw_k &= f_k\left(\cdot, \bar{u}_k, [\bar{u}]_{i_k}, [\underline{u}]_{d_k}\right) - f_k\left(\cdot, \underline{u}_k, [\underline{u}]_{i_k}, [\bar{u}]_{d_k}\right) + M\left(\bar{u}_k - \underline{u}_k\right) \\ &\geqslant f_k\left(\cdot, \bar{u}_k, [\underline{u}]_{i_k}, [\underline{u}]_{d_k}\right) - f_k\left(\cdot, \underline{u}_k, [\underline{u}]_{i_k}, [\underline{u}]_{d_k}\right) + M\left(\bar{u}_k - \underline{u}_k\right) \\ &\geqslant 0 \quad \text{in } Q_T, \\ \mathscr{B}_k w_k &= 0 \quad \text{on } S_T, \\ w_k(x,0) &= 0 \quad \text{in } \Omega. \end{aligned}$$

By the maximum principle, $w_k \geqslant 0$, i.e., $\bar{u}_k^1 \geqslant \underline{u}_k^1$.

Set $z_k = \bar{u}_k^1 - \bar{u}_k^2$. Then z_k satisfies

$$\begin{aligned} \mathscr{L}_k z_k + Mz_k &= f_k\left(\cdot, \bar{u}_k, [\bar{u}]_{i_k}, [\underline{u}]_{d_k}\right) - f_k\left(\cdot, \bar{u}_k^1, [\bar{u}^1]_{i_k}, [\underline{u}^1]_{d_k}\right) + M\left(\bar{u}_k - \bar{u}_k^1\right) \\ &\geqslant f_k\left(\cdot, \bar{u}_k, [\bar{u}^1]_{i_k}, [\underline{u}]_{d_k}\right) - f_k\left(\cdot, \bar{u}_k^1, [\bar{u}^1]_{i_k}, [\underline{u}]_{d_k}\right) + M\left(\bar{u}_k - \bar{u}_k^1\right) \\ &\geqslant 0 \quad \text{in } Q_T, \\ \mathscr{B}_k z_k &= 0 \quad \text{on } S_T, \\ z_k(x,0) &= 0 \quad \text{in } \Omega. \end{aligned}$$

Thus $z_k \geqslant 0$, i.e., $\bar{u}_k^1 \geqslant \bar{u}_k^2$ by the maximum principle. Inductively

$$\underline{u} \leqslant \underline{u}^i \leqslant \bar{u}^i \leqslant \bar{u} \quad \text{for all } i.$$

As a result, point-wise limits $\lim_{i\to\infty} \underline{u}^i = \tilde{u}$ and $\lim_{i\to\infty} \bar{u}^i = \hat{u}$ exist. Evidently, $\tilde{u} \leqslant \hat{u}$.

Step 2. For any given $Q \Subset Q_T$, we shall show that $\underline{u}^i \to \tilde{u}$ and $\bar{u}^i \to \hat{u}$ in $[C^2(\overline{Q})]^m$. Simplify

$$f_{ki}(x,t) := f_k(x,t,\underline{u}_k^i(x,t),[\underline{u}^i]_{\mathrm{i}_k}(x,t),[\bar{u}^i]_{\mathrm{d}_k}(x,t)),$$

$$f_{ki}^*(x,t) := f_k(x,t,\bar{u}_k^i(x,t),[\bar{u}^i]_{\mathrm{i}_k}(x,t),[\underline{u}^i]_{\mathrm{d}_k}(x,t)).$$

As $\underline{u} \leqslant \underline{u}^i, \bar{u}^i \leqslant \bar{u}$, we conclude $\|f_{ki}\|_\infty, \|f_{ki}^*\|_\infty \leqslant C$, and then by the L^p estimate,

$$\|\underline{u}_k^i\|_{W_p^{2,1}(Q_T)} \leqslant C_1(p), \qquad \|\bar{u}_k^i\|_{W_p^{2,1}(Q_T)} \leqslant C_1(p). \tag{4.9}$$

Take $p > n + 2$. Noticing $W_p^{2,1}(Q_T) \overset{C}{\hookrightarrow} C^{1+\alpha,\,(1+\alpha)/2}(\overline{Q}_T)$ for each $0 < \alpha < 1 - (n + 2)/p$, it is deduced that $\underline{u}^i \to \tilde{u}$ and $\bar{u}^i \to \hat{u}$ in $C^{1+\alpha,\,(1+\alpha)/2}(\overline{Q}_T)$. Thereby both $\tilde{u}$ and $\hat{u}$ satisfy initial and boundary conditions of (4.8).

Besides, $f_{ki}, f_{ki}^* \in C^{\alpha,\,\alpha/2}(\overline{Q}_T)$, and

$$|f_{ki}|_{\alpha,\overline{Q}_T} \leqslant C_2, \qquad |f_{ki}^*|_{\alpha,\overline{Q}_T} \leqslant C_2.$$

Applying Theorem 1.19 and (4.9) we have that $\underline{u}_k^i, \bar{u}_k^i \in C^{2+\alpha,\,1+\alpha/2}(Q_T)$, and for each $Q \Subset Q_T$, we can find a constant $C_3 > 0$ such that

$$|\underline{u}_k^i|_{2+\alpha,\overline{Q}} \leqslant C_3, \qquad |\bar{u}_k^i|_{2+\alpha,\overline{Q}} \leqslant C_3 \text{ for all } i.$$

Since $C^{2+\alpha,\,1+\alpha/2}(\overline{Q}) \overset{C}{\hookrightarrow} C^{2,1}(\overline{Q})$, there exist sub-sequences of $\{\underline{u}^i\}$ and $\{\bar{u}^i\}$ converging to $\tilde{u}$ and $\hat{u}$ in $[C^{2,1}(\overline{Q})]^m$, respectively. Because $\{\underline{u}^i\}$ and $\{\bar{u}^i\}$ are monotonic and bounded, we have $\underline{u}^i \to \tilde{u}$ and $\bar{u}^i \to \hat{u}$ in $[C^{2,1}(\overline{Q})]^m$ by the uniqueness of limits.

Step 3. Letting $i \to \infty$ in

$$\mathscr{L}_k \underline{u}_k^{i+1} + M\underline{u}_k^{i+1} = f_k(x,t,\underline{u}_k^i,[\underline{u}^i]_{\mathrm{i}_k},[\bar{u}^i]_{\mathrm{d}_k}) + M\underline{u}_k^i \quad \text{in } Q_T,$$

$$\mathscr{L}_k \bar{u}_k^{i+1} + M\bar{u}_k^{i+1} = f_k(x,t,\bar{u}_k^i,[\bar{u}^i]_{\mathrm{i}_k},[\underline{u}^i]_{\mathrm{d}_k}) + M\bar{u}_k^i \quad \text{in } Q_T,$$

we conclude that $\tilde{u}$ and $\hat{u}$ satisfy

$$\mathscr{L}_k \tilde{u}_k = f_k(x,t,\tilde{u}_k,[\tilde{u}]_{\mathrm{i}_k},[\hat{u}]_{\mathrm{d}_k}) \quad \text{in } Q_T,$$

$$\mathscr{L}_k \hat{u}_k = f_k(x,t,\hat{u}_k,[\hat{u}]_{\mathrm{i}_k},[\tilde{u}]_{\mathrm{d}_k}) \quad \text{in } Q_T.$$

Step 4. Now, we prove $\tilde{u} = \hat{u}$, which implies that $\tilde{u}$ is a solution of (4.8). Let $w = \tilde{u} - \hat{u}$. Recall $\tilde{u} \leqslant \hat{u}$ and (4.6). It deduces that, for $1 \leqslant k \leqslant m$,

$$\mathscr{L}_k w_k = f_k(x,t,\tilde{u}_k,[\tilde{u}]_{\mathrm{i}_k},[\hat{u}]_{\mathrm{d}_k}) - f_k(x,t,\hat{u}_k,[\hat{u}]_{\mathrm{i}_k},[\tilde{u}]_{\mathrm{d}_k})$$

$$\geqslant -M(|\tilde{u}_k - \hat{u}_k| + |[\tilde{u}]_{\mathrm{i}_k} - [\hat{u}]_{\mathrm{i}_k}| + |[\hat{u}]_{\mathrm{d}_k} - [\tilde{u}]_{\mathrm{d}_k}|)$$

$$= Mw_k + M\sum_{j\neq k} w_j \quad \text{in } Q_T,$$

$$\mathscr{B}_k w_k = 0 \text{ on } S_T, \quad w_k(x,0) = 0 \text{ in } \Omega.$$

Theorem 1.31 confirms $w \geqslant 0$, i.e., $\tilde{u} \geqslant \hat{u}$. And then $\tilde{u} = \hat{u}$.

Step 5. At last, we prove the uniqueness. Suppose that $w \in \langle \underline{u}, \bar{u} \rangle$ is a solution of problem (4.8). Then $w = \mathscr{F}(w, w)$. Similar to arguments of monotonic properties of $\{\underline{u}^i\}$ and $\{\bar{u}^i\}$, we can show $\underline{u}^i \leqslant w \leqslant \bar{u}^i$ for all i. This causes $\tilde{u} \leqslant w \leqslant \hat{u}$, and thereby $\tilde{u} = w = \hat{u}$. □

Before ending this section, it is worthy to point out that, similar to Remark 3.4, we can define strong upper and lower solutions for weakly coupled parabolic system, and the upper and lower solutions method is still effective. Moreover, for divergence weakly coupled parabolic system, we can also define weak upper and lower solutions, and conclusions similar to Theorems 3.18 and 3.19 hold.

4.3 APPLICATIONS

In this section, we shall give some examples as applications of the upper and lower solutions method.

4.3.1 A competition model

We first consider a competition model

$$\begin{cases} u_t - \Delta u = u(a - u - hv), & x \in \Omega, \quad t > 0, \\ v_t - \Delta v = v(b - v - ku), & x \in \Omega, \quad t > 0, \\ \mathscr{B}u = \mathscr{B}v = 0, & x \in \partial\Omega, \ t > 0, \\ u = u_0 \geqslant 0, \quad v = v_0 \geqslant 0, & x \in \Omega, \quad t = 0, \end{cases} \tag{4.10}$$

where a, b, h and k are positive constants, Ω is of class C^2, $\mathscr{B}w = w$ or $\mathscr{B}w = \partial_{\boldsymbol{n}}w$, initial data $u_0, v_0 \in W_p^2(\Omega)$ with $p > 1 + n/2$, and

$u_0|_{\partial\Omega} = v_0|_{\partial\Omega} = 0$ when $\mathscr{B}w = w$,

$u_0, v_0 \in C^1(\overline{\Omega})$, $\partial_{\boldsymbol{n}}u_0|_{\partial\Omega} = \partial_{\boldsymbol{n}}v_0|_{\partial\Omega} = 0$ when $\mathscr{B}w = \partial_{\boldsymbol{n}}w$.

Let us denote

$$f_1(u, v) = u(a - u - hv) \quad \text{and} \quad f_2(u, v) = v(b - v - ku).$$

Then (f_1, f_2) is quasi-monotonically decreasing for $u, v \geqslant 0$, and f_1 and f_2 are Lipschitz continuous in the bounded region of u, v.

Let $\bar{u}, \underline{u}, \bar{v}, \underline{v} \geqslant 0$. Then $((\bar{u}, \bar{v}), (\underline{u}, \underline{v}))$ is a pair of coupled upper and lower solutions of (4.10) if

$$\begin{cases} \bar{u}_t - \Delta\bar{u} \geqslant \bar{u}(a - \bar{u} - h\underline{v}), & x \in \Omega, \ t > 0, \\ \underline{v}_t - \Delta\underline{v} \leqslant \underline{v}(b - \underline{v} - k\bar{u}), & x \in \Omega, \ t > 0, \\ \underline{u}_t - \Delta\underline{u} \leqslant \underline{u}(a - \underline{u} - h\bar{v}), & x \in \Omega, \ t > 0, \\ \bar{v}_t - \Delta\bar{v} \geqslant \bar{v}(b - \bar{v} - k\underline{u}), & x \in \Omega, \ t > 0 \end{cases} \tag{4.11}$$

and

$$\mathscr{B}\bar{u} \geqslant 0 \geqslant \mathscr{B}\underline{u}, \quad \mathscr{B}\bar{v} \geqslant 0 \geqslant \mathscr{B}\underline{v}, \quad x \in \partial\Omega, \ t > 0,$$

$$\bar{u} \geqslant u_0 \geqslant \underline{u}, \quad \bar{v} \geqslant v_0 \geqslant \underline{v}, \qquad x \in \Omega, \quad t = 0.$$

In (4.11), $\bar{u}$ and $\underline{v}$ are coupled together, and $\underline{u}$ and $\bar{v}$ are coupled together. These four inequalities can be decomposed into two groups of independent inequalities.

Take $M_1 = \max\{a, \max_{\overline{\Omega}} u_0\}$ and $M_2 = \max\{b, \max_{\overline{\Omega}} v_0\}$. Then any solution (u, v) of (4.10) in the class of $[C^{2,1}(\overline{\Omega} \times (0, T_{\max}))]^2$ satisfies $0 \leqslant u \leqslant M_1$ and $0 \leqslant v \leqslant M_2$ in $\Omega \times (0, T_{\max})$ by the maximum principle, where $T_{\max}$ is the maximal existence time of (u, v). Moreover, if $u_0 \not\equiv 0$ and $v_0 \not\equiv 0$, then $u, v > 0$ in $\Omega \times (0, T_{\max})$ by the positivity lemma. On the other hand, it is easy to check that $(\bar{u}, \bar{v}) = (M_1, M_2)$ and $(\underline{u}, \underline{v}) = (0, 0)$ are coupled upper lower solutions of (4.10). Therefore, (4.10) has a unique global solution (u, v) with $0 \leqslant u \leqslant M_1, 0 \leqslant v \leqslant M_2$ and $u, v \in C^{2+\alpha, 1+\alpha/2}(\overline{\Omega} \times \mathbb{R}_+)$ by the upper and lower solutions method and the interior regularity (cf. Theorem 2.10).

Case 1: $\mathscr{B}w = w$.

Let (λ, ϕ) be the first eigen-pair of (1.33) with $\|\phi\|_\infty = 1$. Then we have the following results.

Theorem 4.10 *If $a \leqslant \lambda$, then $\lim_{t \to \infty} u(x, t) = 0$ uniformly in $\overline{\Omega}$. Analogously, if $b \leqslant \lambda$, then $\lim_{t \to \infty} v(x, t) = 0$ uniformly in $\overline{\Omega}$.*

The proof of this theorem is similar to that of Exercise 4.7.

Theorem 4.11 *Let $\theta_a(x)$ be the unique positive solution of (3.17) with $a > \lambda$.*

(1) *If $a > \lambda$, $b \leqslant \lambda$ and $u_0 \not\equiv 0$, then $\lim_{t \to \infty} u(x, t) = \theta_a(x)$ and $\lim_{t \to \infty} v(x, t) = 0$ uniformly in $\overline{\Omega}$.*

(2) *If $a \leqslant \lambda$, $b > \lambda$ and $v_0 \not\equiv 0$, then $\lim_{t \to \infty} u(x, t) = 0$ and $\lim_{t \to \infty} v(x, t) = \theta_b(x)$ uniformly in $\overline{\Omega}$.*

Proof. We only prove the conclusion (1), as the proof of conclusion (2) is the same. Assume $a > \lambda$ and $b \leqslant \lambda$. Then $\lim_{t \to \infty} v(x, t) = 0$ uniformly in $\overline{\Omega}$ by Theorem 4.10. The comparison principle together with Theorem 3.16 confirm that

$$\limsup_{t \to \infty} u(x, t) \leqslant \theta_a(x) \quad \text{uniformly in } \overline{\Omega}. \tag{4.12}$$

Take $0 < \varepsilon \ll 1$ so that $a - h\varepsilon > \lambda$. Then there exists $T_\varepsilon \gg 1$ such that $v \leqslant \varepsilon$ in $\overline{\Omega} \times (T_\varepsilon, \infty)$. Thereby

$$\begin{cases} u_t - \Delta u \geqslant u(a - h\varepsilon - u), & x \in \Omega, \quad t > T_\varepsilon, \\ u = 0, & x \in \partial\Omega, \ t > T_\varepsilon. \end{cases}$$

The conditions $u_0(x) \geqslant 0, \not\equiv 0$ imply $u(x, T_\varepsilon) > 0$ in Ω. Let u^ε and θ_a^ε be, respectively, the unique positive solutions of

$$\begin{cases} u_t^\varepsilon - \Delta u^\varepsilon = u^\varepsilon(a - h\varepsilon - u^\varepsilon), & x \in \Omega, \quad t > T_\varepsilon, \\ u^\varepsilon = 0, & x \in \partial\Omega, \ t > T_\varepsilon, \\ u^\varepsilon(x, T_\varepsilon) = u(x, T_\varepsilon), & x \in \Omega \end{cases}$$

and

$$\begin{cases} -\Delta\theta_a^\varepsilon = \theta_a^\varepsilon(a - h\varepsilon - \theta_a^\varepsilon), & x \in \Omega, \\ \theta_a^\varepsilon = 0, & x \in \partial\Omega. \end{cases}$$

The comparison principle asserts that $u \geqslant u^\varepsilon$ in $\overline{\Omega} \times (T_\varepsilon, \infty)$. On account of $u(x, T_\varepsilon) > 0$ in $\overline{\Omega}$, we conclude $\lim_{t\to\infty} u^\varepsilon(x, t) = \theta_a^\varepsilon(x)$ uniformly in $\overline{\Omega}$ by Theorem 3.16, which implies

$$\liminf_{t\to\infty} u(x, t) \geqslant \theta_a^\varepsilon(x) \quad \text{uniformly in } \overline{\Omega}. \tag{4.13}$$

By continuity and uniqueness of solutions we can prove $\lim_{\varepsilon\to 0} \theta_a^\varepsilon(x) = \theta_a(x)$ uniformly in $\overline{\Omega}$. Thus, by (4.13), $\liminf_{t\to\infty} u(x, t) \geqslant \theta_a(x)$ uniformly in $\overline{\Omega}$. Recalling (4.12), we find $\lim_{t\to\infty} u(x, t) = \theta_a(x)$ uniformly in $\overline{\Omega}$. □

In the following, let us investigate the case $a, b > \lambda$. We first handle a special case $a = b$ and $h, k < 1$.

Take $\varepsilon, \sigma > 0$ satisfying $h(1 + \varepsilon) < 1 - \sigma$ and $k(1 + \varepsilon) < 1 - \sigma$. Let

$$\bar{u}_s = (1 + \varepsilon)\theta_a, \quad \bar{v}_s = (1 + \varepsilon)\theta_a, \quad \underline{u}_s = \sigma\theta_a \quad \text{and} \quad \underline{v}_s = \sigma\theta_a.$$

After careful calculations, we have

$$\begin{cases} -\Delta\bar{u}_s \geqslant \bar{u}_s(a - \bar{u}_s - h\underline{v}_s) & \text{in } \Omega, \\ -\Delta\underline{v}_s \leqslant \underline{v}_s(a - \underline{v}_s - k\bar{u}_s) & \text{in } \Omega, \\ \bar{u}_s = \underline{v}_s = 0 & \text{on } \partial\Omega \end{cases} \tag{4.14}$$

and

$$\begin{cases} -\Delta\underline{u}_s \leqslant \underline{u}_s(a - \underline{u}_s - h\bar{v}_s) & \text{in } \Omega, \\ -\Delta\bar{v}_s \geqslant \bar{v}_s(a - \bar{v}_s - k\underline{u}_s) & \text{in } \Omega, \\ \underline{u}_s = \bar{v}_s = 0 & \text{on } \partial\Omega. \end{cases} \tag{4.15}$$

Let $(\bar{u}, \underline{v})$ be the unique solution of (4.10) with $(u_0, v_0) = (\bar{u}_s, \underline{v}_s)$, and set $w = \bar{u} - \bar{u}_s$ and $z = \underline{v}_s - \underline{v}$. In view of (4.14), it then follows that

$$\begin{cases} w_t - \Delta w \leqslant A(x, t)w + h\bar{u}_s z, & x \in \Omega, \quad t > 0, \\ z_t - \Delta z \leqslant B(x, t)z + k\underline{v}_s w, & x \in \Omega, \quad t > 0, \\ w = z = 0, & x \in \partial\Omega, \quad t > 0, \\ w(x, 0) = z(x, 0) = 0, & x \in \Omega, \end{cases}$$

where

$$A(x, t) = a - (\bar{u} + \bar{u}_s) - h\underline{v} \quad \text{and} \quad B(x, t) = a - (\underline{v} + \underline{v}_s) - k\bar{u}.$$

Theorem 1.31 gives $w \leqslant 0$ and $z \leqslant 0$, i.e., $\bar{u} \leqslant \bar{u}_s$ and $\underline{v} \geqslant \underline{v}_s$. Using arguments in proving Theorem 3.11, we can show that $\bar{u}$ and $\underline{v}$ are, respectively, monotonically decreasing and monotonically increasing in time t.

Denote $(\underline{u}, \bar{v})$ as the unique solution of (4.10) with $(u_0, v_0) = (\underline{u}_s, \bar{v}_s)$. In the same way, it can be deduced that $\underline{u} \geqslant \underline{u}_s, \bar{v} \leqslant \bar{v}_s$, and $\underline{u}$ and $\bar{v}$ are, before and after respectively, monotonically increasing and monotonically decreasing in time t. Furthermore, since $\bar{u}_s \geqslant \underline{u}_s$ and $\bar{v}_s \geqslant \underline{v}_s$, similarly to the above it can be shown that

$$\underline{u}_s \leqslant \bar{u}, \quad \underline{v} \leqslant \bar{v}_s, \quad \underline{u} \leqslant \bar{u}_s \quad \text{and} \quad \underline{v}_s \leqslant \bar{v}.$$

From the above discussions, we see that point-wise limits

$$\lim_{t \to \infty} \bar{u}(x, t) = u^*(x), \quad \lim_{t \to \infty} \underline{u}(x, t) = u_*(x),$$

$$\lim_{t \to \infty} \bar{v}(x, t) = v^*(x), \quad \lim_{t \to \infty} \underline{v}(x, t) = v_*(x)$$

exist. Clearly, $\underline{u}_s(x) \leqslant u_*(x), u^*(x) \leqslant \bar{u}_s(x)$ and $\underline{v}_s(x) \leqslant v_*(x), v^*(x) \leqslant \bar{v}_s(x)$. Following the proof of Theorem 2.11 step by step, we can show that each one of $\bar{u}, \underline{u}, \bar{v}$ and $\underline{v}$ satisfies the estimate (2.18). Repeating arguments of Theorem 3.12, it can be derived that, as $t \to \infty$,

$$\bar{u}(x, t) \to u^*(x), \quad \underline{u}(x, t) \to u_*(x), \quad \bar{v}(x, t) \to v^*(x) \quad \text{and} \quad \underline{v}(x, t) \to v_*(x) \quad (4.16)$$

in $C^2(\overline{\Omega})$. Also, as $t \to \infty$,

$$\bar{u}_t(x, t) \to 0, \quad \underline{u}_t(x, t) \to 0, \quad \bar{v}_t(x, t) \to 0 \quad \text{and} \quad \underline{v}_t(x, t) \to 0$$

in $C(\overline{\Omega})$. Therefore, both (u^*, v_*) and (u_*, v^*) are solutions of

$$\begin{cases} -\Delta u = u(a - u - hv) & \text{in} \ \Omega, \\ -\Delta v = v(a - v - ku) & \text{in} \ \Omega, \\ u = v = 0 & \text{on} \ \partial\Omega. \end{cases} \quad (4.17)$$

Hence, we have the following theorem.

Theorem 4.12 *Assume that $a = b > \lambda$ and $h, k < 1$. Then problem (4.17) has at least one positive solution. Moreover, if $u_0 \not\equiv 0$ and $v_0 \not\equiv 0$, then the unique solution (u, v) of (4.10) satisfies*

$$\begin{cases} u_*(x) \leqslant \liminf_{t \to \infty} u(x, t) \leqslant \limsup_{t \to \infty} u(x, t) \leqslant u^*(x), \\ v_*(x) \leqslant \liminf_{t \to \infty} v(x, t) \leqslant \limsup_{t \to \infty} v(x, t) \leqslant v^*(x). \end{cases} \quad (4.18)$$

Proof. It is sufficient to prove (4.18). Take $0 < \varepsilon \ll 1$ so that $h(1 + \varepsilon) < 1$ and $k(1 + \varepsilon) < 1$. In consideration of (4.12), there exists $T \gg 1$ such that

$$u(x, T) \leqslant (1 + \varepsilon)\theta_a(x) \quad \text{and} \quad v(x, T) \leqslant (1 + \varepsilon)\theta_a(x) \quad \text{for all} \ x \in \overline{\Omega}.$$

Besides, we can take $0 < \sigma \ll 1$ verifying that $h(1+\varepsilon), k(1+\varepsilon) \leqslant 1 - \sigma$, and $u(x, T)$, $v(x, T) \geqslant \sigma \theta_a(x)$ in Ω. Thus, we have that $\underline{u} \leqslant u \leqslant \bar{u}$ and $\underline{v} \leqslant v \leqslant \bar{v}$ for $t \geqslant T$ and $x \in \Omega$ by the upper and lower solutions method. Recalling (4.16), the conclusion (4.18) is followed. $\square$

For a general case we have the following result.

Theorem 4.13 *Assume $a, b > \lambda$. If $a - \lambda > hb$ and $b - \lambda > ka$, then problem (4.17) has at least one positive solution. Moreover, if $u_0 \not\equiv 0$ and $v_0 \not\equiv 0$, then the unique solution (u, v) of (4.10) satisfies (4.18).*

Proof. Under our assumptions, there exists $\varepsilon > 0$ for which

$$a - \lambda > hb(1 + \varepsilon) \quad \text{and} \quad b - \lambda > ka(1 + \varepsilon).$$

In consideration of (4.12), there exists $T \gg 1$ such that

$$u(x, T) \leqslant (1 + \varepsilon)\theta_a(x) \quad \text{and} \quad v(x, T) \leqslant (1 + \varepsilon)\theta_b(x) \quad \text{for all } x \in \overline{\Omega}.$$

Take $0 < \delta < \min\{a - \lambda - hb(1 + \varepsilon), \, b - \lambda - ka(1 + \varepsilon)\}$ satisfying

$$\delta\phi(x) \leqslant \min\{u(x, T), v(x, T)\} \quad \text{for all } x \in \Omega.$$

Define
$$\bar{u}_s = (1 + \varepsilon)\theta_a, \quad \underline{u}_s = \delta\phi, \quad \bar{v}_s = (1 + \varepsilon)\theta_b \quad \text{and} \quad \underline{v}_s = \delta\phi.$$

Noting $\theta_a(x) < a$ and $\theta_b(x) < b$ in $\overline{\Omega}$, it can be verified that $(\bar{u}_s, \underline{v}_s)$ and $(\underline{u}_s, \bar{v}_s)$ satisfy (4.14) and (4.15), respectively. Following the proof of Theorem 4.12 step by step, the desired result can be obtained. □

Facts (4.14) and (4.15) show that $((\bar{u}_s, \bar{v}_s), (\underline{u}_s, \underline{v}_s))$ is a pair of coupled ordered upper and lower solutions of (4.17). Same as the scalar case, solutions of (4.10) with initial data $(\bar{u}_s, \underline{v}_s)$ and $(\underline{u}_s, \bar{v}_s)$ are monotone in time t. A system with this property is usually called a monotonic system.

Case 2: $\mathscr{B} = \partial_n$.

If parameters satisfy

$$h < a/b < 1/k, \tag{4.19}$$

then (4.10) has a unique positive constant equilibrium solution $\left(\frac{a - bh}{1 - hk}, \frac{b - ak}{1 - hk}\right) =: (A, B)$.

Theorem 4.14 *Assume $\mathscr{B} = \partial_n$ and (4.19) holds. Then (A, B) is globally asymptotically stable, i.e., the unique solution (u, v) of (4.10) satisfies*

$$\lim_{t \to \infty} u(x, t) = A \quad \text{and} \quad \lim_{t \to \infty} v(x, t) = B$$

uniformly in $\overline{\Omega}$ provided $u_0 \not\equiv 0$ and $v_0 \not\equiv 0$.

Proof. Firstly, the condition (4.19) implies $a > hb$ and $b > ka$. Take $\varepsilon > 0$ satisfying

$$a > h(b + \varepsilon) \quad \text{and} \quad b > k(a + \varepsilon).$$

Note that u and v satisfy

$$u_t - \Delta u \leqslant u(a - u) \quad \text{and} \quad v_t - \Delta v \leqslant v(b - v), \quad x \in \Omega, \, t > 0,$$

respectively. The comparison principle and conclusions of Example 3.2 in §3.2 assert $\limsup_{t\to\infty} u(x,t) \leqslant a$ and $\limsup_{t\to\infty} v(x,t) \leqslant b$ uniformly in $\overline{\Omega}$. Then there exists $t_0 \gg 1$ such that

$$u(x,t_0) \leqslant a + \varepsilon \quad \text{and} \quad v(x,t_0) \leqslant b + \varepsilon \quad \text{for all } x \in \overline{\Omega}. \tag{4.20}$$

Clearly, $u(x,t_0) > 0$ and $v(x,t_0) > 0$ in $\overline{\Omega}$ by the positivity lemma. Take $\sigma > 0$ small such that

$$\sigma < \min\left\{ A, B, a - h(b+\varepsilon),\ b - k(a+\varepsilon),\ \min_{\overline{\Omega}} u(x,t_0),\ \min_{\overline{\Omega}} v(x,t_0) \right\}.$$

We choose $0 < q \ll 1$ satisfying

$$q(A - \sigma) < \sigma[a - \sigma - h(b+\varepsilon)] \quad \text{and} \quad q(B - \sigma) < \sigma[b - \sigma - k(a+\varepsilon)],$$

and define

$$\bar{u} = A + (a + \varepsilon - A)\mathrm{e}^{-q(t-t_0)}, \quad \underline{u} = A - (A - \sigma)\mathrm{e}^{-q(t-t_0)}, \quad t \geqslant t_0,$$
$$\bar{v} = B + (b + \varepsilon - B)\mathrm{e}^{-q(t-t_0)}, \quad \underline{v} = B - (B - \sigma)\mathrm{e}^{-q(t-t_0)}, \quad t \geqslant t_0.$$

Careful computations show that $\bar{u}$, $\bar{v}$, $\underline{u}$ and $\underline{v}$ satisfy (4.11) for $t \geqslant t_0$, and

$$\begin{cases} \bar{u}(x,t_0) = a + \varepsilon \geqslant u(x,t_0), \quad \underline{u}(x,t_0) = \sigma \leqslant u(x,t_0), \quad x \in \overline{\Omega}, \\ \bar{v}(x,t_0) = b + \varepsilon \geqslant v(x,t_0), \quad \underline{v}(x,t_0) = \sigma \leqslant v(x,t_0), \quad x \in \overline{\Omega}. \end{cases} \tag{4.21}$$

Obviously, $\partial_n \bar{u} = \partial_n \underline{u} = \partial_n \bar{v} = \partial_n \underline{v} = 0$ on $\partial\Omega$ for all $t \geqslant t_0$. Subsequently the unique solution (u,v) of (4.10) satisfies

$$A - (A - \sigma)\mathrm{e}^{-q(t-t_0)} \leqslant u(x,t) \leqslant A + (a + \varepsilon - A)\mathrm{e}^{-q(t-t_0)},$$
$$B - (B - \sigma)\mathrm{e}^{-q(t-t_0)} \leqslant v(x,t) \leqslant B + (b + \varepsilon - B)\mathrm{e}^{-q(t-t_0)}$$

for all $x \in \overline{\Omega}$ and $t \geqslant t_0$ by the upper and lower solutions method. Since $q > 0$, the desired conclusions are followed. □

If parameters satisfy

$$a > hb \quad \text{and} \quad ka > b, \tag{4.22}$$

then the possible non-negative and non-trivial constant equilibrium solution of (4.10) is $(a,0)$.

Theorem 4.15 *Assume $\mathscr{B} = \partial_n$. If (4.22) holds, then $(a,0)$ is globally asymptotically stable, i.e., the unique solution (u,v) of (4.10) satisfies*

$$\lim_{t\to\infty} u(x,t) = a \quad \text{and} \quad \lim_{t\to\infty} v(x,t) = 0$$

uniformly in $\overline{\Omega}$ provided $u_0 \not\equiv 0$.

Proof. If $v_0 \equiv 0$, then $v \equiv 0$, and the conclusion is obvious. Now, we assume that $v_0 \not\equiv 0$. Same as the proof of Theorem 4.14, we can find $0 < \varepsilon, \sigma \ll 1$, and $t_0 \gg 1$ to guarantee (4.20) and

$$a - \sigma > h(b + \varepsilon), \quad k(a - \sigma) > b + \varepsilon,$$

$$\sigma < \min\left\{ \min_{\overline{\Omega}} u(x, t_0), \, \min_{\overline{\Omega}} v(x, t_0) \right\}.$$

Take $0 < \beta \ll 1$ so that

$$\beta \leqslant k\sigma + \varepsilon \quad \text{and} \quad (a - \sigma)\beta \leqslant \sigma[a - \sigma - h(b + \varepsilon)].$$

Set $q = ka - b + \sigma + k\varepsilon > 0$ and

$$\bar{u} = a + \varepsilon e^{-a(t - t_0)}, \quad \underline{u} = a - (a - \sigma)e^{-\beta(t - t_0)}, \quad t \geqslant t_0,$$

$$\bar{v} = (b + \varepsilon)e^{-\beta(t - t_0)}, \quad \underline{v} = \sigma e^{-q(t - t_0)}, \quad t \geqslant t_0.$$

After careful calculations, we can verify that $\bar{u}, \underline{v}, \underline{u}$, and $\bar{v}$ satisfy (4.11) for $t \geqslant t_0$ and (4.21) holds. Clearly, $\partial_n \bar{u} = \partial_n \underline{u} = \partial_n \bar{v} = \partial_n \underline{v} = 0$ on $\partial\Omega$ for $t \geqslant t_0$. Therefore, the unique solution (u, v) of (4.10) satisfies

$$a - (a - \sigma)e^{-\beta(t - t_0)} \leqslant u(x, t) \leqslant a + \varepsilon e^{-a(t - t_0)},$$

$$\sigma e^{-q(t - t_0)} \leqslant v(x, t) \leqslant (b + \varepsilon)e^{-\beta(t - t_0)}$$

in $\overline{\Omega} \times [t_0, \infty)$ by the upper and lower solutions method. The desired conclusion is followed. □

4.3.2   A prey-predator model

In this subsection we consider a prey-predator model

$$\begin{cases} u_t - \Delta u = u(a - u - hv), & x \in \Omega, \quad t > 0, \\ v_t - \Delta v = v(ku - b), & x \in \Omega, \quad t > 0, \\ u = v = 0, & x \in \partial\Omega, \ t > 0, \\ u = u_0 \geqslant 0, \quad v = v_0 \geqslant 0, & x \in \Omega, \quad t = 0, \end{cases} \tag{4.23}$$

where a, b, h and k are positive constants, Ω is of class C^2, and initial data $u_0, v_0 \in \overset{\circ}{W}^1_p(\Omega) \cap W^2_p(\Omega)$ with $p > 1 + n/2$. This is a mixed quasi-monotonic system for $u, v \geqslant 0$.

Let $\bar{u}, \underline{u}, \bar{v}, \underline{v} \geqslant 0$. Then $((\bar{u}, \bar{v}), (\underline{u}, \underline{v}))$ is a pair of coupled upper and lower solutions of (4.23) if

$$\begin{cases} \bar{u}_t - \Delta \bar{u} \geqslant \bar{u}(a - \bar{u} - h\underline{v}), & x \in \Omega, \ t > 0, \\ \underline{v}_t - \Delta \underline{v} \leqslant \underline{v}(k\underline{u} - b), & x \in \Omega, \ t > 0, \\ \underline{u}_t - \Delta \underline{u} \leqslant \underline{u}(a - \underline{u} - h\bar{v}), & x \in \Omega, \ t > 0, \\ \bar{v}_t - \Delta \bar{v} \geqslant \bar{v}(k\bar{u} - b), & x \in \Omega, \ t > 0 \end{cases} \tag{4.24}$$

and

$$\begin{cases} \bar{u} \geqslant 0 \geqslant \underline{u}, \ \ \bar{v} \geqslant 0 \geqslant \underline{v}, & x \in \partial\Omega, \ t > 0, \\ \bar{u} \geqslant u_0 \geqslant \underline{u}, \ \ \bar{v} \geqslant v_0 \geqslant \underline{v}, & x \in \Omega, \ \ \ t = 0. \end{cases}$$

In (4.24), $\bar{u}, \underline{u}, \bar{v}$ and $\underline{v}$ are coupled together. We cannot decompose (4.24) into two groups of independent inequalities like the competition model.

Let $\underline{u} = \underline{v} = 0$, $\bar{u} = M := \max\{a, \max_{\overline{\Omega}} u_0(x)\}$, and $\bar{v}$ be a solution of

$$\begin{cases} \bar{v}_t - \Delta\bar{v} = \bar{v}(kM - b), & x \in \Omega, \ \ t > 0, \\ \bar{v} = 0, & x \in \partial\Omega, \ t > 0, \\ \bar{v}(x, 0) = v_0(x), & x \in \Omega, \end{cases}$$

which exists uniquely and globally in time t. Evidently, $((\bar{u}, \bar{v}), (\underline{u}, \underline{v}))$ is a pair of coupled upper and lower solutions of (4.23), and then (4.23) has a unique global solution by the upper and lower solutions method.

Theorem 4.16   *Let (λ, ϕ) be the first eigen-pair of (1.33).*

(1) *If $a \leqslant \lambda$, then $\lim_{t\to\infty} u(x, t) = \lim_{t\to\infty} v(x, t) = 0$ uniformly in $\overline{\Omega}$.*

(2) *If $a > \lambda$ and $ka < b$, then $\lim_{t\to\infty} u(x, t) = u_s(x)$ and $\lim_{t\to\infty} v(x, t) = 0$ uniformly in $\overline{\Omega}$ provided $u_0 \not\equiv 0$, where $u_s(x)$ is the unique positive solution of (3.17).*

The proof of Theorem 4.16 is left to readers as an exercise.

4.3.3   The Belousov-Zhabotinskii reaction model

In this subsection, we study the following Belousov-Zhabotinskii reaction model

$$\begin{cases} u_t - d_1\Delta u = u(1 - u - cv), & x \in \Omega, \ \ t > 0, \\ v_t - d_2\Delta v = bw - v - kuv, & x \in \Omega, \ \ t > 0, \\ w_t - d_3\Delta w = h(u - w), & x \in \Omega, \ \ t > 0, \\ u = v = w = 0, & x \in \partial\Omega, \ t > 0, \\ (u, v, w) = (u_0, v_0, w_0), & x \in \Omega, \ \ t = 0, \end{cases} \tag{4.25}$$

where coefficients are all positive constants, Ω is of class C^2, and initial data u_0, v_0, w_0 satisfy

$$u_0, v_0, w_0 \in C_0^1(\overline{\Omega}) \cap W_p^2(\Omega) \text{ with } p > 1 + n/2 \text{ and } u_0, v_0, w_0 \geqslant 0, \not\equiv 0 \text{ in } \Omega.$$

This is a mixed quasi-monotonic system for $u, v, w \geqslant 0$.

Let $\bar{u}, \underline{u}, \bar{v}, \underline{v}, \bar{w}, \underline{w} \geqslant 0$. Then $((\bar{u}, \bar{v}, \bar{w}), (\underline{u}, \underline{v}, \underline{w}))$ is a pair of coupled upper and lower solutions of (4.25) if

$$\begin{cases} \bar{u}_t - d_1\Delta\bar{u} \geqslant \bar{u}(1 - \bar{u} - c\underline{v}), & x \in \Omega, \ t > 0, \\ \underline{v}_t - d_2\Delta\underline{v} \leqslant b\underline{w} - \underline{v} - k\bar{u}\underline{v}, & x \in \Omega, \ t > 0, \\ \underline{w}_t - d_3\Delta\underline{w} \leqslant h(\underline{u} - \underline{w}), & x \in \Omega, \ t > 0, \\ \underline{u}_t - d_1\Delta\underline{u} \leqslant \underline{u}(1 - \underline{u} - c\bar{v}), & x \in \Omega, \ t > 0, \\ \bar{v}_t - d_2\Delta\bar{v} \geqslant b\bar{w} - \bar{v} - k\underline{u}\bar{v}, & x \in \Omega, \ t > 0, \\ \bar{w}_t - d_3\Delta\bar{w} \geqslant h(\bar{u} - \bar{w}), & x \in \Omega, \ t > 0 \end{cases} \tag{4.26}$$

and

$$\begin{cases} \bar{u} \geqslant 0 \geqslant \underline{u}, \ \ \bar{v} \geqslant 0 \geqslant \underline{v}, \ \ \bar{w} \geqslant 0 \geqslant \underline{w}, & x \in \partial\Omega, \ t > 0, \\ \bar{u} \geqslant u_0 \geqslant \underline{u}, \ \ \bar{v} \geqslant v_0 \geqslant \underline{v}, \ \ \bar{w} \geqslant w_0 \geqslant \underline{w}, & x \in \Omega, \ \ t = 0. \end{cases}$$

In (4.26), $\bar{u}, \underline{u}, \bar{v}, \underline{v}, \bar{w}$ and $\underline{w}$ are coupled together, and so (4.26) cannot be decomposed into a simple system of inequalities.

Definition 4.17 *Problem* (4.25) *is referred to as* permanent, *if initial data* $u_0, v_0, w_0 \geqslant 0, \not\equiv 0$ *in* Ω, *then for any* $\Omega_0 \Subset \Omega$ *we have*

$$\liminf_{t\to\infty} \min_{\Omega_0} u(x,t) > 0, \ \ \liminf_{t\to\infty} \min_{\Omega_0} v(x,t) > 0 \ \ and \ \ \liminf_{t\to\infty} \min_{\Omega_0} w(x,t) > 0.$$

Let (λ, ϕ) be the first eigen-pair of (1.33). Normalize ϕ by $\|\phi\|_\infty = 1$.

Theorem 4.18 ([105]) *If* $d_1\lambda < 1 - bc$, *then* (4.25) *is permanent.*

Proof. The proof will be divided into three steps.

Step 1. Owing to $d_1\lambda < 1 - bc < 1$, the elliptic problem

$$\begin{cases} -d_1\Delta u = u(1 - u), & x \in \Omega, \\ u = 0, & x \in \partial\Omega \end{cases}$$

has a unique positive solution $\hat{u}_s(x)$. Because of $h > 0$, the linear problem

$$\begin{cases} -d_3\Delta w + hw = h\hat{u}_s(x), & x \in \Omega, \\ w = 0, & x \in \partial\Omega \end{cases}$$

has a unique positive solution $\hat{w}_s(x)$. Analogously, the linear problem

$$\begin{cases} -d_2\Delta v + v = b\hat{w}_s(x), & x \in \Omega, \\ v = 0, & x \in \partial\Omega \end{cases}$$

admits a unique positive solution $\hat{v}_s(x)$. That is, $(\hat{u}_s, \hat{v}_s, \hat{w}_s)$ is the unique positive solution of

$$\begin{cases} -d_1\Delta u = u(1 - u), & x \in \Omega, \\ -d_2\Delta v = bw - v, & x \in \Omega, \\ -d_3\Delta w = h(u - w), & x \in \Omega, \\ u = v = w = 0, & x \in \partial\Omega. \end{cases} \tag{4.27}$$

Step 2. Since initial data $u_0, v_0, w_0 \in C^1(\overline{\Omega})$, invoking Lemma 3.7, we can find a constant $K \gg 1$ so that

$$(u_0(x), v_0(x), w_0(x)) \leqslant K(\hat{u}_s(x), \hat{v}_s(x), \hat{w}_s(x)) \quad \text{in } \overline{\Omega}.$$

Evidently, $(K\hat{u}_s, K\hat{v}_s, K\hat{w}_s)$ is an upper solution of (4.27). Let $(\hat{u}, \hat{v}, \hat{w})$ be the unique solution of

$$
\begin{cases}
\hat{u}_t - d_1\Delta\hat{u} = \hat{u}(1 - \hat{u}), & x \in \Omega, \quad t > 0, \\
\hat{v}_t - d_2\Delta\hat{v} = b\hat{w} - \hat{v}, & x \in \Omega, \quad t > 0, \\
\hat{w}_t - d_3\Delta\hat{w} = h(\hat{u} - \hat{w}), & x \in \Omega, \quad t > 0, \\
u = v = w = 0, & x \in \partial\Omega, \quad t > 0, \\
(\hat{u}, \hat{v}, \hat{w}) = (K\hat{u}_s, K\hat{v}_s, K\hat{w}_s), & x \in \Omega, \quad t = 0.
\end{cases}
$$

Then $((\hat{u}, \hat{v}, \hat{w}), (0, 0, 0))$ is a pair of coupled upper and lower solutions of (4.25), and therefore

$$0 \leqslant u \leqslant \hat{u}, \quad 0 \leqslant v \leqslant \hat{v}, \quad 0 \leqslant w \leqslant \hat{w}$$

by the upper and lower solutions method. As $u_0, v_0, w_0 \geqslant 0, \not\equiv 0$ in Ω, we have $u, v, w > 0$ in $\Omega \times \mathbb{R}_+$ by the positivity lemma. Using the same discussions as in the proof of Theorem 3.11 we can prove that $(\hat{u}, \hat{v}, \hat{w})$ is monotonically decreasing in time t, and

$$\lim_{t \to \infty}(\hat{u}(x, t), \hat{v}(x, t), \hat{w}(x, t)) = (\hat{u}_s(x), \hat{v}_s(x), \hat{w}_s(x))$$

uniformly in $\overline{\Omega}$. On the other hand, $0 < \hat{u}_s < 1$, $0 < \hat{w}_s < 1$ and $0 < \hat{v}_s < b$ in Ω by the maximum principle. Then there exists $t_0 \gg 1$ such that

$$0 < u < 1, \quad 0 < v < b, \quad 0 < w < 1 \quad \text{in } \Omega \times [t_0, \infty).$$

Furthermore, in consideration of Lemma 3.7, there exists $\varepsilon > 0$ so that

$$u(x, t_0) \geqslant \varepsilon\phi(x), \quad v(x, t_0) \geqslant \varepsilon\phi(x), \quad \text{and} \quad w(x, t_0) \geqslant \varepsilon\phi(x) \quad \text{for all } x \in \Omega.$$

Step 3. Take $0 < \theta_i < \varepsilon$ $(i = 1, 2, 3)$ satisfying

$$\theta_1 \leqslant 1 - bc - d_1\lambda, \quad \theta_3 \leqslant h\theta_1/(h + d_3\lambda) \quad \text{and} \quad \theta_2 \leqslant b\theta_3/(1 + k + d_2\lambda).$$

We can show that $((1, b, 1), (\theta_1\phi, \theta_2\phi, \theta_3\phi))$ is a pair of coupled upper and lower solutions of (4.25) with initial time t_0 and initial datum $(u(\cdot, t_0), v(\cdot, t_0), w(\cdot, t_0))$. Consequently, any solution (u, v, w) of (4.25) satisfies

$$(\theta_1\phi(x), \theta_2\phi(x), \theta_3\phi(x)) \leqslant (u(x, t), v(x, t), w(x, t)) \leqslant (1, b, 1)$$

in $\overline{\Omega} \times [t_0, \infty)$ by the upper and lower solutions method. This indicates that (4.25) is permanent. $\qquad\square$

EXERCISES

4.1 Prove Theorem 4.2.

4.2 Verify that Definition 4.3 and Definition 4.7 are equivalent if f is mixed quasi-monotonous in $\langle \underline{\psi}, \overline{\psi} \rangle$ and satisfies the Lipschitz condition (4.6).

4.3 In the proofs of Theorems 4.14 and 4.15, verify that functions $\overline{u}$, $\underline{v}$, $\underline{u}$ and $\overline{v}$ satisfy (4.11).

4.4 Prove Theorem 4.16.

4.5 Let $u_0, v_0 \in \overset{\circ}{W}{}^1_p(\Omega) \cap W^2_p(\Omega)$ with $p > 1 + n/2$. Use the upper and lower solutions method to prove that there is $T > 0$ such that the problem

$$
\begin{cases}
u_t - d_1 \Delta u = u(a + uv) & \text{in } Q_T, \\
v_t - d_2 \Delta v = v(b + kuv) & \text{in } Q_T, \\
u = v = 0 & \text{on } S_T, \\
u(x,0) = u_0(x), \quad v(x,0) = v_0(x) & \text{in } \Omega
\end{cases}
$$

has a unique solution, where d_1, d_2, a, b and k are positive constants.

4.6 Let $u_0, v_0 \in \overset{\circ}{W}{}^1_p(\Omega) \cap W^2_p(\Omega)$ with $p > 1 + n/2$, and $u_0, v_0 > 0$. Use the upper and lower solutions method to prove that the problem

$$
\begin{cases}
u_t - d_1 \Delta u = u(a - v), & x \in \Omega, \quad t > 0, \\
v_t - d_2 \Delta v = v(b + ku), & x \in \Omega, \quad t > 0, \\
u = v = 0, & x \in \partial\Omega, \ t > 0, \\
u = u_0, \quad v = v_0, & x \in \Omega, \quad t = 0
\end{cases}
$$

has a unique global solution, where d_1, d_2, a, b and k are as above.

4.7 Consider the following initial-boundary value problem

$$
\begin{cases}
u_{it} - d_i \Delta u_i = u_i[a_i + f_i(u_1, u_2)] & \text{in } Q_\infty, \\
u_i = 0 & \text{on } S_\infty, \\
u(x,0) = \varphi_i(x), \quad i = 1, 2 & \text{in } \Omega,
\end{cases}
\tag{P4.1}
$$

where a_i, d_i are positive constants, $\varphi_i \in \overset{\circ}{W}{}^1_p(\Omega) \cap W^2_p(\Omega)$ with $p > 1 + n/2$ and $\varphi_i \geqslant 0, i = 1, 2$. Assume that $f_i : [0, \infty) \times [0, \infty) \to \mathbb{R}$ satisfies

$$
f_i(0,0) = 0, \quad \frac{\partial f_i(u_1, u_2)}{\partial u_j} < 0, \quad i, j = 1, 2
$$

and

$$
a_1 + f_1(c_1, 0) \leqslant 0, \quad a_2 + f_2(0, c_2) \leqslant 0
$$

for some positive constants c_1 and c_2. Let λ be the first eigenvalue of (1.33). Prove that if $a_i < d_i \lambda$ then the solution of (P4.1) satisfies $\lim_{t \to \infty} u_i(x,t) = 0$ uniformly in $\overline{\Omega}$, $i = 1, 2$.

4.8 Verify that $((1, b, 1), (\theta_1\phi, \theta_2\phi, \theta_3\phi))$ is a pair of coupled upper and lower solutions of (4.25) with initial time t_0 and initial datum $(u(\cdot, t_0), v(\cdot, t_0), w(\cdot, t_0))$ in proving Theorem 4.18.

4.9 Use the upper and lower solutions method to investigate existence, uniqueness and longtime behaviours of global solutions of the following problem

$$\begin{cases} u_t - d_1\Delta u = r_1 u \left(1 - \dfrac{u}{a_1 + b_1 v} - c_1 u\right), & x \in \Omega, \quad t > 0, \\ v_t - d_2\Delta v = r_2 v \left(1 - \dfrac{v}{a_2 + b_2 u} - c_2 v\right), & x \in \Omega, \quad t > 0, \\ u = v = 0, & x \in \partial\Omega, \ t > 0, \\ u = u_0 > 0, \quad v = v_0 > 0, & x \in \Omega, \quad t = 0, \end{cases}$$

where d_i, r_i, a_i, b_i and c_i are positive constants, $u_0, v_0 \in \overset{\circ}{W}{}^1_p(\Omega) \cap W^2_p(\Omega)$ with $p > 1 + n/2$.

4.10 Let Γ_1 and Γ_2 be given in (1.3). Consider the problem

$$\begin{cases} \mathscr{L}_k u_k = f_k(x, t, u) & \text{in } Q_T, \\ \mathscr{B}_k u_k = g_k & \text{on } S_T, \\ u_k(x, 0) = \varphi_k(x) & \text{in } \Omega, \\ 1 \leqslant k \leqslant m, \end{cases}$$

where $\mathscr{L}_k$ are as in (4.1), and for each $1 \leqslant k \leqslant m$,

$$\mathscr{B}_k u_k = \partial_n u_k + b_k u_k \ \text{ on } \Gamma_1 \times (0, T], \quad \mathscr{B}_k u_k = u_k \ \text{ on } \Gamma_2 \times (0, T],$$

or

$$\mathscr{B}_k u_k = \partial_n u_k + b_k u_k \ \text{ on } \Gamma_2 \times (0, T], \quad \mathscr{B}_k u_k = u_k \ \text{ on } \Gamma_1 \times (0, T].$$

As Exercise 3.13, discuss the upper and lower solutions method for such a problem.

4.11 Let $\Sigma \subset \mathbb{R}^2$ be a domain, and the system $\{f, g\}$ is quasi-monotonous in Σ (quasi-monotonically increasing or quasi-monotonically decreasing). Consider the following problem

$$\begin{cases} u_t - \Delta u = f(u, v) & \text{in } Q_T, \\ v_t - \Delta v = g(u, v) & \text{in } Q_T, \\ u = v = 0 & \text{on } S_T, \\ u(x, 0) = u_0(x), \ v(x, 0) = v_0(x) & \text{in } \Omega, \end{cases} \tag{P4.2}$$

where $u_0, v_0 \in \overset{\circ}{W}{}^1_p(\Omega) \cap W^2_p(\Omega)$ with $p > 1 + n/2$, and $(u_0(x), v_0(x)) \in \Sigma$ for all $x \in \overline{\Omega}$. Discuss the following issues.

(1) Define the weak solution and coupled weak upper and lower solutions of (P4.2) in $\Sigma^{\dagger}$.

(2) Attach appropriate conditions to the system $\{f, g\}$ and prove the ordering of weak upper and lower solutions.

(3) Let $((\bar{u}, \bar{v}), (\underline{u}, \underline{v}))$ be a pair of coupled weak upper and lower solutions of problem (P4.2). Prove that (P4.2) has a unique weak solution (u, v) between $(\underline{u}, \underline{v})$ and $(\bar{u}, \bar{v})$.

(4) Let $((\bar{u}_1, \bar{v}_1), (\underline{u}_1, \underline{v}_1))$ and $((\bar{u}_2, \bar{v}_2), (\underline{u}_2, \underline{v}_2))$ be two pairs of coupled classical (strong) upper and lower solutions of (P4.2) in Σ. Define

$$\bar{u} = \min\{\bar{u}_1, \bar{u}_2\}, \quad \bar{v} = \min\{\bar{v}_1, \bar{v}_2\}, \quad \underline{u} = \max\{\underline{u}_1, \underline{u}_2\}, \quad \underline{v} = \max\{\underline{v}_1, \underline{v}_2\}.$$

Explore some conditions on the system $\{f, g\}$ and $\bar{u}_i, \bar{v}_i, \underline{u}_i, \underline{v}_i$ $(i = 1, 2)$ to prove that $((\bar{u}, \bar{v}), (\underline{u}, \underline{v}))$ is a pair of coupled weak upper and lower solutions of (P4.2) in Σ.

(5) Give a statement similar to Remark 3.20 for problem (P4.2).

---

†A solution (u, v) of (P4.2) in Σ means that (u, v) is a solution of (P4.2) and the range of (u, v) is in Σ.

CHAPTER 5

Stability Analysis

A basic question in the study of nonlinear parabolic equations is whether a time-dependent solution, as time increases, remains in a neighbourhood of an equilibrium solution and whether it converges to this equilibrium solution as $t \to \infty$. From a practical point of view, it is important to know that, for a given equilibrium solution u_s, whether there is a set of initial data such that their corresponding time-dependent solutions converge to u_s as $t \to \infty$. This leads to a question of stability and asymptotic stability of an equilibrium solution and its stability region.

The stability of an equilibrium solution is not only an important part in the study of partial differential equations, but also has an obvious physical meaning. When an equilibrium solution is globally asymptotically stable, it means that the corresponding physical phenomenon that the equation models is stable and has a unique stable state. In Chapter 3 and Chapter 4, it has been introduced how to deal with the stability of an equilibrium solution by using the upper and lower solutions method. In this chapter, we first include the local stability of equilibrium solutions by using the linearized principal eigenvalue and upper and lower solutions method. The result that linear stability implies local stability is established. Then we introduce three other ways to establish the global stability of equilibrium solutions, namely, *Lyapunov functional method, iteration method* and *average method*.

Throughout this chapter, we assume that Ω is of class $C^{2+\alpha}$ and initial data satisfy the corresponding compatibility conditions. Consider the following problem of *weakly coupled parabolic system*

$$\begin{cases} u_t + L[u] = f(x, u) & \text{in } Q_\infty, \\ B[u] = g(x) & \text{on } S_\infty, \end{cases} \tag{5.1}$$

where $L[u] = (L_1[u_1], \ldots, L_m[u_m])$, $B[u] = (B_1[u_1], \ldots, B_m[u_m])$ and

$$L_k[u_k] = -a_{ij}^k(x)D_{ij}u_k + b_i^k(x)D_i u_k,$$

$B_k[u_k] = u_k$ or $B_k[u_k] = \partial_n u_k + b^k(x)u_k$. We assume that $\partial_t + L_k$ is strongly parabolic in Q_T and $b_i^k \in L^\infty(\Omega)$ for all $1 \leqslant k \leqslant m$ and $1 \leqslant i \leqslant n$.

The equilibrium problem of (5.1) is

$$\begin{cases} L[u] = f(x, u) & \text{in } \Omega, \\ B[u] = g(x) & \text{on } \partial\Omega. \end{cases} \tag{5.2}$$

Let $u(x, t; u_0)$ be a solution of (5.1) with initial datum u_0.

Definition 5.1 *Let X be a Banach space and Y be a subset of X. A solution $u_s \in Y$ of (5.2) is said to be* locally stable *in Y, if for any given $\varepsilon > 0$, there exists a constant $\delta > 0$ such that when $u_0 \in B_\delta(u_s; Y)$, the solution $u(x, t; u_0)$ satisfies*

$$\{u(x, t; u_0)\}_{t \geqslant 0} \subset Y \quad and \quad \|u(x, t; u_0) - u_s(x)\|_X \leqslant \varepsilon \quad for \ all \ t > 0,$$

where $B_\delta(u_s; Y)$ is a neighbourhood of u_s in Y. Otherwise, we call that u_s is unstable *in Y.*

If u_s is locally stable in Y and

$$\lim_{t \to \infty} \|u(x, t; u_0) - u_s(x)\|_X = 0, \tag{5.3}$$

we call that u_s is locally asymptotically stable *in Y. If this limit is uniform with respect to $u_0 \in B_\delta(u_s; Y)$, we say that u_s is* locally uniformly asymptotically stable *in Y.*

If $\{u(x, t; u_0)\}_{t \geqslant 0} \subset Y$ and (5.3) holds for all $u_0 \in Y$, we say that u_s is globally asymptotically stable *in Y.*

Let u_s be a solution of (5.2). The eigenvalue problem of the linearization of (5.2) at u_s is

$$\begin{cases} -L[\phi] + f_u(x, u_s)\phi = \lambda\phi & \text{in } \Omega, \\ B[\phi] = 0 & \text{on } \partial\Omega. \end{cases} \tag{5.4}$$

Definition 5.2 *We say that u_s is* linearly stable *if each eigenvalue λ of (5.4) has a negative real part. If problem (5.4) has an eigenvalue whose real part is positive, then we say that u_s is* not linearly stable. *We say that u_s is* degenerate *if $\lambda = 0$ is an eigenvalue of (5.4). Otherwise, u_s is* non-degenerate.

5.1 LOCAL STABILITY DETERMINED BY THE PRINCIPAL EIGENVALUE

For scalar case, i.e., $m = 1$, we intend to use the upper and lower solutions method to prove that the local stability of u_s can be determined by the sign of the principal eigenvalue of (5.4), namely, the linear stability implies the local stability. Let $\lambda_1(u_s)$ be the principal eigenvalue of (5.4) and ϕ be the corresponding positive eigenfunction with $\|\phi\|_\infty = 1$. It is well-known that $\lambda_1(u_s)$ is real.

Theorem 5.3 *Let $f \in C^{\alpha,1}(\overline{\Omega} \times \mathbb{R})$ and $f_u(x, \cdot) \in C(\mathbb{R})$ be uniformly with respect to $x \in \overline{\Omega}$. Let u be the unique solution of (5.1) with initial datum u_0.*

(1) *If $\lambda_1(u_s) < 0$, then there exist constants $0 < \rho, k \ll 1$ such that*

$$|u(x, t) - u_s(x)| \leqslant \rho e^{-kt}\phi(x) \quad for \ all \ t > 0, \ x \in \overline{\Omega}$$

provided that it holds at $t = 0$.

(2) *If $\lambda_1(u_s) > 0$, then there exists $\rho > 0$ such that for any $0 < \sigma < 1$ we can find $0 < k \ll 1$ for which*

$$u(x,t) \geqslant u_s(x) + \rho\left(1 - \sigma e^{-kt}\right)\phi(x) \quad \text{for all } t > 0, \ x \in \overline{\Omega}$$

provided that it holds at $t = 0$.

(3) *In the case $\lambda_1(u_s) = 0$, we assume that f_{uu} is continuous. If $f_{uu}(x, u_s(x)) > 0$ for all $x \in \overline{\Omega}$, then u_s is not locally asymptotically stable.*

Proof. We first prove the conclusion (1). Let $\bar{u} = u_s(x) + \rho e^{-kt}\phi(x)$. Then we have

$$\bar{u}_t + L[\bar{u}] - f(x, \bar{u}) = L[u_s] + \rho e^{-kt}L[\phi] - k\rho e^{-kt}\phi - f(x, \bar{u})$$
$$= f(x, u_s) - f(x, \bar{u}) + [f_u(x, u_s) - \lambda_1(u_s) - k]\rho e^{-kt}\phi.$$

Using $\lambda_1(u_s) < 0$ we see that when $\rho, k > 0$ are small enough,

$$f(x, u_s) - f(x, \bar{u}) + [f_u(x, u_s) - \lambda_1(u_s) - k]\rho e^{-kt}\phi$$
$$= [f_u(x, u_s) - f_u(x, u_s + \theta\phi) - \lambda_1(u_s) - k]\rho e^{-kt}\phi > 0, \quad 0 \leqslant \theta \leqslant \rho,$$

as $f_u(x, u_s) - f_u(x, u_s + \theta\phi) \to 0$ when $\rho \to 0$. Therefore, $\bar{u}$ is an upper solution provided $u_0(x) \leqslant u_s(x) + \rho\phi(x)$.

Analogously, $\underline{u} = u_s(x) - \rho e^{-kt}\phi(x)$ is a lower solution provided that $0 < \rho, k \ll 1$ and $u_0(x) \geqslant u_s(x) - \rho\phi(x)$.

By the upper and lower solutions method, the unique solution u of (5.1) with initial datum u_0 satisfies

$$u_s(x) - \rho e^{-kt}\phi(x) \leqslant u(x,t) \leqslant u_s(x) + \rho e^{-kt}\phi(x) \quad \text{for all } x \in \overline{\Omega}, \ t \geqslant 0,$$

if $u_s(x) - \rho\phi(x) \leqslant u_0(x) \leqslant u_s(x) + \rho\phi(x)$.

The proof of (2) is left to readers as an exercise.

Now we prove the conclusion (3). Assume $\lambda_1(u_s) = 0$ and $f_{uu}(x, u_s(x)) > 0$ for all $x \in \overline{\Omega}$. Let $\varepsilon > 0$ and $\bar{u} = u_s + \varepsilon\phi$. Then we have, by using Taylor's formula,

$$L[\bar{u}] - f(x, \bar{u}) = L[u_s] + \varepsilon L[\phi] - f(x, u_s + \varepsilon\phi)$$
$$= f(x, u_s) + \varepsilon f_u(x, u_s)\phi - f(x, u_s + \varepsilon\phi)$$
$$= -\frac{1}{2}\varepsilon^2\phi^2 f_{uu}(x, u_s + \tau\phi) \quad \text{with } 0 \leqslant \tau \leqslant \varepsilon$$
$$< 0$$

when $0 < \varepsilon \ll 1$ by the continuity of f_{uu}. This shows that $u_s(x) + \varepsilon\phi(x)$ is a lower solution of (5.2). Take $u_0(x) = u_s(x) + \varepsilon\phi(x)$. Then the solution u of (5.1) with initial datum u_0 is strictly monotone increasing by Theorem 3.11. And so, $u > u_s(x) + \varepsilon\phi(x)$ in $\Omega \times \mathbb{R}_+$. Which implies that u_s is not locally asymptotically stable. $\square$

Example 5.1 *Consider the problem*

$$\begin{cases} -\Delta u = (a - u)^2(1 - u) =: f(u) & \text{in } \Omega, \\ \partial_{\boldsymbol{n}}u = 0 & \text{on } \partial\Omega, \end{cases}$$

where $0 < a < 1$ is a constant.

It has two positive constant equilibria a and 1. From relations

$$f'(u) = (u-a)(2+a-3u) \quad \text{and} \quad f''(u) = 2 + 4a - 6u,$$

we see that

$$f'(a) = 0, \quad f'(1) < 0 \quad \text{and} \quad f''(a) > 0.$$

Thus, $\lambda_1(a) = 0$ and $\lambda_1(1) = f'(1) < 0$. It follows from Theorem 5.3 that $u_s = a$ is not locally asymptotically stable and $u_s = 1$ is locally stable.

5.2 LYAPUNOV FUNCTIONAL METHOD

The Lyapunov function (functional) method is a powerful tool in the study of stability of equilibrium point (solution) of differential equations, especially parabolic partial differential equations. In this section we shall introduce the *Lyapunov functional method* to study global stability of the unique equilibrium solution of reaction-diffusion systems including the system coupled by reaction-diffusion equations and ordinary differential equations.

5.2.1 Abstract results

Let $\tau_1, \ldots, \tau_m$ be non-negative constants, and $u_1(x,t), \ldots, u_m(x,t)$ be functions of (x,t). We define $\tau = (\tau_1, \ldots, \tau_m)$, $u(x,t) = (u_1(x,t), \ldots, u_m(x,t))$ and $u^\tau(x,t) = (u_1(x, t-\tau_1), \ldots, u_m(x, t-\tau_m))$. Let l, k and m be positive integers with $1 \leqslant l < k < m$. Consider the following initial-boundary value problem

$$\begin{cases} u_{it} - d_i \Delta u_i = f_i(x, u, u^\tau), & x \in \Omega, \quad t > 0, & 1 \leqslant i \leqslant k, \\ u_{it} = f_i(x, u, u^\tau), & x \in \Omega, \quad t > 0, & k+1 \leqslant i \leqslant m, \\ u_i = 0, & x \in \partial\Omega, \ t > 0, & 1 \leqslant i \leqslant l, \\ \partial_{\boldsymbol{n}} u_i = 0, & x \in \partial\Omega, \ t > 0, & l+1 \leqslant i \leqslant k, \\ u_i(x,t) = u_{i0}(x,t), & x \in \Omega, \quad -\tau_i \leqslant t \leqslant 0, & 1 \leqslant i \leqslant m, \end{cases} \quad (5.5)$$

where Ω is of class C^2, $d_i > 0$ are constants $(1 \leqslant i \leqslant k)$, and $u_{i0}(\cdot, t) \in C^\alpha(\overline{\Omega})$ uniformly with respect to $t \in [-\tau_i, 0]$, $i = 1, \ldots, m$. This is an initial-boundary value problem of system coupled by reaction-diffusion equations and ordinary differential equations.

Define $d_i = 0$ for $k+1 \leqslant i \leqslant m$. We make the following assumptions.

(H1) The corresponding equilibrium problem of (5.5)

$$\begin{cases} -d_i \Delta u_i = f_i(x, u, u) & \text{in } \Omega, \quad 1 \leqslant i \leqslant m, \\ u_i = 0 & \text{on } \partial\Omega, \ 1 \leqslant i \leqslant l, \\ \partial_{\boldsymbol{n}} u_i = 0 & \text{on } \partial\Omega, \ l+1 \leqslant i \leqslant k \end{cases}$$

has a unique solution $\tilde{u}(x) = (\tilde{u}_1(x), \ldots, \tilde{u}_m(x))$.

(H2) Problem (5.5) has a unique global solution u. Moreover, there exist positive constants α and ρ with $0 < \alpha < 1$, such that, for all $1 \leqslant i \leqslant m$,

$$u_i - \tilde{u}_i, \, d_i \nabla(u_i - \tilde{u}_i) \in C^{\alpha}([\rho, \infty), L^2(\Omega))$$

and

$$\|u_i - \tilde{u}_i\|_{C^{\alpha}([\rho,\infty), L^2(\Omega))} + d_i \|\nabla(u_i - \tilde{u}_i)\|_{C^{\alpha}([\rho,\infty), L^2(\Omega))} \leqslant C,$$

where C is a constant depending only on α and ρ. Here, function $u \in C^{\alpha}([\rho, \infty), L^2(\Omega))$ means that

$$\sup_{t,s \geqslant \rho, \, t \neq s} \frac{\|u(\cdot,t) - u(\cdot,s)\|_{2,\Omega}}{|t-s|^{\alpha}} < \infty.$$

We mention that, in general, the regularity and estimate in **(H2)** can be obtained by Theorems 2.11 and 2.13.

Theorem 5.4 (Lyapunov functional method [114]) *Let conditions* **(H1)** *and* **(H2)** *hold. Assume that there exists a function $E(u, \tilde{u})$ such that, along the orbit u of (5.5), the functional*

$$F(t) = \int_{\Omega} E(u(x,t), \tilde{u}(x))$$

is non-negative and satisfies

$$F'(t) \leqslant -A \sum_{i=1}^{k} \int_{\Omega} |\nabla u_i(x,t) - \nabla \tilde{u}_i(x)|^2 - B \sum_{i=l+1}^{m} \int_{\Omega} |u_i(x,t) - \tilde{u}_i(x)|^2$$

for some positive constants A, B. Then

$$\lim_{t \to \infty} \|u_i(\cdot, t) - \tilde{u}_i(\cdot)\|_{H^1(\Omega)} = 0 \quad for \ \ 1 \leqslant i \leqslant k,$$

$$\lim_{t \to \infty} \|u_i(\cdot, t) - \tilde{u}_i(\cdot)\|_{2,\Omega} = 0 \quad for \ \ k+1 \leqslant i \leqslant m.$$

Theorem 5.4 can be deduced from the following lemma directly, and the proof is omitted here.

Lemma 5.5 *Let a and b be positive constants, $\varphi \in C^1([a, \infty))$ and φ be bounded from below. Assume that $\psi \geqslant 0$, $\int_a^{\infty} h(t) \mathrm{d}t < \infty$ and*

$$\varphi'(t) \leqslant -b\psi(t) + h(t). \tag{5.6}$$

If either $\psi \in C^1([a, \infty))$ and $\psi'(t) \leqslant k$ in $[a, \infty)$ for some constant $k > 0$, or ψ is uniform continuous in $[a, \infty)$, then $\lim_{t \to \infty} \psi(t) = 0$.

Proof. We only deal with the case that $\psi \in C^1([a, \infty))$ and $\psi'(t) \leqslant k$ in $[a, \infty)$. Assume that the conclusion were not valid. Then there exist $\varepsilon > 0$ and $t_i \nearrow \infty$ such that $\psi(t_i) > \varepsilon$ and $t_i - t_{i-1} > \varepsilon/(2k)$ for all i. Take $0 < \sigma < \varepsilon/(2k)$. The condition $\psi'(t) \leqslant k$ guarantees

$$\psi(t_i) - \psi(t) \leqslant k(t_i - t) \quad \text{for all } t \in [t_i - \sigma, t_i],$$

and then

$$\psi(t) \geqslant \psi(t_i) - k(t_i - t) > \varepsilon - k\sigma > \varepsilon/2 \quad \text{for all } t \in [t_i - \sigma, t_i].$$

Integrating (5.6) from a to $t \geqslant t_N$ with large N and using the condition $\psi \geqslant 0$, we obtain immediately

$$\varphi(t) \leqslant \varphi(a) - b\int_a^t \psi(s)\mathrm{d}s + \int_a^t h(s)\mathrm{d}s$$

$$\leqslant \varphi(a) - b\sum_{i=1}^N \int_{t_i-\sigma}^{t_i} \psi(s)\mathrm{d}s + \int_a^t h(s)\mathrm{d}s$$

$$\leqslant \varphi(a) - Nb\varepsilon\sigma/2 + \int_a^t h(s)\mathrm{d}s \quad \text{for all } t \geqslant t_N.$$

It follows that $\varphi(t) \to -\infty$ as $t \to \infty$ due to the fact that $\int_a^\infty h(t)\mathrm{d}t < \infty$. A contradiction is attained, and the proof is complete. $\qquad\square$

If $f_i(x, u, u^\tau) = f_i(u, u^\tau)$ does not depend on x, $i = 1, \ldots, m$, then the ordinary differential system corresponding to (5.5) is

$$\begin{cases} y_{it} = f_i(y, y^\tau), & t > 0, \\ y_i(t) = y_{i0}(t), & -\tau_i \leqslant t \leqslant 0, \\ i = 1, \ldots, m. \end{cases} \tag{5.7}$$

When we find a Lyapunov function of (5.7), a natural question is that, does it have to be a Lyapunov functional of (5.5)? In the following we will answer this question in some sense. For simplicity, we only consider the case $\tau_i = 0$ and write $f_i(u, u^\tau) = f_i(u)$, $1 \leqslant i \leqslant m$.

Let $\Sigma \subset \mathbb{R}^m$ be an invariant set with respect to both (5.7) and (5.5), i.e., $y(0) \in \Sigma$ implies $y(t) \in \Sigma$ for all $t \geqslant 0$, and $u(x, 0) \in \Sigma$ for all $x \in \overline{\Omega}$ implies $u(x, t) \in \Sigma$ for all $x \in \overline{\Omega}$ and $t \geqslant 0$. Let $y^* \in \Sigma$ be the unique equilibrium point of (5.7).

Lemma 5.6 *Suppose $\tau_i = 0$ and $f_i = f_i(u)$, $1 \leqslant i \leqslant m$. Assume that $y(0) \in \Sigma$, $u(x, 0) \in \Sigma$ for all $x \in \overline{\Omega}$, and (5.7) has a Lyapunov function $V(y)$, i.e.,*

$$\sum_{i=1}^m V_{y_i}(y)f_i(y) \leqslant 0 \quad \text{for all } y \in \Sigma.$$

Denote $D = \mathrm{diag}(d_1, \ldots, d_m)$. Suppose that along the orbit of (5.5),

$$\sum_{i,j=1}^m \nabla u_i d_i V_{u_i u_j}(u) \cdot \nabla u_j \geqslant 0, \quad \int_{\partial\Omega} \sum_{i=1}^l d_i V_{u_i}(u)\big|_{u_i=0} \partial_{\boldsymbol{n}} u_i \mathrm{d}S \leqslant 0.$$

Then functional

$$F(t) = \int_\Omega V(u(x,t))$$

satisfies

$$F'(t) \leqslant 0 \quad \text{for all } t > 0.$$

Proof. A straightforward computation shows that along the orbit of (5.5),

$$F'(t) = \int_\Omega \sum_{i=1}^m V_{u_i}(u)u_{it}$$

$$= \int_\Omega \sum_{i=1}^m d_i V_{u_i}(u)\Delta u_i + \int_\Omega \sum_{i=1}^m V_{u_i}(u)f_i(u)$$

$$= \int_{\partial\Omega} \sum_{i=1}^l d_i V_{u_i}(u)\big|_{u_i=0}\partial_{\boldsymbol{n}} u_i \mathrm{d}S - \int_\Omega \sum_{i,j=1}^m \nabla u_i d_i V_{u_i u_j}(u) \cdot \nabla u_j$$

$$+ \int_\Omega \sum_{i=1}^m V_{u_i}(u)f_i(u)$$

$$\leqslant 0 \quad \text{for all } t > 0.$$

The proof is complete. □

Example 5.2 *We consider an initial-boundary value problem with homogeneous Dirichlet boundary condition of scalar equation*

$$\begin{cases} u_t - \Delta u = kf(u), & x \in \Omega, \quad t > 0, \\ u = 0, & x \in \partial\Omega, \; t > 0, \\ u(x,0) = u_0(x), & x \in \Omega, \end{cases} \tag{5.8}$$

where Ω is of class $C^{2+\alpha}$, $k \geqslant 0$ is a constant, $u_0 \in W_p^2(\Omega)$ with $p > 1$, and $f \in C^1(\mathbb{R})$ satisfies

$$f(0) = 0, \quad f'(0) > 0 \quad \text{and} \quad f'(v) \leqslant f'(0) \quad \text{for all } v \in \mathbb{R}.$$

Using the interior estimate in t direction we can show that $\nabla u_t \in L^2(\Omega)$ for $t > 0$.

It is clear that $u_s = 0$ is a equilibrium solution of (5.8). The eigenvalue problem of linearization of (5.8) at $u_s = 0$ is

$$\begin{cases} \Delta\phi + kf'(0)\phi = \mu\phi & \text{in } \Omega, \\ \phi = 0 & \text{on } \partial\Omega. \end{cases} \tag{5.9}$$

Let λ be the first eigenvalue of (1.33). Then the first eigenvalue of (5.9) is $\mu_1 = kf'(0) - \lambda$. If $kf'(0) > \lambda$, then $\mu_1 > 0$, and so $u_s = 0$ is not linearly stable. If $kf'(0) < \lambda$, then $\mu_1 < 0$, and so every eigenvalue μ of (5.9) is negative. Consequently, $u_s = 0$ is linearly stable and is also locally asymptotically stable. Fortunately, the following theorem shows that $u_s = 0$ is also globally asymptotically stable.

Theorem 5.7 *If $kf'(0) < \lambda$, then $u_s = 0$ is globally asymptotically stable with respect to problem* (5.8).

Proof. Define

$$W(t) = \int_\Omega |\nabla u(x,t)|^2 \mathrm{d}x.$$

Recall that

$$\|\nabla v\|_2^2 \leqslant \frac{1}{\lambda}\|\Delta v\|_2^2, \quad \forall\, v \in H_0^1(\Omega) \cap H^2(\Omega).$$

Then, along the orbit of (5.8), we have

$$\begin{aligned}
W'(t) &= 2\int_\Omega \nabla u \cdot \nabla u_t \mathrm{d}x \\
&= -2\int_\Omega \Delta u u_t \mathrm{d}x \\
&= -2\int_\Omega |\Delta u|^2 \mathrm{d}x - 2\int_\Omega kf(u)\Delta u \mathrm{d}x \\
&= 2k\int_\Omega f'(u)|\nabla u|^2 \mathrm{d}x - 2\int_\Omega |\Delta u|^2 \mathrm{d}x \\
&\leqslant 2k\int_\Omega f'(u)|\nabla u|^2 \mathrm{d}x - 2\lambda\int_\Omega |\nabla u|^2 \mathrm{d}x.
\end{aligned}$$

By the assumption that $f'(v) \leqslant f'(0)$ for all $v \in \mathbb{R}$, it follows that

$$W'(t) \leqslant 2(kf'(0) - \lambda)\int_\Omega |\nabla u|^2 \mathrm{d}x = -2(\lambda - kf'(0))W(t).$$

Thanks to $\lambda - kf'(0) > 0$ and $W(t) \geqslant 0$, we have $\lim_{t\to\infty} W(t) = 0$, i.e., $\lim_{t\to\infty} u = 0$ in $H_0^1(\Omega)$. By Theorem 2.11,

$$\|u\|_{C^{2+\alpha,\,1+\alpha/2}(\overline\Omega \times [1,\infty))} \leqslant C.$$

We can use this fact and the compactness arguments to deduce that $\lim_{t\to\infty} u = 0$ in $C^2(\overline\Omega)$. $\qquad\square$

In the following three subsections we will use the Lyapunov functional to study the global asymptotic stability of the constant equilibrium solution in conjunction with specific examples. For the convenience, we shall use the notation

$$F(x,y) = x - y - y\ln\frac{x}{y} \quad \text{with} \quad x,y > 0.$$

Recently, Ni et al. [86] made an extension of Lyapunov functional method to study the global asymptotic stability of non-constant equilibrium solution in heterogeneous environment. The models discussed in [86] include the scalar parabolic equations

$$\begin{cases}
u_t = d(x)\Delta u + uf(x,u), & x \in \Omega, \quad t > 0, \\
\partial_{\boldsymbol n} u = 0, & x \in \partial\Omega, \; t > 0, \\
u(x,0) = \varphi(x) \geqslant 0, \not\equiv 0, & x \in \Omega
\end{cases}$$

and

$$\begin{cases} u_t = d(x)\mathrm{div}[a(x)\nabla\Phi(x,u)] + f(x,u), & x \in \Omega, \ t > 0, \\ \partial_{\boldsymbol{n}}\Phi(x,u) = g(x,u), & x \in \partial\Omega, \ t > 0, \\ u(x,0) = \varphi(x) \geqslant 0, \not\equiv 0, & x \in \Omega, \end{cases}$$

as well as the Lotka-Volterra competition model with k species

$$\begin{cases} u_{it} = d_i(x)\Delta u_i + u_i\Big(m_i(x) - \sum_{j=1}^k a_{ij}(x)u_j\Big), & x \in \Omega, \ t > 0, \ 1 \leqslant i \leqslant k, \\ \partial_{\boldsymbol{n}} u_i = 0, & x \in \partial\Omega, \ t > 0, \ 1 \leqslant i \leqslant k, \\ u_i(x,0) = \varphi_i(x) \geqslant 0, \not\equiv 0, & x \in \Omega, \ 1 \leqslant i \leqslant k, \end{cases}$$

where for each i, either $d_i(x) > 0$ or $d_i(x) \equiv 0$ in $\overline{\Omega}$.

5.2.2   A variable-territory prey-predator model

A so-called variable territory model posits that the self-limitation of predator depends inversely on the availability of the prey. Under suitable re-scaling, a variable-territory prey-predator model can be written as ([117])

$$\begin{cases} u_t - d_1\Delta u = u(\beta - au - bv), & x \in \Omega, \quad t > 0, \\ v_t - d_2\Delta v = v(u - 1 - v/u), & x \in \Omega, \quad t > 0, \\ \partial_{\boldsymbol{n}} u = \partial_{\boldsymbol{n}} v = 0, & x \in \partial\Omega, \ t > 0, \\ u(x,0) = u_0(x), \ v(x,0) = v_0(x), & x \in \Omega, \end{cases} \tag{5.10}$$

where $a, b, \beta > 0$ are constants, $u_0, v_0 \in C^2(\overline{\Omega})$ and $u_0, v_0 > 0$ in $\overline{\Omega}$. Consider the initial value problem of ODE system

$$\begin{cases} \bar{u}'(t) = \bar{u}(\beta - a\bar{u} - b\underline{v}), \\ \underline{u}'(t) = \underline{u}(\beta - a\underline{u} - b\bar{v}), \\ \bar{v}'(t) = \bar{v}(\bar{u} - 1 - \bar{v}/\bar{u}), \\ \underline{v}'(t) = \underline{v}(\underline{u} - 1 - \underline{v}/\underline{u}), \\ \bar{u}(0) = \max_{\overline{\Omega}} u_0(x), \\ \underline{u}(0) = \min_{\overline{\Omega}} u_0(x), \\ \bar{v}(0) = \max_{\overline{\Omega}} v_0(x), \\ \underline{v}(0) = \min_{\overline{\Omega}} v_0(x). \end{cases}$$

Clearly, $((\bar{u},\bar{v}),(\underline{u},\underline{v}))$ is a pair of coupled upper and lower solutions of (5.10). In order to show the global existence of $(\bar{u},\underline{u},\bar{v},\underline{v})$, we set $w = \bar{v}/\bar{u}$ and $z = \underline{v}/\underline{u}$. Then

$(\bar{u}, \underline{u}, \bar{v}, \underline{v}, w, z)$ satisfies

$$
\begin{cases}
\bar{u}'(t) = \bar{u}(\beta - a\bar{u} - b\underline{v}), \\
\underline{u}'(t) = \underline{u}(\beta - a\underline{u} - b\bar{v}), \\
\bar{v}'(t) = \bar{v}(\bar{u} - 1 - w), \\
\underline{v}'(t) = \underline{v}(\underline{u} - 1 - z), \\
w'(t) = w[(a+1)\bar{u} + b\underline{v} - (1+\beta) - w], \\
z'(t) = z[(a+1)\underline{u} + b\bar{v} - (1+\beta) - z]
\end{cases}
$$

with positive initial data

$$
\bar{u}(0) = \max_{\overline{\Omega}} u_0(x), \quad \underline{u}(0) = \min_{\overline{\Omega}} u_0(x), \quad \bar{v}(0) = \max_{\overline{\Omega}} v_0(x),
$$

$$
\underline{v}(0) = \min_{\overline{\Omega}} v_0(x), \quad w(0) = \bar{v}(0)/\bar{u}(0), \quad z(0) = \underline{v}(0)/\underline{u}(0).
$$

Clearly, this problem has a unique global positive solution $(\bar{u}, \underline{u}, \bar{v}, \underline{v}, w, z)$. Therefore, $(\bar{u}, \underline{u}, \bar{v}, \underline{v})$ exists globally. By the upper and lower solutions method, (5.10) has a unique global solution (u, v) which satisfies

$$
\underline{u}(t) \leqslant u(x, t) \leqslant \bar{u}(t) \quad \text{and} \quad \underline{v}(t) \leqslant v(x, t) \leqslant \bar{v}(t).
$$

Moreover, it is clear that

$$
\bar{u}(t) \leqslant \max\{\beta/a, \max_{\overline{\Omega}} u_0(x)\} =: A \quad \text{and} \quad \bar{v}(t) \leqslant \max\{A^2, \max_{\overline{\Omega}} v_0(x)\} =: B,
$$

and thereby, the solution (u, v) of (5.10) satisfies

$$
0 < u \leqslant A \quad \text{and} \quad 0 < v \leqslant B \quad \text{in} \quad \overline{\Omega} \times [0, \infty). \tag{5.11}
$$

The system (5.10) has a positive constant equilibrium solution if and only if $\beta > a$, and in that case, the positive constant equilibrium solution $(\tilde{u}, \tilde{v})$ is uniquely given by

$$
\tilde{u} = \frac{b - a + [(a-b)^2 + 4b\beta]^{1/2}}{2b} \quad \text{and} \quad \tilde{v} = \frac{\beta - a\tilde{u}}{b}.
$$

Obviously, $\tilde{u} > 1$. We point out that contents of this part take from [117], and the choice of $k(s)$ in proving Theorem 5.8 is inspired by [24].

Theorem 5.8 *If $\beta > a$, then $(\tilde{u}, \tilde{v})$ is globally asymptotically stable for problem (5.10).*

Proof. Step 1. We define

$$
f_1(u, v) = u(\beta - au - bv), \quad f_2(u, v) = v(u - 1 - v/u)
$$

and

$$k(s) = \frac{2a + b(s + \tilde{u} - 1)}{b^2}, \tag{5.12}$$

$$V(u, v) = \int_{\tilde{u}}^{u} k(s) \frac{s - \tilde{u}}{s^2} ds + F(v, \tilde{v}).$$

After careful computations we have

$$V_u(u, v) f_1(u, v) + V_v(u, v) f_2(u, v)$$

$$= k(u) \frac{u - \tilde{u}}{u} (\beta - au - bv) + (v - \tilde{v}) \frac{u^2 - u - v}{u}$$

$$= -\frac{k(u)}{u} [a(u - \tilde{u})^2 + b(u - \tilde{u})(v - \tilde{v})] + \frac{v - \tilde{v}}{u} [(u + \tilde{u} - 1)(u - \tilde{u}) - (v - \tilde{v})]$$

$$= -\frac{ak(u)(u - \tilde{u})^2 - [u + \tilde{u} - 1 - bk(u)](u - \tilde{u})(v - \tilde{v}) + (v - \tilde{v})^2}{u}. \tag{5.13}$$

It is well known that if

$$(u + \tilde{u} - 1 - bk(u))^2 < 4ak(u), \tag{5.14}$$

then the quadratic form

$$ak(u)\xi^2 - [u + \tilde{u} - 1 - bk(u)]\xi\eta + \eta^2$$

is positive unless $\xi = \eta = 0$. The inequality (5.14) can be rewritten as

$$b^2 k^2(u) - k(u)[4a + 2b(u + \tilde{u} - 1)] + (u + \tilde{u} - 1)^2 < 0. \tag{5.15}$$

Obviously, the equation

$$b^2 k^2 - k[4a + 2b(u + \tilde{u} - 1)] + (u + \tilde{u} - 1)^2 = 0$$

has two positive roots:

$$k_1 = k_1(u) = \frac{1}{b^2} \left(2a + b(u + \tilde{u} - 1) - 2\sqrt{a^2 + ab(u + \tilde{u} - 1)} \right),$$

$$k_2 = k_2(u) = \frac{1}{b^2} \left(2a + b(u + \tilde{u} - 1) + 2\sqrt{a^2 + ab(u + \tilde{u} - 1)} \right),$$

and (5.15) holds for all $k_1(u) < k(u) < k_2(u)$. It is easy to see that the function $k(u)$ given by (5.12) satisfies $k_1(u) < k(u) < k_2(u)$. Accordingly, the inequality (5.14) holds. Furthermore, as $0 \leqslant u \leqslant A$, we declare

$$ak(u)\xi^2 - [u + \tilde{u} - 1 - bk(u)]\xi\eta + \eta^2 \geqslant \varepsilon(|\xi|^2 + |\eta|^2)$$

for some $\varepsilon > 0$. This, together with (5.13), derives

$$V_u(u, v) f_1(u, v) + V_v(u, v) f_2(u, v) \leqslant -C \left(|u - \tilde{u}|^2 + |v - \tilde{v}|^2 \right). \tag{5.16}$$

Step 2. Let (u, v) denote the solution of (5.10), and define

$$W(t) = \int_\Omega V(u(x,t), v(x,t)), \quad f(u) = k(u)\frac{u - \tilde{u}}{u^2}.$$

Then we have

$$W'(t) = \int_\Omega \left(-d_1 f'(u)|\nabla u|^2 - d_2 \frac{\tilde{v}}{v^2}|\nabla v|^2\right)$$

$$+ \int_\Omega \left[V_u(u,v)f_1(u,v) + V_v(u,v)f_2(u,v)\right]. \quad (5.17)$$

Now we calculate $f'(u)$. Firstly, rewrite $f(u)$ as

$$f(u) = \frac{2a + b(\tilde{u} - 1)}{b^2}\left(\frac{1}{u} - \frac{\tilde{u}}{u^2}\right) + \frac{u - \tilde{u}}{bu}.$$

In view of $\tilde{u}(\tilde{u} - 1) = \tilde{v}$ and $\beta = a\tilde{u} + b\tilde{v}$, it is easy to calculate

$$f'(u) = \frac{2a + b(\tilde{u} - 1)}{b^2}\frac{2\tilde{u} - u}{u^3} + \frac{\tilde{u}}{bu^2}$$

$$= \frac{1}{b^2 u^3}(4a\tilde{u} - 2au + 2b\tilde{v} + bu)$$

$$> \frac{2}{b^2 u^3}(2a\tilde{u} + b\tilde{v} - au)$$

$$= \frac{2a}{b^2 u^3}\left(\tilde{u} + \frac{\beta}{a} - u\right). \quad (5.18)$$

On the other hand, the first equation of (5.10) implies $u_t - d_1\Delta u \leqslant u(\beta - au)$, and so $\limsup_{t\to\infty} \max_{\overline{\Omega}} u(\cdot, t) \leqslant \beta/a$ by Lemma 5.14. Then one can take $T \gg 1$ such that

$$u \leqslant \tilde{u}/2 + \beta/a \quad \text{in } \overline{\Omega} \times [T, \infty).$$

This estimate combined with (5.18) allows us to produce

$$f'(u) \geqslant \frac{a\tilde{u}}{b^2 u^3} \geqslant \frac{a\tilde{u}}{b^2 A^3} \quad \text{in } \overline{\Omega} \times [T, \infty). \quad (5.19)$$

Step 3. Since v is bounded, it follows from (5.16), (5.17) and (5.19) that

$$W'(t) \leqslant -C\int_\Omega \left(|\nabla u|^2 + |u - \tilde{u}|^2\right) - C\int_\Omega \left(|\nabla v|^2 + |v - \tilde{v}|^2\right)$$

for $t \geqslant T$. Certainly,

$$W'(t) \leqslant -C\int_\Omega |u - \tilde{u}|^2 =: -\psi(t) \quad \text{for all } t \geqslant T. \quad (5.20)$$

Using the equation of u and the estimate (5.11), we have

$$\begin{aligned}
\psi'(t) &= 2C \int_\Omega u_t(u - \tilde{u}) \\
&= 2C \int_\Omega (d_1 \Delta u + \beta u - au^2 - buv)(u - \tilde{u}) \\
&= -2Cd_1 \int_\Omega |\nabla u|^2 + 2C \int_\Omega (\beta u - au^2 - buv)(u - \tilde{u}) \\
&\leqslant 2C \int_\Omega (\beta u - au^2 - buv)(u - \tilde{u}) \\
&\leqslant k
\end{aligned}$$

for some constant $k > 0$. Applying Lemma 5.5 to (5.20) we conclude $\psi(t) \to 0$, and then $u \to \tilde{u}$ in $L^2(\Omega)$ as $t \to \infty$. Recalling (5.11) and Theorem 2.13, it achieves

$$\|u(\cdot, t)\|_{C^{1+\alpha}(\overline{\Omega})} \leqslant M \quad \text{for all } t \geqslant T + 1,$$

and then

$$\lim_{t \to \infty} \|u(\cdot, t) - \tilde{u}\|_{C^1(\overline{\Omega})} = 0.$$

As a result, $u \geqslant \tilde{u}/2$ in $\overline{\Omega} \times [T_1, \infty)$ for some $T_1 > T$. Hence $u - 1 - v/u$ is bounded in $\overline{\Omega} \times [T_1, \infty)$. Similar to discussions about u, it can be deduced that

$$\lim_{t \to \infty} \|v(\cdot, t) - \tilde{v}\|_{C^1(\overline{\Omega})} = 0.$$

Step 4. Above arguments show that $\tilde{u}/2 \leqslant u \leqslant A$ and $0 < v \leqslant B$ in $\overline{\Omega} \times [T_0, \infty)$ with $T_0 \gg 1$. Invoking Theorem 2.11 we can get further that

$$\|u(\cdot, t)\|_{C^{2+\alpha}(\overline{\Omega})} \leqslant C, \quad \|v(\cdot, t)\|_{C^{2+\alpha}(\overline{\Omega})} \leqslant C \quad \text{for all } t \geqslant T_0,$$

which implies

$$\lim_{t \to \infty} \|u(\cdot, t) - \tilde{u}\|_{C^2(\overline{\Omega})} = \lim_{t \to \infty} \|v(\cdot, t) - \tilde{v}\|_{C^2(\overline{\Omega})} = 0.$$

The proof is complete. $\qquad \square$

5.2.3 A prey-predator model with trophic interactions

In this part we treat the following prey-predator model with trophic interactions of three levels ([68]):

$$\begin{cases}
u_t - d_1 \Delta u = u f_1(u, v), & x \in \Omega, \quad t > 0, \\
v_t - d_2 \Delta v = v f_2(u, v, w), & x \in \Omega, \quad t > 0, \\
w_t - d_3 \Delta w = w f_3(v, w), & x \in \Omega, \quad t > 0, \\
\partial_{\boldsymbol{n}} u = \partial_{\boldsymbol{n}} v = \partial_{\boldsymbol{n}} w = 0, & x \in \partial\Omega, t > 0, \\
(u, v, w) = (u_0, v_0, w_0) > 0, & x \in \Omega, \quad t = 0,
\end{cases} \tag{5.21}$$

where

$$f_1(u, v) = 1 - u - \frac{av}{v + \rho},$$

$$f_2(u, v, w) = \frac{bu}{v + \rho} - \gamma - \frac{cw}{w + \sigma},$$

$$f_3(v, w) = \frac{mv}{w + \sigma} - \beta,$$

u, v and w are population densities of resource (level 0 species), consumer (level 1 species) and top predator (level 2 species) respectively. We assume $u_0, v_0, w_0 \in W_p^2(\Omega)$. Then problem (5.21) has a unique global solution (u, v, w). The content of this part is due to [68].

We first analyze positive solutions of the algebraic system

$$f_1(u, v) = 0, \quad f_2(u, v, w) = 0, \quad f_3(v, w) = 0. \tag{5.22}$$

Let $(\tilde{u}, \tilde{v}, \tilde{w})$ be a positive solution of (5.22). Then $(\tilde{u}, \tilde{v}, \tilde{w})$ satisfies

$$\begin{cases} \tilde{u} = \dfrac{\rho + (1 - a)\tilde{v}}{\tilde{v} + \rho}, \\[2mm] \tilde{w} = \dfrac{m\tilde{v} - \beta\sigma}{\beta}, \\[2mm] \dfrac{b\tilde{v}[\rho + (1 - a)\tilde{v}]}{(\tilde{v} + \rho)^2} = (\gamma + c)\tilde{v} - \dfrac{c\beta\sigma}{m}. \end{cases} \tag{5.23}$$

Let us define

$$\underline{v} = \beta\sigma/m, \quad f(v) = (\gamma + c)v(v + \rho)^2 - c\underline{v}(v + \rho)^2 - b\rho v - b(1 - a)v^2.$$

Then $(\tilde{u}, \tilde{v}, \tilde{w})$ is a positive solution of (5.23) if and only if $f(v) = 0$ has a positive root $\tilde{v}$ and

(1) $\tilde{v} > \underline{v}$ when $a \leqslant 1$;

(2) $\underline{v} < \tilde{v} < \rho/(a - 1) =: \bar{v}$ when $a > 1$.

Notice

$$f(\pm\infty) = \pm\infty, \quad f(-\rho) = ab\rho^2 > 0 \text{ and } f(0) = -c\rho^2\underline{v} < 0.$$

There exist three numbers $v_1, v_2, \tilde{v}$ satisfying $-\infty < v_1 < v_2 < 0 < \tilde{v}$ such that

$$f(v_1) = f(v_2) = f(\tilde{v}) = 0, \quad f'(v_1) > 0, \quad f'(v_2) < 0, \quad f'(\tilde{v}) > 0,$$

$$f(v) < 0 \text{ in } (-\infty, v_1) \cup (v_2, \tilde{v}), \quad f(v) > 0 \text{ in } (v_1, v_2) \cup (\tilde{v}, \infty).$$

This indicates that the system (5.23) has at most one positive solution, and

(1) when $a \leqslant 1$, system (5.23) has a positive solution if and only if $f(\underline{v}) < 0$;

(2) when $a > 1$, system (5.23) has a positive solution if and only if $f(\underline{v}) < 0 < f(\bar{v})$.

A direct calculation gives

$$f(\underline{v}) = (\gamma + c)\underline{v}(\underline{v} + \rho)^2 - c\underline{v}(\underline{v} + \rho)^2 - b\rho\underline{v} - b(1-a)\underline{v}^2$$
$$= \underline{v}[\gamma(\underline{v} + \rho)^2 - b\rho - b(1-a)\underline{v}].$$

It follows that, when $a \leqslant 1$, the system (5.23) has a positive solution if and only if

$$\gamma(\underline{v} + \rho)^2 - b\rho - b(1-a)\underline{v} < 0. \tag{5.24}$$

When $a > 1$ and $\beta\sigma/m = \underline{v} < \bar{v} = \rho/(a-1)$, a direct computation gives

$$f(\bar{v}) = (\gamma + c)\bar{v}(\bar{v} + \rho)^2 - c\underline{v}(\bar{v} + \rho)^2 - b\rho\bar{v} - b(1-a)\bar{v}^2$$
$$= (\bar{v} + \rho)^2[(\gamma + c)\bar{v} - c\underline{v}] > 0.$$

As a result, when $a > 1$, (5.23) has a positive solution if and only if (5.24) holds and

$$\beta\sigma/m = \underline{v} < \bar{v} = \rho/(a-1). \tag{5.25}$$

So far, we get

Theorem 5.9 *When $a \leqslant 1$, the system (5.22) has a positive solution if and only if (5.24) holds. When $a > 1$, (5.22) has a positive solution if and only if (5.24) and (5.25) hold. Furthermore, in any case, (5.22) has at most one positive solution.*

Theorem 5.10 *If (5.22) has a positive solution $(\tilde{u}, \tilde{v}, \tilde{w})$, then, as the equilibrium solution of (5.21), it is globally asymptotically stable.*

Proof. Define

$$h = \frac{b(\tilde{v} + \rho)}{a\rho}, \quad k = \frac{c\sigma}{m(\tilde{w} + \sigma)}$$

and

$$W(t) = \int_\Omega [hF(u, \tilde{u}) + F(v, \tilde{v}) + kF(w, \tilde{w})].$$

After careful calculations, we get

$$h(u - \tilde{u})f_1(u, v) + (v - \tilde{v})f_2(u, v, w) + k(w - \tilde{w})f_3(v, w)$$
$$= -\left(h(u - \tilde{u})^2 + \frac{b\tilde{u}(v - \tilde{v})^2}{(\tilde{v} + \rho)(v + \rho)} + \frac{mk\tilde{v}(w - \tilde{w})^2}{(\tilde{w} + \sigma)(w + \sigma)} \right),$$

and then

$$W'(t) = \int_\Omega \left(\frac{h(u - \tilde{u})}{u}u_t + \frac{v - \tilde{v}}{v}v_t + \frac{k(w - \tilde{w})}{w}w_t \right)$$
$$= -\int_\Omega \left(d_1 h\tilde{u}\frac{|\nabla u|^2}{u^2} + d_2\tilde{v}\frac{|\nabla v|^2}{v^2} + d_3 k\tilde{w}\frac{|\nabla w|^2}{w^2} \right)$$
$$- \int_\Omega \left(h(u - \tilde{u})^2 + \frac{b\tilde{u}(v - \tilde{v})^2}{(\tilde{v} + \rho)(v + \rho)} + \frac{mk\tilde{v}(w - \tilde{w})^2}{(\tilde{w} + \sigma)(w + \sigma)} \right).$$

Repeating the last part in proving Theorem 5.8, we can show that

$$\lim_{t\to\infty} \|u - \tilde{u}\|_{C^2(\overline{\Omega})} = \lim_{t\to\infty} \|v - \tilde{v}\|_{C^2(\overline{\Omega})} = \lim_{t\to\infty} \|w - \tilde{w}\|_{C^2(\overline{\Omega})} = 0.$$

The proof is complete. □

Now, suppose that (5.22) has no positive solution. Then, possible non-negative solutions of (5.22) are $(0,0,0)$, $(1,0,0)$ and $(u^*, v^*, 0)$, where (u^*, v^*) is a possible positive solution of

$$u^* = 1 - \frac{av^*}{v^* + \rho}, \quad \frac{bu^*}{v^* + \rho} = \gamma. \tag{5.26}$$

Hence, (u^*, v^*) exists if and only if $\rho < b/\gamma$. And in this case, v^* is uniquely given by

$$v^* = \frac{1}{2}\left(\frac{b(1-a)}{\gamma} - 2\rho + \sqrt{\frac{b^2(a-1)^2}{\gamma^2} + \frac{4\rho ab}{\gamma}}\right).$$

Theorem 5.11 *Suppose that (5.22) has no positive solution and (5.26) has a positive solution (u^*, v^*). Then the non-negative equilibrium solution $(u^*, v^*, 0)$ of (5.21) is globally asymptotically stable.*

Proof. Step 1. We first verify that v^* satisfies

$$mv^* \leqslant \beta\sigma. \tag{5.27}$$

When $a \leqslant 1$, owing to that (5.22) has no positive solution, we must have

$$\gamma(\underline{v} + \rho)^2 - b\rho - b(1-a)\underline{v} \geqslant 0$$

by Theorem 5.9. Since v^* satisfies

$$\gamma(v^* + \rho)^2 - b\rho - b(1-a)v^* = 0,$$

it is deduced that $v^* \leqslant \underline{v}$, i.e., (5.27) holds.

When $a > 1$, we infer that at least one of (5.24) and (5.25) does not hold by Theorem 5.9. If (5.24) is not valid, same as the case $a \leqslant 1$ we can get (5.27). If (5.25) is not valid, then

$$\beta\sigma/m \geqslant \rho/(a-1).$$

Now, if (5.27) does not hold, then $v^* > \beta\sigma/m \geqslant \rho/(a-1)$ and accordingly

$$\sqrt{\frac{b^2(a-1)^2}{\gamma^2} + \frac{4\rho ab}{\gamma}} > \frac{2\rho}{a-1} + 2\rho + \frac{b(a-1)}{\gamma} = \frac{2a\rho}{a-1} + \frac{b(a-1)}{\gamma},$$

which implies $2a\rho/(a-1) < 0$. This is a contradiction, and then (5.27) holds.

Step 2. Similar to the proof of Theorem 5.10, we take

$$p = \frac{b(v^* + \rho)}{a\rho}, \quad q = \frac{c}{m}$$

and

$$U(t) = \int_\Omega \left[pF(u, u^*) + F(v, v^*) + qw \right].$$

Notice (5.27) and $u, v, w > 0$. Careful calculations lead to

$$\begin{aligned}
U'(t) &= \int_\Omega \frac{p(u - u^*)}{u} u_t + \int_\Omega \frac{v - v^*}{v} v_t + \int_\Omega q w_t \\
&= -d_1 p u^* \int_\Omega \frac{|\nabla u|^2}{u^2} - d_2 v^* \int_\Omega \frac{|\nabla v|^2}{v^2} \\
&\quad + \int_\Omega \left[p(u - u^*) \left(u^* - u + \frac{av^*}{v^* + \rho} - \frac{av}{v + \rho} \right) \right. \\
&\quad \left. + (v - v^*) \left(\frac{bu}{v + \rho} - \frac{bu^*}{v^* + \rho} - \frac{cw}{w + \sigma} \right) + qw \left(\frac{mv}{w + \sigma} - \beta \right) \right] \\
&= -d_1 p u^* \int_\Omega \frac{|\nabla u|^2}{u^2} - d_2 v^* \int_\Omega \frac{|\nabla v|^2}{v^2} - p \int_\Omega (u - u^*)^2 \\
&\quad - \int_\Omega \left[\frac{bu^*(v - v^*)^2}{(v^* + \rho)(v + \rho)} + \frac{\beta q w^2}{w + \sigma} + q(\beta \sigma - mv^*) \frac{w}{w + \sigma} \right].
\end{aligned}$$

Repeating the last part in proving Theorem 5.8, we can show that

$$\lim_{t \to \infty} \|u - u^*\|_{C^2(\overline{\Omega})} = \lim_{t \to \infty} \|v - v^*\|_{C^2(\overline{\Omega})} = \lim_{t \to \infty} \|w\|_{C^2(\overline{\Omega})} = 0.$$

The proof is complete. □

Similarly, we can prove the following theorem.

Theorem 5.12 *If both (5.22) and (5.26) do not have any positive solution, then the equilibrium solution $(1, 0, 0)$ of (5.21) is globally asymptotically stable.*

Above results provide a complete qualitative description of dynamical behaviour of (5.21). In particular, we note that in consideration of asymptotic behaviour, either populations of three species tend to a (constant) positive equilibrium, or one or more of predator species must become extinct. Further, it elaborates that, if conditions are favorable for the positive equilibrium $(\tilde{u}, \tilde{v}, \tilde{w})$ to exist, then its existence dominates the entire dynamical behaviour in the sense that all positive solutions are globally attracted to this equilibrium. If this positive equilibrium does not exist, but the semi-positive equilibrium $(u^*, v^*, 0)$ exists, then this semi-positive equilibrium dominates the dynamical behaviour in the above sense. In the absence of these equilibria, we see the extinction of two predator species. Accordingly, one can discern a hierarchy of dynamical behaviour that is characterized by the preservation of the maximum number of species in the system.

5.2.4 A system coupled by two PDEs and one ODE

A two-predators-one-prey system with strategy may be written as an ODE system ([87]) with $u = (u_1, u_2, u_3)$ that satisfies

$$
\begin{cases}
\dfrac{du_1}{dt} = u_1\left(\dfrac{u_2 u_3}{u_1 + u_2} - 1\right) =: f_1(u), \\[2mm]
\dfrac{du_2}{dt} = u_2\left(\dfrac{b u_1 u_3}{u_1 + u_2} - a\right) =: f_2(u), \\[2mm]
\dfrac{du_3}{dt} = u_3\left(r - u_3 - \dfrac{(1+b)u_1 u_2}{u_1 + u_2}\right) =: f_3(u),
\end{cases}
\tag{5.28}
$$

where d_1, d_2, a, b, r are positive constants, u_1, u_2 and u_3 are population densities of two predator species and a prey species, respectively.

It is easy to see that (5.28) has a positive constant equilibrium solution if and only if

$$rb > a + b.$$

In this case, the positive constant equilibrium solution is uniquely given by

$$
\tilde{u}_1 = (a+b)\frac{rb - (a+b)}{b^2(1+b)}, \quad
\tilde{u}_2 = (a+b)\frac{rb - (a+b)}{ab(1+b)}, \quad
\tilde{u}_3 = \frac{a+b}{b}.
$$

To take into account the in-homogeneous distribution of predators and prey in different spatial locations within a fixed bounded domain $\Omega \subset \mathbb{R}^n$ at any given time, and the natural tendency of each species to diffuse to areas of smaller population concentration, we have to pay attention to the role of diffusion. Let us consider that two predators are birds and one prey is worm. It is well known that diffusions of birds are very fast, while that of worm is so slow that we can omit its diffusion. Based on these reasons, we study the following system coupled by two reaction-diffusion equations and one ordinary differential equation

$$
\begin{cases}
u_{1t} - d_1 \Delta u_1 = f_1(u), & x \in \Omega, \quad t > 0, \\
u_{2t} - d_2 \Delta u_2 = f_2(u), & x \in \Omega, \quad t > 0, \\
u_{3t} = f_3(u), & x \in \Omega, \quad t > 0, \\
\partial_{\boldsymbol{n}} u_1 = \partial_{\boldsymbol{n}} u_2 = 0, & x \in \partial\Omega, \ t > 0, \\
u_i(x, 0) = u_{i0}(x), & i = 1,2,3, \ x \in \Omega,
\end{cases}
\tag{5.29}
$$

where $f_1(u)$, $f_2(u)$ and $f_3(u)$ are given in (5.28), $u_{i0} \in W_p^2(\Omega)$ for some $p > 1 + n/2$ and $i = 1,2$, $u_{30} \in C^\gamma(\overline{\Omega})$ for some $0 < \gamma < 1$ and $u_{10}(x) > 0$, $u_{20}(x) > 0$ and $u_{30}(x) \geqslant 0$ in $\overline{\Omega}$.

Same as Theorem 4.1, we can show that problem (5.29) has a unique local solution (u_1, u_2, u_3) satisfying $u_1, u_2 \in W_p^{2,1}(Q_T)$ and $u_3 \in C^{\alpha,1}(\overline{Q}_T)$ for some $0 < T < \infty$ and $0 < \alpha < \min\{\gamma, 2 - (n+2)/p\}$. We leave details to readers as an exercise (Exercise 5.9).

Let $T_{\max}$ be the maximal existence time of (5.29). It is easy to see from the third equation of (5.29) that $u_3 \leqslant \max\{r, \max_{\overline{\Omega}} u_{30}(x)\}$. Integrating equations of (5.29) over Ω and adding the result, we have

$$\frac{\mathrm{d}}{\mathrm{d}t} \int_\Omega (u_1 + u_2 + u_3) \leqslant - \int_\Omega (u_1 + au_2 + u_3) + C.$$

It follows that $\|u_i(\cdot, t)\|_{1,\Omega}$ is bounded in $[0, T_{\max})$, $i = 1, 2$. Using Theorem 2.14, we can obtain that $\|u_i(\cdot, t)\|_{\infty,\Omega}$ is also bounded in $[0, T_{\max})$, $i = 1, 2$. Therefore, $T_{\max} = \infty$ and (u_1, u_2, u_3) is bounded.

Theorem 5.13 ([114]) *The positive constant equilibrium solution $(\tilde{u}_1, \tilde{u}_2, \tilde{u}_3)$ of (5.29) is globally asymptotically stable in the sense that*

$$\lim_{t\to\infty} \|u_i(\cdot, t) - \tilde{u}_i\|_{C^1(\overline{\Omega})} = 0, \quad i = 1, 2; \quad \lim_{t\to\infty} \|u_3(\cdot, t) - \tilde{u}_3\|_{2,\Omega} = 0.$$

Proof. Applying Theorem 2.13 to equations of u_1 and u_2, we have, for some $0 < \alpha < 1$,

$$\|\nabla u_i\|_{C^{\alpha/2}([1,\infty), C(\overline{\Omega}))} + \|u_i\|_{C^{(1+\alpha)/2}([1,\infty), C(\overline{\Omega}))} \leqslant C, \tag{5.30}$$

$i = 1, 2$. It then follows from the differential equation of u_3 that

$$\|u_3\|_{C^1([1,\infty), C(\overline{\Omega}))} \leqslant C. \tag{5.31}$$

We set $p = \dfrac{b^2}{a}$, $q = \dfrac{b(a+b)}{a(1+b)}$, and define

$$V(t) = \int_\Omega [pF(u_1, \tilde{u}_1) + F(u_2, \tilde{u}_2) + qF(u_3, \tilde{u}_3)].$$

Then $V(t) \geqslant 0$ for all $t \geqslant 0$. Using (5.29) and integrating by parts, we obtain

$$V'(t) = - \int_\Omega \left(\frac{pd_1\tilde{u}_1}{u_1^2} |\nabla u_1|^2 + \frac{d_2\tilde{u}_2}{u_2^2} |\nabla u_2|^2 \right)$$
$$+ \int_\Omega \left(p(u_1 - \tilde{u}_1)\frac{f_1(u)}{u_1} + (u_2 - \tilde{u}_2)\frac{f_2(u)}{u_2} + q(u_3 - \tilde{u}_3)\frac{f_3(u)}{u_3} \right).$$

Through careful calculations we can get

$$p(u_1 - \tilde{u}_1)\frac{f_1(u)}{u_1} + (u_2 - \tilde{u}_2)\frac{f_2(u)}{u_2} + q(u_3 - \tilde{u}_3)\frac{f_3(u)}{u_3}$$
$$= -\frac{(bu_1 - au_2)^2}{a(u_1 + u_2)} - \frac{a+b}{ab(1+b)}(a + b - bu_3)^2.$$

Therefore,

$$V'(t) = - \int_\Omega \left(\frac{pd_1\tilde{u}_1}{u_1^2} |\nabla u_1|^2 + \frac{d_2\tilde{u}_2}{u_2^2} |\nabla u_2|^2 \right)$$
$$- \int_\Omega \left(\frac{(bu_1 - au_2)^2}{a(u_1 + u_2)} + \frac{a+b}{ab(1+b)}(a + b - bu_3)^2 \right).$$

As $u_i^2 \leqslant C$ for $i = 1, 2, 3$, it follows that

$$V'(t) \leqslant -\frac{C'}{C} \int_\Omega (|\nabla u_1|^2 + |\nabla u_2|^2) - \int_\Omega \left(\frac{(bu_1 - au_2)^2}{a(u_1 + u_2)} + \frac{a+b}{ab(1+b)}(a+b-bu_3)^2 \right)$$

$$=: -\psi_1(t) - \psi_2(t). \tag{5.32}$$

According to (5.30) and (5.31), we have $\|\psi_1\|_{C^{\alpha/2}([1,\infty))} \leqslant C$ and $\|\psi_2\|_{C^1([1,\infty))} \leqslant C$. Applying Lemma 5.5 to (5.32) we conclude that $\psi_1(t), \psi_2(t) \to 0$ as $t \to \infty$. Therefore,

$$\lim_{t \to \infty} \int_\Omega (|\nabla u_1|^2 + |\nabla u_2|^2) = 0, \tag{5.33}$$

$$\lim_{t \to \infty} \int_\Omega \frac{(bu_1 - au_2)^2}{(u_1 + u_2)} = \lim_{t \to \infty} \int_\Omega (a + b - bu_3)^2 = 0.$$

The latter demonstrates that

$$\lim_{t \to \infty} \int_\Omega |au_2 - bu_1| = \lim_{t \to \infty} \int_\Omega (u_3 - \tilde{u}_3)^2 = 0. \tag{5.34}$$

From (5.33) and the Poincaré inequality, we deduce

$$\lim_{t \to \infty} \int_\Omega [(u_1 - \bar{u}_1)^2 + (u_2 - \bar{u}_2)^2] = 0, \tag{5.35}$$

where $\bar{f} = \frac{1}{|\Omega|} \int_\Omega f$ for $f \in L^1(\Omega)$. The second limit of (5.34) implies $\bar{u}_3(t) \to \tilde{u}_3$. Noticing that $\bar{u}_1(t)$ and $\bar{u}_2(t)$ are bounded, and using the first limit of (5.34), we infer that there exist a sequence $\{t_k\}$ with $t_k \to \infty$ and a non-negative constant $\hat{u}_1$ such that

$$\bar{u}_3'(t_k) \to 0, \quad \bar{u}_1(t_k) \to \hat{u}_1 \quad \text{and} \quad \bar{u}_2(t_k) \to \hat{u}_1 b/a. \tag{5.36}$$

At $t = t_k$, we write

$$|\Omega|\bar{u}_3' = \int_\Omega (u_3 - \tilde{u}_3) \left(r - u_3 - \tilde{u}_3 - (1+b)\frac{u_1 u_2}{u_1 + u_2} \right)$$

$$+ \tilde{u}_3 \int_\Omega \left(r - \tilde{u}_3 - \frac{b(1+b)}{a+b} u_1 \right) + \frac{1+b}{b} \int_\Omega \frac{u_1(bu_1 - au_2)}{u_1 + u_2}. \tag{5.37}$$

Noticing (5.34), (5.36) and $r > \tilde{u}_3$, it follows from (5.37) that $\hat{u}_1 \neq 0$ and $r - \tilde{u}_3 - \hat{u}_1 b(1+b)/(a+b) = 0$, and hence $\hat{u}_1 = \tilde{u}_1$. Consequently,

$$\lim_{k \to \infty} \bar{u}_i(t_k) = \tilde{u}_i, \quad i = 1, 2. \tag{5.38}$$

Since $\|u_i(\cdot, t_k)\|_{C^{1+\alpha}(\overline{\Omega})} \leqslant C$, there exist a sub-sequence of $\{t_k\}$, still denoted by itself, and non-negative functions $w_i \in C^1(\overline{\Omega})$, such that

$$\lim_{k \to \infty} \|u_i(\cdot, t_k) - w_i(\cdot)\|_{C^1(\overline{\Omega})} = 0, \quad i = 1, 2.$$

This combined with (5.35) and (5.38) allows us to derive $w_i \equiv \tilde{u}_i$. Therefore,

$$\lim_{t \to \infty} \|u_i(\cdot, t) - \tilde{u}_i\|_{C^1(\overline{\Omega})} = 0, \quad i = 1, 2.$$

Theorem 5.13 is thus proved. $\qquad \square$

5.3 ITERATION METHOD

The idea of *iteration method* is to construct *iterative sequences* and to prove these sequences converge to the unique equilibrium solution. We only take the following ratio-dependent prey-predator model ([116]) as an example to introduce this method:

$$
\begin{cases}
u_t - d_1 \Delta u = u\left(1 - ku - \dfrac{v}{cu+v}\right), & x \in \Omega, \quad t > 0, \\[2mm]
v_t - d_2 \Delta v = v\left(-a - bv + \dfrac{u}{cu+v}\right), & x \in \Omega, \quad t > 0, \\[2mm]
\partial_{\boldsymbol{n}} u = \partial_{\boldsymbol{n}} v = 0, & x \in \partial\Omega, \ t > 0, \\[2mm]
u(x,0) = u_0(x), \ v(x,0) = v_0(x), & x \in \Omega,
\end{cases}
\tag{5.39}
$$

where d_1, d_2, c, k, a and b are positive constants, $u_0, v_0 \in W_p^2(\Omega)$ and $u_0, v_0 > 0$ in $\overline{\Omega}$.

By the upper and lower solutions method, (5.39) has a unique global solution (u, v) and $0 < u, v \leqslant C$ in $\overline{\Omega} \times [0, \infty)$ for some constant C.

Problem (5.39) has a positive constant equilibrium solution $(\tilde{u}, \tilde{v})$ if and only if $ac < 1$, and in this case it is uniquely given by

$$
\tilde{u} = \frac{c(a + b\tilde{v})}{k}, \quad \tilde{v} = \frac{c(1 - ac)}{k + bc^2}.
$$

We first state a lemma which plays an important role in the construction of iterative sequences and can be proved by the comparison principle and known results of ODEs. Details will be omitted.

Lemma 5.14 *Let* $f \in C^1([0, \infty))$ *with* $f > 0$, *and* d, β, γ *be constants with* $d > 0$ *and* $\beta \geqslant 0$. *Assume that* $T \in [0, \infty)$ *and* $w \in C^{2,1}(\Omega \times (T, \infty)) \cap C^{1,0}(\overline{\Omega} \times [T, \infty))$ *is a positive function.*

(1) *If* $\gamma > 0$ *and* w *satisfies*

$$
\begin{cases}
w_t - d\Delta w \leqslant (\geqslant) \, w^{1+\beta} f(w)(\gamma - w) & in \ \ \Omega \times (T, \infty), \\
\partial_{\boldsymbol{n}} w = 0 & on \ \ \partial\Omega \times [T, \infty),
\end{cases}
$$

then $\limsup\limits_{t \to \infty} \max\limits_{\overline{\Omega}} w(\cdot, t) \leqslant \gamma \ \left(\liminf\limits_{t \to \infty} \min\limits_{\overline{\Omega}} w(\cdot, t) \geqslant \gamma \right)$.

(2) *If* $\gamma \leqslant 0$ *and* w *satisfies*

$$
\begin{cases}
w_t - d\Delta w \leqslant w^{1+\beta} f(w)(\gamma - w) & in \ \ \Omega \times (T, \infty), \\
\partial_{\boldsymbol{n}} w = 0 & on \ \ \partial\Omega \times [T, \infty),
\end{cases}
$$

then $\limsup\limits_{t \to \infty} \max\limits_{\overline{\Omega}} w(\cdot, t) \leqslant 0$.

Theorem 5.15 ([116]) *If parameters appearing in (5.39) satisfy* $ac < 1$ *and*

$$
k < 2bc^2 + 2ack, \tag{5.40}
$$

then $(\tilde{u}, \tilde{v})$ *is globally asymptotically stable for problem (5.39).*

Proof. Our intention is to use Lemma 5.14 to construct *iterative sequences* and to prove that these sequences are monotonic and converge to $(\tilde{u}, \tilde{v})$.

For $s > 0$, we define

$$h(v; s) = -bv^2 - (a + bcs)v + (1 - ac)s.$$

Then $h(v; s) = 0$ has two roots:

$$\psi(s) = \frac{-(a + bcs) + \sqrt{(a + bcs)^2 + 4b(1 - ac)s}}{2b} > 0,$$

$$\psi_-(s) = \frac{-(a + bcs) - \sqrt{(a + bcs)^2 + 4b(1 - ac)s}}{2b} < 0.$$

Therefore,

$$h(v; s) = b(v - \psi_-(s))(\psi(s) - v).$$

Step 1. It follows from the first equation of (5.39) that $u_t - d_1 \Delta u \leqslant u(1 - ku)$. Then by Lemma 5.14,

$$\limsup_{t \to \infty} \max_{\overline{\Omega}} u(\cdot, t) \leqslant 1/k =: \bar{u}_1.$$

Step 2. For any given $\varepsilon > 0$, we can find $T_1^\varepsilon \gg 1$ so that $u \leqslant \bar{u}_1 + \varepsilon$ in $\overline{\Omega} \times [T_1^\varepsilon, \infty)$. Then we have that, for $x \in \Omega$ and $t \geqslant T_1^\varepsilon$,

$$v_t - d_2 \Delta v \leqslant -av - bv^2 + \frac{(\bar{u}_1 + \varepsilon)v}{c(\bar{u}_1 + \varepsilon) + v}$$

$$= \frac{v}{c(\bar{u}_1 + \varepsilon) + v} h(v; \bar{u}_1 + \varepsilon)$$

$$= \frac{bv}{c(\bar{u}_1 + \varepsilon) + v} (v - \psi_-(\bar{u}_1 + \varepsilon))(\psi(\bar{u}_1 + \varepsilon) - v)$$

by the second equation of (5.39). Hence $\limsup_{t \to \infty} \max_{\overline{\Omega}} v(\cdot, t) \leqslant \psi(\bar{u}_1 + \varepsilon)$ by Lemma 5.14, and the arbitrariness of ε gives

$$\limsup_{t \to \infty} \max_{\overline{\Omega}} v(\cdot, t) \leqslant \psi(\bar{u}_1) =: \bar{v}_1.$$

Step 3. A straightforward computation shows that

$$k\bar{v}_1 < c \quad \text{if and only if} \quad k < 2bc^2 + 2ack.$$

Thanks to the condition (5.40), one can choose $\varepsilon_0 > 0$ so small that

$$c > k(\bar{v}_1 + \varepsilon) \quad \text{for all } 0 < \varepsilon \leqslant \varepsilon_0.$$

Take $T_2^\varepsilon \gg 1$ such that $v \leqslant \bar{v}_1 + \varepsilon$ in $\overline{\Omega} \times [T_2^\varepsilon, \infty)$. It then follows from the equation of u that

$$u_t - d_1 \Delta u \geqslant u^2 \left(\frac{c}{cu + \bar{v}_1 + \varepsilon} - k \right) = u^2 \frac{c - k(\bar{v}_1 + \varepsilon) - cku}{cu + \bar{v}_1 + \varepsilon} \quad \text{in } \Omega \times [T_2^\varepsilon, \infty).$$

Using Lemma 5.14, we have $\liminf\limits_{t\to\infty}\min\limits_{\overline{\Omega}} u(\cdot,t)\geqslant [c-k(\bar{v}_1+\varepsilon)]/(ck)$, and then the arbitrariness of ε informs

$$\liminf_{t\to\infty}\min_{\overline{\Omega}} u(\cdot,t)\geqslant (c-k\bar{v}_1)/(ck)=:\underline{u}_1>0.$$

Step 4. For any given $0<\varepsilon<\underline{u}_1$, we choose $T_3^\varepsilon\gg 1$ such that $u\geqslant\underline{u}_1-\varepsilon$ in $\overline{\Omega}\times[T_3^\varepsilon,\infty)$. Making use of the second equation of (5.39) and careful calculations, we obtain that, for $x\in\Omega$ and $t\geqslant T_3^\varepsilon$,

$$v_t-d_2\Delta v\geqslant -av-bv^2+\frac{(\underline{u}_1-\varepsilon)v}{c(\underline{u}_1-\varepsilon)+v}$$
$$=\frac{v}{c(\underline{u}_1-\varepsilon)+v}h(v;\underline{u}_1-\varepsilon)$$
$$=\frac{bv}{c(\underline{u}_1-\varepsilon)+v}(v-\psi_-(\underline{u}_1-\varepsilon))(\psi(\underline{u}_1-\varepsilon)-v).$$

Lemma 5.14 gives $\liminf\limits_{t\to\infty}\min\limits_{\overline{\Omega}} v(\cdot,t)\geqslant\psi(\underline{u}_1-\varepsilon)$, and hence, by the arbitrariness of ε,

$$\liminf_{t\to\infty}\min_{\overline{\Omega}} v(\cdot,t)\geqslant\psi(\underline{u}_1)=:\underline{v}_1.$$

Step 5. For any given $0<\varepsilon<\underline{v}_1$, we take $T_4^\varepsilon\gg 1$ so that $v\geqslant\underline{v}_1-\varepsilon$ in $\overline{\Omega}\times[T_4^\varepsilon,\infty)$. Consequently, for $x\in\Omega$ and $t\geqslant T_4^\varepsilon$,

$$u_t-d_1\Delta u\leqslant u^2\left(\frac{c}{cu+\underline{v}_1-\varepsilon}-k\right)=u^2\frac{c-k(\underline{v}_1-\varepsilon)-cku}{cu+\underline{v}_1-\varepsilon}.$$

Same as above,

$$\limsup_{t\to\infty}\max_{\overline{\Omega}} u(\cdot,t)\leqslant (c-k\underline{v}_1)/(ck)=:\bar{u}_2.$$

Step 6. Define $\varphi(s)=(c-ks)/(ck)$. Then $\varphi'(s)<0$, $\psi'(s)>0$, and constants $\bar{u}_1,\bar{v}_1,\underline{u}_1,\underline{v}_1$ and $\bar{u}_2$ constructed above satisfy

$$\underline{v}_1=\psi(\underline{u}_1)<\psi(\bar{u}_1)=\bar{v}_1,\quad \underline{u}_1=\varphi(\bar{v}_1)<\varphi(\underline{v}_1)=\bar{u}_2\leqslant\bar{u}_1,\qquad(5.41)$$

$$\underline{u}_1\leqslant\liminf_{t\to\infty}\min_{\overline{\Omega}} u(\cdot,t)\leqslant\limsup_{t\to\infty}\max_{\overline{\Omega}} u(\cdot,t)\leqslant\bar{u}_2,$$

$$\underline{v}_1\leqslant\liminf_{t\to\infty}\min_{\overline{\Omega}} v(\cdot,t)\leqslant\limsup_{t\to\infty}\max_{\overline{\Omega}} v(\cdot,t)\leqslant\bar{v}_1.$$

Step 7. Applying inductive method, we can construct four sequences $\{\underline{u}_i\}$, $\{\underline{v}_i\}$, $\{\bar{u}_i\}$ and $\{\bar{v}_i\}$ by

$$\bar{v}_i=\psi(\bar{u}_i),\quad \underline{u}_i=\varphi(\bar{v}_i),\quad \underline{v}_i=\psi(\underline{u}_i),\quad \bar{u}_{i+1}=\varphi(\underline{v}_i),\qquad(5.42)$$

and $\underline{u}_i,\underline{v}_i,\bar{u}_i,\bar{v}_i$ satisfy

$$\begin{cases}\underline{u}_i\leqslant\liminf\limits_{t\to\infty}\min\limits_{\overline{\Omega}} u(\cdot,t)\leqslant\limsup\limits_{t\to\infty}\max\limits_{\overline{\Omega}} u(\cdot,t)\leqslant\bar{u}_i,\\[2mm]\underline{v}_i\leqslant\liminf\limits_{t\to\infty}\min\limits_{\overline{\Omega}} v(\cdot,t)\leqslant\limsup\limits_{t\to\infty}\max\limits_{\overline{\Omega}} v(\cdot,t)\leqslant\bar{v}_i.\end{cases}\qquad(5.43)$$

On account that $\varphi'(s) < 0$ and $\psi'(s) > 0$ for $s > 0$, it follows from (5.41) and (5.42) that

$$\underline{v}_{i-1} < \underline{v}_i = \psi(\underline{u}_i) < \psi(\bar{u}_i) = \bar{v}_i < \bar{v}_{i-1},$$
$$\underline{u}_{i-1} < \underline{u}_i = \varphi(\bar{v}_i) < \varphi(\underline{v}_i) = \bar{u}_{i+1} < \bar{u}_i$$

inductively. Therefore, sequences $\{\underline{u}_i\}, \{\underline{v}_i\}, \{\bar{u}_i\}$ and $\{\bar{v}_i\}$ have point-wise limits:

$$\lim_{i\to\infty} \underline{u}_i = \underline{u}, \quad \lim_{i\to\infty} \underline{v}_i = \underline{v}, \quad \lim_{i\to\infty} \bar{u}_i = \bar{u} \quad \text{and} \quad \lim_{i\to\infty} \bar{v}_i = \bar{v}. \tag{5.44}$$

Obviously, $0 < \underline{u} \leqslant \bar{u}$ and $0 < \underline{v} \leqslant \bar{v}$, and $\underline{u}, \underline{v}, \bar{u}, \bar{v}$ satisfy

$$\underline{u} = \varphi(\bar{v}), \quad \bar{u} = \varphi(\underline{v}), \quad \underline{v} = \psi(\underline{u}) \quad \text{and} \quad \bar{v} = \psi(\bar{u}). \tag{5.45}$$

By a direct calculation we see that (5.45) is equivalent to

$$\underline{u} = (c - k\bar{v})/(ck), \quad \bar{u} = (c - k\underline{v})/(ck), \tag{5.46}$$

$$a + b\underline{v} - \underline{u}/(c\underline{u} + \underline{v}) = 0, \quad a + b\bar{v} - \bar{u}/(c\bar{u} + \bar{v}) = 0. \tag{5.47}$$

It follows from (5.46) that

$$c\underline{u} + \bar{v} = c/k = c\bar{u} + \underline{v}, \quad \text{which implies} \quad c(\bar{u} - \underline{u}) = \bar{v} - \underline{v}. \tag{5.48}$$

Step 8. Now, we prove $\bar{v} = \underline{v}$, and then get $\bar{u} = \underline{u}$ by (5.48). Substituting the second equation of (5.46) into that of (5.47), we produce

$$a + b\bar{v} - \frac{c - k\underline{v}}{c(c - k\underline{v}) + ck\bar{v}} = 0,$$

that is,

$$ac(c - k\underline{v}) + ack\bar{v} + bc(c - k\underline{v})\bar{v} + bck\bar{v}^2 = c - k\underline{v}. \tag{5.49}$$

Similarly, by the first equation of (5.46) and that of (5.47), we have

$$ac(c - k\bar{v}) + ack\underline{v} + bc(c - k\bar{v})\underline{v} + bck\underline{v}^2 = c - k\bar{v}. \tag{5.50}$$

Subtracting (5.50) into (5.49) results in

$$bck(\bar{v}^2 - \underline{v}^2) + bc^2(\bar{v} - \underline{v}) + 2ack(\bar{v} - \underline{v}) = k(\bar{v} - \underline{v}). \tag{5.51}$$

The proof will be accomplished by a contradiction argument. Suppose otherwise that $\bar{v} \neq \underline{v}$. Then $\bar{v} > \underline{v}$ and

$$k = bck(\bar{v} + \underline{v}) + bc^2 + 2ack \tag{5.52}$$

in accordance with (5.51). The condition (5.40) asserts $k < 2bc^2 + 2ack$. This together with (5.52) alleges $bc^2 > bck(\bar{v} + \underline{v})$, i.e., $\bar{v} + \underline{v} < c/k$. Hence, by (5.48),

$$c\underline{u} + \bar{v} = c/k > \bar{v} + \underline{v} \quad \Longrightarrow \quad \underline{v} < c\underline{u}. \tag{5.53}$$

Owing to (5.47) we have as well

$$b(\bar{v} - \underline{v}) = \frac{\bar{u}}{c\bar{u} + \bar{v}} - \frac{\underline{u}}{c\underline{u} + \underline{v}} = \frac{(\bar{u} - \underline{u})\underline{v} - \underline{u}(\bar{v} - \underline{v})}{(c\bar{u} + \bar{v})(c\underline{u} + \underline{v})}.$$

It follows from (5.48) immediately that

$$b(\bar{v} - \underline{v}) = \frac{\frac{1}{c}(\bar{v} - \underline{v})\underline{v} - \underline{u}(\bar{v} - \underline{v})}{(c\bar{u} + \bar{v})(c\underline{u} + \underline{v})} = \frac{(\bar{v} - \underline{v})(\underline{v} - c\underline{u})}{c(c\bar{u} + \bar{v})(c\underline{u} + \underline{v})},$$

which implies $\underline{v} > c\underline{u}$ since $\bar{v} - \underline{v} > 0$. This contradicts (5.53). And subsequently, $\bar{v} = \underline{v}$ and $\bar{u} = \underline{u}$, which in turn yield $\bar{u} = \underline{u} = \tilde{u}$ and $\bar{v} = \underline{v} = \tilde{v}$. Using (5.43) and (5.44), we finally get

$$\lim_{t \to \infty} (u(x,t), v(x,t)) = (\tilde{u}, \tilde{v})$$

uniformly in $\overline{\Omega}$. □

5.4 AVERAGE METHOD

Let u denote the unique solution of the following initial-boundary value problem

$$\begin{cases} u_{it} - d_i \Delta u_i = f_i(x,t,u), & x \in \Omega, \quad t > 0, \\ \partial_n u_i = 0, & x \in \partial\Omega, \ t > 0, \\ u_i(x,0) = u_{i0}(x), & x \in \Omega, \\ i = 1, \ldots, m, \end{cases} \tag{5.54}$$

where $d_i > 0$. Define

$$\bar{u}_i(t) = \frac{1}{|\Omega|} \int_\Omega u_i(x,t), \quad \bar{u}(t) = (\bar{u}_1(t), \ldots, \bar{u}_m(t)).$$

Then $\bar{u}_i$ satisfies

$$\bar{u}_i'(t) = \frac{1}{|\Omega|} \int_\Omega f_i(x,t,u(x,t)), \quad t > 0; \quad \bar{u}_i(0) = \frac{1}{|\Omega|} \int_\Omega u_{i0}(x).$$

This is an initial value problem of the first order ODEs. In some cases, it is possible to obtain limit of $\bar{u}_i(t)$ as $t \to \infty$. The *average method*, just as what its name suggests, is to determine the limit of $u_i(x,t)$ by that of $\bar{u}_i(t)$.

5.4.1 Abstract results

Let u be the unique global solution of (5.54) and $u(x,t) \in \Sigma$ for some bounded domain $\Sigma \subset \mathbb{R}^m$ and all $(x,t) \in Q_\infty$. We define

$$K = \sup\{|f_u(x,t,u)| : (x,t) \in Q_\infty, \ u \in \Sigma\},$$

$$F(t) = \sup\{|\nabla_x f(x,t,u)| : x \in \overline{\Omega}, \ u \in \Sigma\}.$$

Let $\eta > 0$ be the second eigenvalue of $-\Delta$ in Ω with the homogeneous Neumann boundary condition and $d = \min\{d_1, \ldots, d_m\}$.

Theorem 5.16 *Let f be bounded in $Q_\infty \times \Sigma$. If $\sigma = d\eta - K > 0$, then*

$$\|\nabla u(\cdot,t)\|_2^2 \leqslant e^{-\sigma t}\left(\|\nabla u_0\|_2^2 + \frac{|\Omega|}{\sigma}\int_0^t e^{\sigma s}F^2(s)\mathrm{d}s\right), \quad t \geqslant 0, \qquad (5.55)$$

$$\|u(\cdot,t) - \bar{u}(t)\|_2^2 \leqslant \frac{1}{\eta}e^{-\sigma t}\left(\|\nabla u_0\|_2^2 + \frac{|\Omega|}{\sigma}\int_0^t e^{\sigma s}F^2(s)\mathrm{d}s\right), \quad t \geqslant 0. \qquad (5.56)$$

If we further assume that $\lim_{t\to\infty} F(t) = 0$, then, for any $0 < \alpha < 1$,

$$\lim_{t\to\infty} \|u(\cdot,t) - \bar{u}(t)\|_{C^\alpha(\overline{\Omega})} = 0. \qquad (5.57)$$

Proof. Firstly, by the Poincaré inequality,

$$\|\nabla w\|_2^2 \geqslant \eta\|w - \bar{w}\|_2^2, \quad \|\Delta w\|_2^2 \geqslant \eta\|\nabla w\|_2^2 \qquad (5.58)$$

for all $w \in H^2(\Omega)$ with $\partial_{\boldsymbol{n}}w = 0$ on $\partial\Omega$. Denote

$$v(t) = \|\nabla u(\cdot,t)\|_2^2, \quad D = \mathrm{diag}(d_1,\ldots,d_m).$$

Then, by the second inequality of (5.58) and the Young inequality, it derives

$$\begin{aligned}
v'(t) &= 2\int_\Omega \langle \nabla u, \nabla u_t\rangle \\
&= 2\int_\Omega \langle \nabla u, \nabla(D\Delta u + f)\rangle \\
&= -2\int_\Omega \langle \Delta u, D\Delta u\rangle + 2\int_\Omega \langle \nabla u, f_u \cdot \nabla u + \nabla_x f\rangle \\
&\leqslant -2d\eta\int_\Omega |\nabla u|^2 + 2K\int_\Omega |\nabla u|^2 + \sigma\int_\Omega |\nabla u|^2 + \frac{|\Omega|}{\sigma}F^2(t) \\
&= -\sigma v(t) + \frac{|\Omega|}{\sigma}F^2(t),
\end{aligned}$$

which implies

$$v(t) \leqslant v(0)e^{-\sigma t} + \frac{|\Omega|}{\sigma}e^{-\sigma t}\int_0^t e^{\sigma s}F^2(s)\mathrm{d}s.$$

This is exactly (5.55). Using this fact and the first inequality of (5.58) we get (5.56) immediately.

If in addition $\lim_{t\to\infty} F(t) = 0$, it then follows from (5.55) and (5.56) that

$$\lim_{t\to\infty} \|\nabla u(\cdot,t)\|_2 = \lim_{t\to\infty} \|u(\cdot,t) - \bar{u}(t)\|_2 = 0. \qquad (5.59)$$

Since f is bounded in $Q_\infty \times \Sigma$ and $u(x,t) \in \Sigma$ for all $(x,t) \in Q_\infty$, we have that $|\nabla u|$ is bounded in $\Omega \times [1,\infty)$ by Theorem 2.13. Take $p > \max\{n,2\}$ satisfying $\alpha < 1 - n/p$. Then we have

$$\|u(\cdot,t) - \bar{u}(t)\|_{C^\alpha(\overline{\Omega})} \leqslant C\|u(\cdot,t) - \bar{u}(t)\|_{W_p^1(\Omega)} \leqslant C'\|u(\cdot,t) - \bar{u}(t)\|_{H^1(\Omega)}^{2/p}, \quad t \geqslant 1.$$

This, together with (5.59), leads to (5.57). $\qquad\square$

5.4.2 A special diffusive competition model

To understand effects of migration and spatial heterogeneity, Dockery et al. ([21]) and Lou ([80]) studied the following initial-boundary value problem of a diffusive competition model

$$\begin{cases} u_t - d_1\Delta u = u(m(x) - u - v), & x \in \Omega, \ t > 0, \\ v_t - d_2\Delta v = v(m(x) - u - v), & x \in \Omega, \ t > 0, \\ \partial_{\boldsymbol{n}} u = \partial_{\boldsymbol{n}} v = 0, & x \in \partial\Omega, \ t \geqslant 0, \\ u(x,0) = u_0(x), \ v(x,0) = v_0(x), & x \in \Omega, \end{cases}$$

and established various interesting results about equilibrium solutions and their stabilities. Here, we deal with a special case $m(x) \equiv 1$, i.e.,

$$\begin{cases} u_t - d_1\Delta u = u(1 - u - v), & x \in \Omega, \quad t > 0, \\ v_t - d_2\Delta v = v(1 - u - v), & x \in \Omega, \quad t > 0, \\ \partial_{\boldsymbol{n}} u = \partial_{\boldsymbol{n}} v = 0, & x \in \partial\Omega, \ t \geqslant 0, \\ u(x,0) = u_0(x), \ v(x,0) = v_0(x), & x \in \Omega. \end{cases} \tag{5.60}$$

Obviously, for each $s \in (0,1)$, $(s, 1 - s)$ is a positive equilibrium solution of (5.60). On the other hand, it is easily shown that any positive equilibrium solution of (5.60) must be constant and has form $(s, 1 - s)$ for some $s \in (0,1)$. That is, segment $\{(a,b) : a,b \in (0,1), a + b = 1\}$ is the set of positive equilibrium solutions of (5.60). Our main aim is to investigate longtime behaviour of solutions of (5.60). Contents of this part are due to [85].

In this subsection, we assume $u_0, v_0 \in W_p^2(\Omega) \cap C(\overline{\Omega})$ and $u_0, v_0 \geqslant 0, \not\equiv 0$.

Theorem 5.17 (1) *Problem* (5.60) *has a unique bounded non-negative global solution* (u, v). *Moreover,*

$$\frac{\underline{u}_0}{\underline{u}_0 + \bar{v}_0} \leqslant \liminf_{t\to\infty} \min_{x\in\overline{\Omega}} u(x,t) \leqslant \limsup_{t\to\infty} \max_{x\in\overline{\Omega}} u(x,t) \leqslant \frac{\bar{u}_0}{\bar{u}_0 + \underline{v}_0},$$

$$\frac{\underline{v}_0}{\bar{u}_0 + \underline{v}_0} \leqslant \liminf_{t\to\infty} \min_{x\in\overline{\Omega}} v(x,t) \leqslant \limsup_{t\to\infty} \max_{x\in\overline{\Omega}} v(x,t) \leqslant \frac{\bar{v}_0}{\underline{u}_0 + \bar{v}_0},$$

where $\bar{u}_0 = \max_{\overline{\Omega}} u_0(x)$, $\underline{u}_0 = \min_{\overline{\Omega}} u_0(x)$, $\bar{v}_0 = \max_{\overline{\Omega}} v_0(x)$ *and* $\underline{v}_0 = \min_{\overline{\Omega}} v_0(x)$.

(2) *There exists a constant* $C > 0$ *such that*

$$\|u(\cdot,t)\|_{C^{2+\alpha}(\overline{\Omega})}, \ \|v(\cdot,t)\|_{C^{2+\alpha}(\overline{\Omega})} \leqslant C \ \text{ for all } t \geqslant 1. \tag{5.61}$$

Proof. Since problem (5.60) is a quasi-monotonically decreasing system for $u, v \geqslant 0$, the conclusion (1) can be derived by the upper and lower solutions method directly. The conclusion (2) is a consequence of Remark 2.12. $\qquad\square$

Lemma 5.18 *Let (u, v) be the unique solution of (5.60) and $w = u + v$. Then*

$$\limsup_{t \to \infty} \bar{w}(t) \leqslant 1. \tag{5.62}$$

Moreover, we have that either

$$\bar{w}(T) \leqslant 1 \quad \text{for some } T \geqslant 0, \quad \text{which implies } \bar{w}(t) \leqslant 1 \quad \text{for all } t \geqslant T, \tag{5.63}$$

or

$$\bar{w}(t) > 1 \quad \text{for all } t \geqslant 0. \tag{5.64}$$

Proof. It is easy to see that

$$\bar{w}'(t) = \frac{1}{|\Omega|} \int_\Omega w(1 - w).$$

By using the Hölder inequality we have $\bar{w}'(t) \leqslant \bar{w}(1 - \bar{w})$. The desired conclusions can be derived from this inequality directly. □

Recalling Theorem 5.17 (1), we may assume $u, v \leqslant 1$ in the following discussions. Set

$$E(t) = \int_\Omega (u - \bar{u})^2 + \int_\Omega (v - \bar{v})^2.$$

Similar to the discussion in proving Theorem 5.16, we can find a constant $k > 0$ such that

$$E'(t) \leqslant -2 \int_\Omega (d_1 |\nabla u|^2 + d_2 |\nabla v|^2) + kE(t). \tag{5.65}$$

Lemma 5.19 *If $d_1, d_2 > 2/\eta$, then*

$$\lim_{t \to \infty} \int_\Omega (u - \bar{u})^2 = \lim_{t \to \infty} \int_\Omega (v - \bar{v})^2 = 0. \tag{5.66}$$

Proof. Set $f(u, v) = u(1 - u - v)$ and $g(u, v) = v(1 - u - v)$. Then $|f_u, f_v, g_u, g_v| \leqslant 2$. Invoking notations in §5.4.1, we have $K \leqslant 2$ and $F(t) \equiv 0$. If $d_1, d_2 > 2/\eta$, then the limit (5.66) can be followed by Theorem 5.16. □

Theorem 5.20 *If $d_1, d_2 \geqslant \dfrac{k+2}{2\eta}$, then there is a constant $0 \leqslant s \leqslant 1$ such that* $\lim_{t \to \infty} u(x, t) = s$ *and* $\lim_{t \to \infty} v(x, t) = 1 - s$ *uniformly in* $\overline{\Omega}$.

Proof. Let ε be a constant to be determined, and set

$$Q(t) = \int_\Omega u + \varepsilon E(t).$$

Using (5.65), we have

$$E'(t) \leqslant -2\eta \frac{k+2}{2\eta} E(t) + kE(t) \leqslant -2E(t). \tag{5.67}$$

When (5.63) holds, we have, by taking $\varepsilon = -1$ and using (5.67), that

$$Q'(t) \geqslant \int_\Omega \{u(1-w) + 2[(u-\bar{u})^2 + (v-\bar{v})^2]\}$$

$$\geqslant \int_\Omega \{2[(u-\bar{u})^2 + (v-\bar{v})^2] - (u-\bar{u})^2 - (u-\bar{u})(v-\bar{v})\}$$

$$\geqslant \frac{1}{2} \int_\Omega [(u-\bar{u})^2 + 3(v-\bar{v})^2] \geqslant 0, \quad t > T.$$

This implies that $Q(t)$ is convergent as $t \to \infty$ since $Q(t)$ is bounded. By (5.66), there exists a constant $0 \leqslant s_1 \leqslant 1$ such that $\bar{u}$ converges to s_1 as $t \to \infty$. Using (5.66) again we conclude $\lim_{t\to\infty} \|u(\cdot,t) - s_1\|_{2,\Omega} = 0$.

When (5.64) holds, taking $\varepsilon = 1/2$ and using (5.67) one has

$$Q'(t) \leqslant \int_\Omega \{u(1-w) - [(u-\bar{u})^2 + (v-\bar{v})^2]\}$$

$$\leqslant -\int_\Omega \{2(u-\bar{u})^2 + (u-\bar{u})(v-\bar{v}) + (v-\bar{v})^2\}$$

$$\leqslant -\frac{3}{2} \int_\Omega (u-\bar{u})^2 - \frac{1}{2} \int_\Omega (v-\bar{v})^2 \leqslant 0.$$

Then, arguing as above, we likewise deduce $\lim_{t\to\infty} \|u(\cdot,t) - s_1\|_{2,\Omega} = 0$.

Similarly, there exists a constant $0 \leqslant s_2 \leqslant 1$ such that $\lim_{t\to\infty} \|v(\cdot,t) - s_2\|_{2,\Omega} = 0$. Recalling (5.61), it derives that

$$\lim_{t\to\infty} \|u(\cdot,t) - s_1\|_{C(\overline{\Omega})} = \lim_{t\to\infty} \|v(\cdot,t) - s_2\|_{C(\overline{\Omega})} = 0.$$

As a result, $\lim_{t\to\infty} \|w(\cdot,t) - s_1 - s_2\|_{C(\overline{\Omega})} = 0$. In addition, $\lim_{t\to\infty} \bar{w}(t) = s_1 + s_2 \leqslant 1$ by (5.62). We claim that $s_1 + s_2 = 1$. If $s_1 + s_2 < 1$, then there is $T \gg 1$ such that $w < \delta := (1 + s_1 + s_2)/2 < 1$ in $\overline{\Omega} \times [T, \infty)$. As a consequence,

$$\frac{d}{dt} \int_\Omega u = \int_\Omega u(1-w) \geqslant (1-\delta) \int_\Omega u \quad \text{for } t \geqslant T,$$

from which it yields that $\lim_{t\to\infty} \bar{u}(t) = \infty$ since $\bar{u}(T) > 0$. It is a contradiction. $\quad\square$

In the following, we consider a special case $d_1 = d_2 =: d$. Then $w = u + v$ satisfies

$$\begin{cases} w_t - d\Delta w = w(1-w), & x \in \Omega, \quad t > 0, \\ \partial_{\boldsymbol{n}} w = 0, & x \in \partial\Omega, \ t \geqslant 0, \\ w(x,0) = w_0(x), & x \in \Omega \end{cases}$$

with $w_0(x) = u_0(x) + v_0(x)$. Clearly,

$$\lim_{t\to\infty} w(x,t) = 1 \quad \text{uniformly in } \overline{\Omega}. \tag{5.68}$$

Without loss of generality, we assume $w_0 > 0$ in $\overline{\Omega}$. Take $m = \min_{\overline{\Omega}} w_0(x)$ and $M = \max_{\overline{\Omega}} w_0(x)$. Then $m > 0$ and

$$\frac{me^t}{1 - m + me^t} \leqslant w(x,t) \leqslant \frac{Me^t}{1 - M + Me^t}.$$

Problem (5.60) reduces to

$$\begin{cases} u_t - d\Delta u = u(1 - w(x,t)), & x \in \Omega, \quad t > 0, \\ v_t - d\Delta v = v(1 - w(x,t)), & x \in \Omega, \quad t > 0, \\ \partial_{\boldsymbol{n}} u = \partial_{\boldsymbol{n}} v = 0, & x \in \partial\Omega, \ t \geqslant 0, \\ u(x,0) = u_0(x), \ v(x,0) = v_0(x), & x \in \Omega. \end{cases}$$

It will be shown that

$$\lim_{t\to\infty} u(x,t) = s, \quad \lim_{t\to\infty} v(x,t) = 1 - s \quad \text{uniformly in } \overline{\Omega}$$

for some $s \in [0,1]$. For this motivation, let us first study a new problem

$$\begin{cases} \varphi_t - d\Delta\varphi = \varphi(1 - w(x,t)), & x \in \Omega, \quad t > 0, \\ \partial_{\boldsymbol{n}}\varphi = 0, & x \in \partial\Omega, \ t \geqslant 0, \\ \varphi(x,0) = \varphi_0(x) \in C(\overline{\Omega}), & x \in \Omega. \end{cases} \tag{5.69}$$

In the sequel, we will prove that the unique solution φ of (5.69) converges to a constant uniformly in $\overline{\Omega}$ as $t \to \infty$.

Let ψ denote the unique solution of

$$\begin{cases} \psi_t - d\Delta\psi = 0, & x \in \Omega, \quad t > 0, \\ \partial_{\boldsymbol{n}}\psi = 0, & x \in \partial\Omega, \ t \geqslant 0, \\ \psi(x,0) = \varphi_0(x), & x \in \Omega. \end{cases}$$

Then $\lim_{t\to\infty} \psi(x,t) = \bar{\varphi}_0$ uniformly in $\overline{\Omega}$. Moreover, the upper and lower solutions method gives

$$\psi(x,t)\frac{e^t}{1 - M + Me^t} \leqslant \varphi(x,t) \leqslant \psi(x,t)\frac{e^t}{1 - m + me^t}, \tag{5.70}$$

which implies

$$\frac{1}{M}\bar{\varphi}_0 \leqslant \liminf_{t\to\infty} \varphi(x,t) \leqslant \limsup_{t\to\infty} \varphi(x,t) \leqslant \frac{1}{m}\bar{\varphi}_0. \tag{5.71}$$

Theorem 5.21 *Let $d_1 = d_2 = d$, and let (u,v) be the unique solution of (5.60) and φ be the unique solution of (5.69) with $w(x,t) = u(x,t) + v(x,t)$. If $\varphi_0 \geqslant 0$, then φ converges to a constant uniformly in $\overline{\Omega}$ as $t \to \infty$.*

Proof. For any given $\delta > 0$, let us denote

$$m_\delta = \min_{\overline{\Omega}} w(x,\delta), \quad M_\delta = \max_{\overline{\Omega}} w(x,\delta), \quad \varphi_\delta(x) = \varphi(x,\delta).$$

It follows from (5.68) that

$$\lim_{\delta\to\infty} m_\delta = \lim_{\delta\to\infty} M_\delta = 1. \tag{5.72}$$

Similar to derivations of (5.70) and (5.71), we have

$$\frac{1}{M_\delta}\bar{\varphi}_\delta \leqslant \liminf_{t\to\infty} \varphi(x, t+\delta) = \liminf_{t\to\infty} \varphi(x, t)$$

$$\leqslant \limsup_{t\to\infty} \varphi(x, t) = \limsup_{t\to\infty} \varphi(x, t+\delta) \leqslant \frac{1}{m_\delta}\bar{\varphi}_\delta, \quad (5.73)$$

where $\bar{\varphi}_\delta = \frac{1}{|\Omega|}\int_\Omega \varphi_\delta(x)$. In accordance with (5.72) and the boundedness of $\bar{\varphi}_\delta$ in δ, it then follows that

$$0 \leqslant \limsup_{t\to\infty} \varphi(x, t) - \liminf_{t\to\infty} \varphi(x, t) \leqslant \left(\frac{1}{m_\delta} - \frac{1}{M_\delta}\right)\bar{\varphi}_\delta \to 0$$

as $\delta \to \infty$. This implies that φ is uniformly convergent in $\overline{\Omega}$ as $t \to \infty$, and then $\bar{\varphi}_\delta$ converges to a constant as $\delta \to \infty$. Recalling (5.73), we assert that φ converges to a constant uniformly in $\overline{\Omega}$ as $t \to \infty$. □

Let $\varphi(x, t; \phi)$ be the unique solution of (5.69) with initial function $\varphi_0(x) = \phi(x)$. Then, for any $k_1, k_2 \in \mathbb{R}$, there holds

$$\varphi(x, t; k_1\phi_1 + k_2\phi_2) = k_1\varphi(x, t; \phi_1) + k_2\varphi(x, t; \phi_2). \quad (5.74)$$

Noticing $0 \leqslant u_0 \leqslant w_0$ and $0 \leqslant v_0 \leqslant w_0$ in $\overline{\Omega}$, the following theorem can be derived from Theorem 5.21 and the limit (5.68).

Theorem 5.22   *There is $0 \leqslant s \leqslant 1$ for which $\lim_{t\to\infty}(u(x,t), v(x,t)) = (s, 1-s)$ uniformly in $\overline{\Omega}$.*

Especially, we can get more accurate results under some requirements on initial densities.

Theorem 5.23   (1) *If $w_0 \equiv C$ for some constant $C > 0$, then*

$$\lim_{t\to\infty} u(x,t) = \bar{u}_0/C \quad and \quad \lim_{t\to\infty} v(x,t) = \bar{v}_0/C \ \text{uniformly in } \overline{\Omega}.$$

In particular, for the corresponding ODE problem of (5.60), if $u_0 + v_0 > 0$ then

$$\lim_{t\to\infty} u(t) = u_0/(u_0 + v_0) \quad and \quad \lim_{t\to\infty} v(t) = v_0/(u_0 + v_0).$$

(2) *If $u_0 \equiv kv_0$ for some constant $k \geqslant 0$, then*

$$\lim_{t\to\infty} u(x,t) = k/(1+k) \quad and \quad \lim_{t\to\infty} v(x,t) = 1/(1+k) \ \text{uniformly in } \overline{\Omega}.$$

Proof. Conclusions can be derived from (5.68), (5.71) and (5.74). □

Remark 5.24   *For the case $w_0(x) \equiv C > 0$, or for the corresponding ODE problem of (5.60), we can get the exact limits of u and v. Unfortunately, for the general case, we can only prove that u and v have limits as $t \to \infty$ and cannot give the exact values of them.*

EXERCISES

5.1 Prove conclusion (2) of Theorem 5.3.

5.2 Let L and B be as in (3.10) and λ_1 be the first eigenvalue of

$$\begin{cases} L[\phi] = \lambda\phi & \text{in } \Omega, \\ B[\phi] = 0 & \text{on } \partial\Omega. \end{cases}$$

Let $f \in C^1(\overline{\Omega} \times \mathbb{R})$ and $u_s(x)$ be a solution of

$$\begin{cases} L[u] = f(x, u) & \text{in } \Omega, \\ B[u] = 0 & \text{on } \partial\Omega. \end{cases}$$

By using the upper and lower solutions method, prove the following conclusions.

(1) If $f_u(x, u_s(x)) < \lambda_1$ for all $x \in \overline{\Omega}$, then $u_s(x)$ is locally asymptotically stable.

(2) If $f_u(x, u_s(x)) > \lambda_1$ for all $x \in \overline{\Omega}$, then $u_s(x)$ is unstable.

5.3 Let λ be the first eigenvalue of (1.33). Prove that

$$\|u\|_2^2 \leqslant \frac{1}{\lambda}\|\nabla u\|_2^2 \text{ for all } u \in H_0^1(\Omega),$$

$$\|\nabla u\|_2^2 \leqslant \frac{1}{\lambda}\|\Delta u\|_2^2 \text{ for all } u \in H_0^1(\Omega) \cap H^2(\Omega),$$

and $1/\lambda$ is the best constant.

5.4 Consider problem

$$\begin{cases} -\Delta u = u(a - u)^2(1 - u) & \text{in } \Omega, \\ \partial_{\boldsymbol{n}} u = 0 & \text{on } \partial\Omega, \end{cases}$$

where $0 < a < 1$ is a constant. Discuss local stabilities of 0, a and 1.

5.5 Assume that $\dfrac{\partial f_1}{\partial u_2} \geqslant 0$ and $\dfrac{\partial f_2}{\partial u_1} \geqslant 0$. Let (u_{1s}, u_{2s}) be a solution of

$$\begin{cases} -\Delta u_1 = f_1(u_1, u_2) & \text{in } \Omega, \\ -\Delta u_2 = f_2(u_1, u_2) & \text{in } \Omega, \\ u_1 = u_2 = 0 & \text{on } \partial\Omega. \end{cases}$$

Denote

$$a_{ij} = \frac{\partial f_i}{\partial u_j}(u_{1s}, u_{2s}), \quad i, j = 1, 2.$$

Let λ_0 be the principle eigenvalue of

$$\begin{cases} \Delta\phi_1 + a_{11}\phi_1 + a_{12}\phi_2 = \lambda\phi_1 & \text{in } \Omega, \\ \Delta\phi_2 + a_{21}\phi_1 + a_{22}\phi_2 = \lambda\phi_2 & \text{in } \Omega, \\ \phi_1 = \phi_2 = 0 & \text{on } \partial\Omega, \end{cases}$$

and $\phi = (\phi_1, \phi_2)$ be the corresponding eigenfunction with $\phi_1, \phi_2 > 0$ in Ω.

(1) Prove that (u_{1s}, u_{2s}) is locally stable if $\lambda_0 < 0$, while it is unstable if $\lambda_0 > 0$.

(2) When $\dfrac{\partial f_1}{\partial u_2} \leqslant 0$ and $\dfrac{\partial f_2}{\partial u_1} \leqslant 0$, what kind of conclusion can we draw?

5.6 Prove Lemma 5.5 under the condition that ψ is uniformly continuous in $[a, \infty)$.

5.7 Let $h : [0, \infty) \to \mathbb{R}$ be uniformly continuous, and $\lim\limits_{t \to \infty} \int_0^t h(s)ds$ exist. Prove $\lim\limits_{t \to \infty} h(t) = 0$.

5.8 Prove Theorem 5.12.

5.9 Assume $u_{i0} \in W_p^2(\Omega)$ for some $p > 1 + n/2$ and $i = 1, 2$, $u_{30} \in C^\gamma(\overline{\Omega})$ for some $0 < \gamma < 1$, and $u_{10}, u_{20} > 0$, $u_{30} \geqslant 0$ in $\overline{\Omega}$. Using the space $C(\overline{Q_{T_0}})$ instead of $L^p(Q_{T_0})$, and imitating Theorem 4.1 to prove that problem (5.29) has a unique local solution (u_1, u_2, u_3) satisfying $u_1, u_2 \in W_p^{2,1}(Q_T)$ and $u_3 \in C^{\alpha, 1}(\overline{Q_T})$ for some $0 < T < \infty$ and $0 < \alpha < \min\{\gamma, 2 - (n+2)/p\}$.

5.10 Continue to study problem (5.39) and assume $ac < 1$. Prove that if $k < bc^2 + 2ack$ then the positive constant equilibrium solution $(\tilde{u}, \tilde{v}) = \left(\dfrac{c(a + b\tilde{v})}{k}, \dfrac{c(1 - ac)}{k + bc^2}\right)$ is globally asymptotically stable by using the Lyapunov functional method.

5.11 Give the proof of Lemma 5.14.

5.12 Consider a prey-predator model with a transmissible disease in the prey population

$$
\begin{cases}
u_t - d_1\Delta u = au(1 - u - v) - uv - muw, & x \in \Omega, \ t > 0, \\
v_t - d_2\Delta v = uv - bvw - cv, & x \in \Omega, \ t > 0, \\
w_t - d_3\Delta w = \theta muw - \theta bvw - kw, & x \in \Omega, \ t > 0, \qquad \text{(P5.1)} \\
\partial_{\boldsymbol{n}}u = \partial_{\boldsymbol{n}}v = \partial_{\boldsymbol{n}}w = 0, & x \in \partial\Omega, \ t > 0, \\
u = u_0(x), \ v = v_0(x), \ w = w_0(x), & x \in \Omega, \ t = 0,
\end{cases}
$$

where coefficients are positive constants, $u_0, v_0, w_0 \in W_p^2(\Omega)$ and $u_0, v_0, w_0 \geqslant 0, \not\equiv 0$.

(1) Prove that if $c < 1$ then (P5.1) has a non-negative constant equilibrium solution $E^* = \left(c, \dfrac{a(1 - c)}{a + 1}, 0\right)$ which is called the predator-free equilibrium solution.

(2) Assume $c < 1$ and $mc + ab(1 - c) < k/\theta$. Take advantage of the Lyapunov functional method to prove that the predator-free equilibrium solution E^* is globally asymptotically stable.

5.13 Give details of the proof of Theorem 5.23.

5.14 Consider a Lotka-Volterra model of a two-predator and one-prey system

$$
\begin{cases}
u_t - d_1 \Delta u = u\left(-1 + \dfrac{vw}{u+v}\right), & x \in \Omega, \quad t > 0, \\[3mm]
v_t - d_2 \Delta v = v\left(-a + \dfrac{buw}{u+v}\right), & x \in \Omega, \quad t > 0, \\[3mm]
w_t - d_3 \Delta w = w\left(r - w - \dfrac{(1+b)uv}{u+v}\right), & x \in \Omega, \quad t > 0, \\[2mm]
\partial_n u = \partial_n v = \partial_n w = 0, & x \in \partial\Omega, t > 0, \\[2mm]
u = u_0(x),\ v = v_0(x),\ w = w_0(x), & x \in \Omega, \quad t = 0,
\end{cases}
\tag{P5.2}
$$

where coefficients are positive constants, $u_0, v_0, w_0 \in W_p^2(\Omega)$ and $u_0, v_0, w_0 > 0$ in $\overline{\Omega}$.

(1) Utilize Theorem 2.14 to prove that the solution (u, v, w) of (P5.2) exists globally and is uniformly bounded.

(2) Prove that (P5.2) has a positive constant equilibrium solution if and only if $rb > a + b$, and in this case the positive constant equilibrium solution is uniquely given by

$$
\tilde{u} = (a+b)\frac{rb - (a+b)}{b^2(1+b)}, \quad \tilde{v} = (a+b)\frac{rb - (a+b)}{ab(1+b)}, \quad \tilde{w} = \frac{a+b}{b}.
$$

(3) Use the Lyapunov functional method to prove that $(\tilde{u}, \tilde{v}, \tilde{w})$ is globally asymptotically stable.

5.15 Consider a predator-prey model with hyperbolic mortality and nonlinear prey harvesting

$$
\begin{cases}
u_t - d_1 \Delta u = u(1 - u) - \dfrac{suv}{a+u} - \dfrac{bu}{\rho+u}, & x \in \Omega, \quad t > 0, \\[3mm]
v_t - d_2 \Delta v = \dfrac{uv}{a+u} - \dfrac{rv^2}{1+rv}, & x \in \Omega, \quad t > 0, \\[2mm]
\partial_n u = \partial_n v = 0, & x \in \partial\Omega, t > 0, \\[2mm]
u(x,0) = u_0(x),\ v(x,0) = v_0(x), & x \in \Omega,
\end{cases}
\tag{P5.3}
$$

where s, a, b, ρ and r are positive constants, $u_0, v_0 \in W_p^2(\Omega)$ and $u_0, v_0 \geqslant 0, \not\equiv 0$.

(1) Prove that if $b < \rho < a$ then (P5.3) has a unique positive constant equilibrium solution (u^*, v^*).

(2) Assume $b < \rho < a$ and

$$
s < ra^2,\ b < \rho(1 - s/(ra^2)),\ \rho > 1,\ a(1 - \rho) + \rho < b.
$$

Use the iteration method to prove that (u^*, v^*) is globally asymptotically stable.

CHAPTER 6

Global Existence and Finite Time Blowup

Blowup phenomena occur in various types of nonlinear evolution equations such as the Schrödinger equations, hyperbolic equations as well as parabolic equations. The phenomenon of finite time blowup has important applications in many engineering fields.

For global existence and finite time blowup of solutions, there are some natural questions:

- will the blowup occur in finite time?

- how to determine conditions for the global existence and finite time blowup of solutions?

- is there a critical exponent and how to determine it?

- when the blowup occurs in finite time, where are blowup points, and can the blowup occur in a region?

- when the blowup occurs in finite time, how large is the blowup time?

- what is the profile of solution near blowup time?

The global existence and finite time blowup of solutions for nonlinear parabolic equations and systems have been studied in numerous papers. Survey papers [63, 33, 11, 18] are the good starting point to get a sense of current development of global existence and finite time blowup theory. There are several books (e.g., [88, 96, 37, 93, 48]) with very detailed and extensive study on global existence and finite time blowup theory. These are very nice reference books for experts and researchers who want to quote results in this field.

In this chapter, we shall introduce some basic concepts and methods to deal with global existence and finite time blowup of solutions to parabolic differential equations. We first present simple comparison methods. Some early papers also appeared in the 1970s, e.g., [101, 45, 102, 10, 58, 9]. Then we introduce a simple eigenvalue method from [55, 32]. We also cover the energy method from [15] and concavity method from [64, 65, 62]. Finally, illustrated by specific examples, we systematically present critical exponents, boundary blowup and estimates of blowup rates.

Throughout this chapter, we always assume that Ω is of class C^2, and initial data $u_0, v_0 \in W_p^2(\Omega) \cap C(\overline{\Omega})$ with $p > 1$ satisfy the corresponding compatibility conditions.

6.1   SOME BASIC METHODS

Consider the problem

$$\begin{cases} u_t - \Delta u = f(u), & x \in \Omega, \ t > 0, \\ a\partial_{\boldsymbol{n}} u + bu = 0, & x \in \partial\Omega, \ t > 0, \\ u(x,0) = u_0(x), & x \in \Omega, \end{cases} \tag{6.1}$$

where $a, b \geqslant 0$ are constants with $a + b > 0$, $f \in C^1$ and $u_0 > 0$ in $\overline{\Omega}$.

We take

$$\sigma < \min\left\{0, \inf_{\Omega} u_0(x)\right\}, \quad A > \max\left\{0, \sup_{\Omega} u_0(x)\right\},$$

$$\beta < \min\left\{0, \min_{2\sigma \leqslant u \leqslant 2A} f(u)\right\}, \quad B > \max\left\{0, \max_{2\sigma \leqslant u \leqslant 2A} f(u)\right\}.$$

Then $\bar{u} = Bt + A$ and $\underline{u} = \beta t + \sigma$ are, respectively, an upper solution and a lower solution of (6.1) in $[0,T]$ so long as $T \leqslant \min\{|\sigma/\beta|, A/B\}$. Thus problem (6.1) has a unique solution in $\Omega \times [0,T]$ by the upper and lower solutions method. Repeating this process we can prove that there exists $0 < T_{\max} \leqslant \infty$ for which (6.1) has a unique solution u defined in $\Omega \times [0, T_{\max})$, and $T_{\max} < \infty$ implies $\limsup_{t \to T_{\max}^-} \max_{\overline{\Omega}} |u(\cdot, t)| = \infty$. Such a $T_{\max}$ is called the *maximal existence time*.

When $T_{\max} = \infty$, we call u a *global solution* of (6.1), or we say that solution u of (6.1) exists globally; when $T_{\max} < \infty$, we say that u *blows up in finite time* and such a finite time $T_{\max}$ is called the *blowup time*. This phenomenon is also called *blowup in finite time*.

In this section we shall introduce four methods dealing with the finite time blowup.

6.1.1   Comparison method

Let us start with a *comparison method*. We first study problem (3.5). For convenience, we rewrite it here

$$\begin{cases} u_t - \Delta u = c(e^{au} - 1), & x \in \Omega, \ t > 0, \\ \partial_{\boldsymbol{n}} u + bu = 0, & x \in \partial\Omega, \ t > 0, \\ u(x,0) = u_0(x), & x \in \Omega, \end{cases} \tag{6.2}$$

where $u_0 > 0$ in $\overline{\Omega}$, a, c and b are constants with $a, c > 0$ and $b \geqslant 0$.

Theorem 6.1 ([124]) *Let* (λ_1, ϕ_1) *be the first eigen-pair of problem* (3.6). *If*

$$ac \geqslant \lambda_1, \quad u_0(x) \geqslant \delta\phi_1(x) \ \ in \ \Omega$$

for some constant $\delta > 0$, then there exists $0 < T_{\max} < \infty$ for which (6.2) has a unique solution u defined in $\overline{\Omega} \times [0, T_{\max})$, and

$$T_{\max} \leqslant \frac{1}{\gamma}, \quad u(x,t) \geqslant \delta(1 - \gamma t)^{-1}\phi_1(x), \quad and \quad \lim_{t \to T_{\max}^-} \max_{\overline{\Omega}} u(\cdot, t) = \infty,$$

where

$$\gamma = \frac{1}{2}c\delta a^2 \min_{\overline{\Omega}} \phi_1(x).$$

This indicates that u blows up in finite time.

Proof. We first look for a lower solution which blows up in finite time. Take $\underline{u} = p(t)\phi_1(x)$ with $p(t) > 0$. To our purpose, we require

$$(\partial_t - \Delta)p(t)\phi_1(x) = [p'(t) + \lambda_1 p(t)]\phi_1(x) \leqslant c(e^{ap(t)\phi_1(x)} - 1),$$

$$p(0) \leqslant \delta.$$

Since

$$c(e^{ap\phi_1} - 1) \geqslant cap\phi_1 + \frac{1}{2}ca^2p^2\phi_1^2 \geqslant [cap + \frac{1}{2}ca^2p^2 \min_{\overline{\Omega}} \phi_1(x)]\phi_1,$$

it is enough to demand that $p(t) > 0$ satisfies

$$p'(t) + (\lambda_1 - ca)p \leqslant \frac{1}{2}ca^2p^2 \min_{\overline{\Omega}} \phi_1(x), \quad p(0) \leqslant \delta.$$

As $\lambda_1 - ca \leqslant 0$, it suffices to take $p(t)$ as the unique solution of

$$p' = \frac{1}{2}ca^2p^2 \min_{\overline{\Omega}} \phi_1(x), \quad p(0) = \delta,$$

that is,

$$p(t) = \frac{\delta}{1 - \gamma t}, \quad t \in [0, 1/\gamma).$$

Therefore, $\underline{u} = p(t)\phi_1(x)$ is a lower solution of (6.2) which blows up in finite time, and so does the solution of (6.2) by the comparison principle. $\square$

In the following, we study an initial-boundary value problem of semilinear equation with nonlinear boundary condition

$$\begin{cases} u_t - \Delta u = u^p, & x \in \Omega, \ t > 0, \\ \partial_n u = u^q, & x \in \partial\Omega, \ t > 0, \\ u(x, 0) = u_0(x), & x \in \Omega, \end{cases} \tag{6.3}$$

where p, q are positive constants, and $u_0 > 0$ in $\overline{\Omega}$.

Theorem 6.2 ([105]) *Assume $\max\{p, q\} \leqslant 1$. Then every solution of (6.3) exists globally.*

Proof. We seek a global upper solution $\bar{u}$ of (6.3). Let $h(x)$ be a solution of

$$\Delta h = k = \frac{|\partial\Omega|}{|\Omega|} \quad \text{in } \Omega, \quad \partial_n h = 1 \quad \text{on } \partial\Omega.$$

Without loss of generality, we assume $h(x) > 0$ in $\overline{\Omega}$. Define

$$\ell = \max_{\overline{\Omega}} |\nabla h(x)|^2, \quad r = 1 + \max_{\overline{\Omega}} u_0(x), \quad \sigma = r(k + r\ell + 1),$$

and set

$$\bar{u} = re^{\sigma t + rh(x)}.$$

We then find that $\bar{u}$ satisfies

$$\begin{cases} \bar{u}_t \geqslant \Delta\bar{u} + \bar{u}^p, & x \in \Omega, \ t > 0, \\ \partial_n\bar{u} \geqslant \bar{u}^q, & x \in \partial\Omega, \ t > 0, \\ \bar{u}(x,0) \geqslant u_0(x), & x \in \Omega, \end{cases}$$

and certainly $\bar{u}$ is a desired upper solution which exists globally in time t. $\quad\square$

Theorem 6.3 ([105]) *If* $\max\{p,q\} > 1$, *then the solution u of (6.3) blows up in finite time.*

Proof. We first consider the case $p > 1$. Take $a > 0$ so large that $a^{-1/(p-1)} \leqslant \min_{\overline{\Omega}} u_0(x)$, and set

$$\underline{u} = [a - (p-1)t]^{-1/(p-1)}.$$

Then $\underline{u}$ satisfies

$$\begin{cases} \underline{u}_t = \Delta\underline{u} + \underline{u}^p, & x \in \Omega, \ t > 0, \\ \partial_n\underline{u} \leqslant \underline{u}^q, & x \in \partial\Omega, \ t > 0, \\ \underline{u}(x,0) \leqslant u_0(x), & x \in \Omega. \end{cases}$$

Therefore, $u \geqslant \underline{u}$ and u blows up in finite time.

Now, we handle the case $q > 1$. Because Ω is bounded, we can take $d > 0$ suitable large such that $|x_1| < d$ for all $x \in \overline{\Omega}$. Set

$$\min_{\overline{\Omega}} u_0(x) = 2\varepsilon > 0, \quad g(s) = [\varepsilon^{1-q} - (q-1)s]^{-1/(q-1)}.$$

Then $g'(s) = g^q(s)$ and $g(0) = \varepsilon$. Choose $s_0 > 0$ satisfying $g(s_0) = 2\varepsilon$. Denote $c = \min\{1, s_0/(2d)\}$, $\sigma = c^2 q\varepsilon^{q-1}$ and

$$\underline{u}(x,t) = g(\sigma t + c(x_1 + d)).$$

Then $\underline{u}$ blows up in finite time, and

$$\underline{u}(x,0) = g(c(x_1 + d)) \leqslant g(2cd) \leqslant g(s_0) = 2\varepsilon \leqslant u_0(x), \quad x \in \overline{\Omega}.$$

A straightforward computation yields

$$\underline{u}_t = \sigma g', \quad \underline{u}_{x_1} = cg', \quad \underline{u}_{x_1 x_1} = c^2 g'' = c^2 q g^{q-1} g'$$

and $\underline{u}_{x_j} = \underline{u}_{x_j x_j} = 0$ when $j > 1$. Thus we have

$$\partial_n \underline{u} \leqslant cg' = cg^q = c\underline{u}^q \leqslant \underline{u}^q, \quad x \in \partial\Omega, \ t \geqslant 0,$$

$$\underline{u}_t - \Delta\underline{u} = \sigma g' - c^2 q g^{q-1} g' \leqslant g'(\sigma - c^2 q \varepsilon^{q-1}) = 0, \quad x \in \Omega, \ t \geqslant 0.$$

Above arguments show that $\underline{u}$ is a lower solution of (6.3), and the proof is complete. □

6.1.2 Kaplan's first eigenvalue method

Here we are going to introduce the *first eigenvalue method* which was introduced in 1963 by Kaplan [55]. As one will see from the proof, it is a very simple method and yet applied to a large class of equations.

We first study problem (6.2). Let (λ_1, ϕ_1) be the first eigen-pair of eigenvalue problem (3.6). From Theorem 6.1, we have known that if $ac \geqslant \lambda_1$ and $u_0(x) \geqslant 0, \not\equiv 0$, then the unique solution of (6.2) blows up in finite time. In this part, we will prove that even if $ac < \lambda_1$, its solution may blow up in finite time.

Theorem 6.4 *Assume $ac < \lambda_1$. Then solution u of (6.2) will blow up in finite time for large initial datum $u_0(x)$.*

Proof. Normalize ϕ_1 by $\int_\Omega \phi_1(x) = 1$. Multiplying the first equation of (6.2) by ϕ_1 and integrating the result over Ω, we get

$$\int_\Omega u_t \phi_1 + \lambda_1 \int_\Omega u\phi_1 = c \int_\Omega (e^{au} - 1)\phi_1 \geqslant c \int_\Omega \left(au + \frac{a^2 u^2}{2} \right) \phi_1.$$

Set $h(t) = \int_\Omega u\phi_1$. Then, in view of the Hölder inequality, it can be inferred that $h(t)$ satisfies

$$h'(t) + \lambda_1 h(t) \geqslant ach(t) + \frac{ca^2}{2} h^2(t), \quad t > 0,$$

$$h(0) = \int_\Omega u_0(x)\phi_1 > 0.$$

It is not hard to show that $h(t)$ blows up in finite time provided that $h(0)$ is suitably large, and so does u. □

Next, we consider a special case of (6.1) in the form of

$$\begin{cases} u_t - \Delta u = f(u), & x \in \Omega, \quad t > 0, \\ u = 0, & x \in \partial\Omega, \ t > 0, \\ u(x, 0) = u_0(x), & x \in \Omega, \end{cases} \tag{6.4}$$

where $f \in C^1$ and $u_0 \geqslant 0$ in $\overline{\Omega}$. Let (λ, ϕ) be the first eigen-pair of the eigenvalue problem (1.33) with $\int_\Omega \phi(x) = 1$. Then $\lambda > 0$ and $\phi > 0$ in Ω.

Theorem 6.5 *Assume that there exist positive constants a, b and $p > 1$ such that $f(s) \geqslant as^p - bs$ for all $s \geqslant 0$. Let u be the unique solution of (6.4). If*

$$\int_\Omega u_0(x)\phi(x) > \left(\frac{\lambda + b}{a}\right)^{1/(p-1)},$$

then u blows up in finite time.

Proof. Firstly, $u \geqslant 0$. The function

$$g(t) = \int_\Omega u(x, t)\phi(x) \geqslant 0$$

and satisfies

$$\begin{aligned}
g'(t) &= \int_\Omega u_t\phi = \int_\Omega [\Delta u + f(u)]\phi \\
&= \int_\Omega u\Delta\phi + \int_\Omega f(u)\phi \\
&= -\lambda g(t) + \int_\Omega f(u)\phi \\
&\geqslant -\lambda g(t) + \int_\Omega [au^p - bu]\phi \\
&= -(\lambda + b)g(t) + a\int_\Omega \phi u^p.
\end{aligned}$$

It follows that

$$g(t) \leqslant \left(\int_\Omega \phi u^p\right)^{1/p} \left(\int_\Omega \phi\right)^{(p-1)/p} = \left(\int_\Omega \phi u^p\right)^{1/p}$$

according to the Hölder inequality, and hence

$$g'(t) \geqslant g(t)[ag^{p-1}(t) - (\lambda + b)].$$

Since $ag^{p-1}(0) > \lambda + b$, the above inequality implies $\lim_{t\to T_{\max}^-} g(t) = \infty$ for some $0 < T_{\max} < \infty$. Consequently, u blows up in finite time. ◻

6.1.3 Energy method

In this subsection, we continue to study problem (6.4). Set $F(u) = \int_0^u f(s)\mathrm{d}s$, and define an energy functional with the form

$$E(u(t)) = \frac{1}{2}\int_\Omega |\nabla u(x, t)|^2 - \int_\Omega F(u(x, t)).$$

Lemma 6.6 *Let u be the unique solution of (6.4) and $T > 0$. If*

$$u \in C((0, T), C(\overline{\Omega})) \cap C((0, T), H_0^1(\Omega) \cap H^2(\Omega)) \cap C^1((0, T), L^2(\Omega)),$$

then

$$\int_s^t \int_\Omega (\Delta u + f(u))u_\tau + E(u(t)) = E(u(s)) \quad \text{for all } 0 < s < t < T.$$

Proof. Based on the denseness, it can be assumed that

$$u \in C^1([s,t], H_0^1(\Omega) \cap H^2(\Omega)).$$

Accordingly, we have

$$\int_\Omega (\Delta u + f(u))u_t = \int_\Omega (-\nabla u \cdot \nabla u_t + f(u)u_t) = -\frac{\mathrm{d}}{\mathrm{d}t} E(u(t)).$$

Integration of the above equation from s to t derives the desired result. □

Lemma 6.7 *Let u be the unique solution of (6.4) defined in $(0,T)$. Then*

$$\frac{\mathrm{d}}{\mathrm{d}t} \int_\Omega u^2 + 2\int_\Omega |\nabla u|^2 = 2\int_\Omega uf(u), \qquad (6.5)$$

$$\int_s^t \int_\Omega u_\tau^2 + E(u(t)) = E(u(s)) \quad \text{for all } 0 < s < t < T. \qquad (6.6)$$

Proof. Multiplying the first equation of (6.4) by u and integrating the result over Ω, we obtain (6.5) immediately. The equation (6.6) can be deduced from Lemma 6.6 and (6.4). □

Theorem 6.8 ([15]) *Suppose that there are positive constants K and ε such that*

$$uf(u) \geqslant (2+\varepsilon)F(u) \quad \text{for all } |u| \geqslant K.$$

Set $\eta = \min_{|u| \leqslant K} uf(u)$ and $\gamma = \max_{|u| \leqslant K} F(u)$. If $u_0 \in H_0^1(\Omega)$ and

$$(2+\varepsilon)E(u_0) < |\Omega|[\eta - (2+\varepsilon)\gamma], \qquad (6.7)$$

then the unique solution u of (6.4) blows up in finite time.

Proof. Write $g(t) = \int_\Omega u^2(x,t)\mathrm{d}x$. Then by (6.5),

$$g'(t) = -2\int_\Omega |\nabla u|^2 + 2\int_{\{|u| \leqslant K\}} uf(u) + 2\int_{\{|u| > K\}} uf(u)$$

$$\geqslant -2\int_\Omega |\nabla u|^2 + 2\int_{\{|u| \leqslant K\}} uf(u) + 2(2+\varepsilon)\int_{\{|u| > K\}} F(u). \qquad (6.8)$$

As $u_0 \in H_0^1(\Omega)$, we can take $s = 0$ in (6.6) and get

$$\int_{\{|u| > K\}} F(u) = -\int_{\{|u| \leqslant K\}} F(u) + \frac{1}{2}\int_\Omega |\nabla u|^2 - E(u_0) + \int_0^t \int_\Omega u_t^2. \qquad (6.9)$$

As $\eta \leqslant 0 = uf(u)|_{u=0} = F(0) \leqslant \gamma$, by the definitions of η and γ one has

$$2\int_{\{|u| \leqslant K\}} uf(u) - 2(2+\varepsilon)\int_{\{|u| \leqslant K\}} F(u) \geqslant 2|\Omega|[\eta - (2+\varepsilon)\gamma]. \qquad (6.10)$$

It follows from (6.8)–(6.10) that

$$g'(t) \geq 2(2+\varepsilon) \int_0^t \int_\Omega u_t^2 + \varepsilon \int_\Omega |\nabla u|^2 + 2\{|\Omega|[\eta - (2+\varepsilon)\gamma] - (2+\varepsilon)E(u_0)\}.$$

This, together with the assumption (6.7), deduces

$$g'(t) > 2(2+\varepsilon) \int_0^t \int_\Omega u_t^2 + \varepsilon \int_\Omega |\nabla u|^2 \geq 2(2+\varepsilon) \int_0^t \int_\Omega u_t^2. \tag{6.11}$$

Set $h(t) = \int_0^t g(s)\mathrm{d}s$. By using the Cauchy-Schwarz inequality we have

$$h'(t) - h'(0) = \int_0^t g'(s) = 2 \int_0^t \int_\Omega uu_t$$

$$\leq 2 \left(\int_0^t \int_\Omega u^2 \right)^{1/2} \left(\int_0^t \int_\Omega u_t^2 \right)^{1/2}$$

$$= 2h^{1/2}(t) \left(\int_0^t \int_\Omega u_t^2 \right)^{1/2}.$$

This combined with (6.11) allows us to yield

$$h(t)h''(t) \geq 2(2+\varepsilon)h(t) \int_0^t \int_\Omega u_t^2 \geq (1+\varepsilon/2)(h'(t) - h'(0))^2. \tag{6.12}$$

On the contrary, we assume that u exists globally. It follows from (6.11) that $h''(t) = g'(t) > 0$. Consequently, $h'(t) > h'(t_1) > h'(0)$ for every $t > t_1 > 0$, and then $h'(0) < (1-\sigma)h'(t)$ for all $t > t_1$ and $h(t) \to \infty$ as $t \to \infty$, where $0 < \sigma < 1$. In view of (6.12), it declares that $h(t)h''(t) \geq (1+\varepsilon/2)\sigma^2(h'(t))^2$ for all $t > t_1$, i.e.,

$$\frac{h''(t)}{h'(t)} \geq \left(1+\frac{\varepsilon}{2}\right)\sigma^2 \frac{h'(t)}{h(t)}, \quad t > t_1.$$

It yields that $h'(t) \to \infty$ since $h(t) \to \infty$. Hence, there exists $t_2 > t_1$ so that

$$\left(1+\frac{\varepsilon}{2}\right)[h'(t) - h'(0)]^2 \geq \left(1+\frac{\varepsilon}{4}\right)(h'(t))^2 \quad \text{for } t \geq t_2.$$

Combining this with (6.12), we get

$$h(t)h''(t) \geq \left(1+\frac{\varepsilon}{4}\right)(h'(t))^2, \quad t \geq t_2.$$

Let $\eta(t) = h^{-\varepsilon/4}(t)$, then

$$\eta''(t) \leq 0 \quad \text{for all } t \geq t_2. \tag{6.13}$$

Obviously, $\lim_{t\to\infty} \eta(t) = 0$ since $\lim_{t\to\infty} h(t) = \infty$. Then there exists $t_0 > t_2$ for which $\eta'(t_0) < 0$. According to (6.13), we have $\eta'(t) \leq \eta'(t_0) < 0$ for all $t \geq t_0$, and furthermore $\eta(t) \to -\infty$ as $t \to \infty$. This is a contradiction. Therefore, u must blow up in finite time. $\qquad\square$

6.1.4 Concavity method

We first state a lemma which is a basic conclusion of convex functions and its proof will be omitted.

Lemma 6.9 *Let $J \in C^2([0,T)) \cap C([0,T])$, $J(0) > 0$, $J'(0) < 0$, and $J''(t) \leqslant 0$ in $[0,T)$. If*

$$T \geqslant -J(0)/J'(0),$$

then there exists $T_* : 0 < T_* \leqslant -J(0)/J'(0)$ such that $J(T_*) = 0$.

Theorem 6.10 ([105]) *Suppose $f \in C^1(\mathbb{R})$ and $f(u) \geqslant 0$ for $u \geqslant 0$. Let $u_0 \in H_0^1(\Omega) \cap H^2(\Omega)$, and $u_0 \geqslant 0, \not\equiv 0$ and satisfy*

$$\Delta u_0 + f(u_0) \geqslant 0, \not\equiv 0 \quad in \ \Omega. \tag{6.14}$$

We further assume that there exists $\psi \in C^1([0,\infty)) \cap C^3(\mathbb{R}_+)$ such that

(i) $\psi(0) = \psi'(0) = 0, \psi(u) \geqslant 0, \not\equiv 0, \psi''(u) \geqslant 0, \psi'''(u) \geqslant 0$ *and $\psi'(u) > 0$ for all $u > 0$;*

(ii) $\psi'(u)f'(u) - \psi''(u)f(u) \geqslant 0$ *for all $u \geqslant 0$;*

(iii) *there exists $\theta > 0$ so that $\psi(u)\psi''(u) \geqslant \frac{\theta+1}{2}(\psi'(u))^2$ for $u \geqslant 0$.*

If u is a solution of (6.4) and satisfies

$$u \in C^1((0, T_{\max}), H_0^1(\Omega) \cap H^2(\Omega)) \cap C([0, T_{\max}), C(\overline{\Omega})),$$

where $T_{\max} \leqslant \infty$ is the maximal existence time of u, then $T_{\max} < \infty$.

Proof. Firstly, by (6.14) and the maximum principle we see that u is strictly monotone increasing in t, and $u_t(\cdot, t) \geqslant 0, \not\equiv 0$ for $t \in (0, T_{\max})$. Thus $u \geqslant 0$, and $u > 0$ in $\Omega \times (0, T_{\max})$. Set

$$I(t) = \int_\Omega \psi(u(x,t)).$$

Then $I(t) > 0$ for all $t \in [0, T_{\max})$, and

$$I'(t) = \int_\Omega \psi'(u(x,t))u_t(x,t) > 0 \quad \text{for all } 0 < t < T_{\max}.$$

Without loss of generality, we assume $I'(0) > 0$. As $u = 0$ on $\partial\Omega$, a direct calculation causes

$$I'(t) = \int_\Omega \psi'(u)u_t = \int_\Omega \psi'(u)(\Delta u + f(u))$$

$$= -\int_\Omega \psi''(u)|\nabla u|^2 + \int_\Omega \psi'(u)f(u)$$

$$= 2\int_\Omega \psi'(u)u_t + \int_\Omega \psi''(u)|\nabla u|^2 - \int_\Omega \psi'(u)f(u), \quad 0 \leqslant t < T_{\max},$$

$$I''(t) = \int_\Omega \left\{ 2\psi''(u)u_t^2 + [\psi'(u)f'(u) - \psi''(u)f(u)]u_t + \psi'''(u)|\nabla u|^2 u_t \right\}$$

$$\geqslant 2\int_\Omega \psi''(u)u_t^2, \quad 0 \leqslant t < T_{\max}.$$

Define $J(t) = I^{-\theta}(t)$ for $t \in [0, T_{\max})$, where θ is given by the condition (iii). Clearly,

$$J(0) = I^{-\theta}(0) > 0, \quad J'(0) = -\theta I^{-(\theta+1)}(0)I'(0) < 0,$$

and

$$J''(t) = \theta I^{-(\theta+2)}(t)[(\theta+1)(I'(t))^2 - I(t)I''(t)], \quad 0 \leqslant t < T_{\max}.$$

Applying the Schwarz inequality and the condition (iii), we obtain

$$I(t)I''(t) \geqslant \int_\Omega \psi(u) \int_\Omega 2\psi''(u)u_t^2$$

$$\geqslant \left(\int_\Omega \sqrt{2\psi(u)\psi''(u)}u_t \right)^2$$

$$\geqslant (\theta+1)(I'(t))^2, \quad 0 \leqslant t < T_{\max}.$$

Hence, $J''(t) \leqslant 0$ for all $0 \leqslant t < T_{\max}$. By means of Lemma 6.9, if

$$T_{\max} > -J(0)/J'(0) = I(0)/(\theta I'(0)), \tag{6.15}$$

then there exists $T_* \leqslant I(0)/(\theta I'(0))$ such that $J(T_*) = 0$, i.e.,

$$\lim_{t \to T_*^-} \int_\Omega \psi(u(x,t)) = \infty,$$

which implies $\lim_{t \to T_*^-} \|u(\cdot, t)\|_\infty = \infty$. This contradicts (6.15), and so $T_{\max} \leqslant I(0)/(\theta I'(0)) < \infty$. □

6.2 A SYSTEM WITH DIRICHLET BOUNDARY CONDITIONS

By Theorem 6.5 we see that the unique solution of

$$\begin{cases} u_t - \Delta u = u^p, & x \in \Omega, \ t > 0, \\ u = 0, & x \in \partial\Omega, \ t > 0, \\ u(x,0) = u_0(x) > 0, & x \in \Omega \end{cases} \tag{6.16}$$

blows up in finite time when $u_0(x)$ is suitably large.

However, the solution u of (6.16) exists globally and $\lim_{t \to \infty} u(x,t) = 0$ uniformly in $\overline{\Omega}$ when $u_0(x)$ is suitably small. In fact, let (λ, ϕ) be the first eigen-pair of problem (1.33). Take $0 < \varepsilon \ll 1$ satisfying $\varepsilon^{p-1} \max_{\overline{\Omega}} \phi^{p-1}(x) \leqslant \lambda/2$. Then $\bar{u} = \varepsilon e^{-\lambda t/2}\phi(x)$ is an upper solution of (6.16) when $u_0(x) \leqslant \varepsilon\phi(x)$, and as a consequence, solution u of (6.16) exists globally and $\lim_{t \to \infty} u(x,t) = 0$ uniformly in $\overline{\Omega}$.

In this section we discuss the following problem

$$\begin{cases} u_t - \Delta u = u^\rho v^p, & x \in \Omega, \quad t > 0, \\ v_t - \Delta v = u^q v^\beta, & x \in \Omega, \quad t > 0, \\ u = v = 0, & x \in \partial\Omega, \ t > 0, \\ u = u_0(x), \quad v = v_0(x), & x \in \Omega, \quad t = 0, \end{cases} \tag{6.17}$$

where p, q, ρ, β are non-negative constants, $u_0, v_0 \geqslant 0$ in Ω. It will be shown that (6.17) exhibits very different properties from (6.16).

6.2.1 Critical exponents

The content of this part refers to [106].

Theorem 6.11 (Critical exponents) *Let (λ, ϕ) be the first eigen-pair of (1.33).*

(1) *Assume*

$$\rho, \beta \leqslant 1 \quad \text{and} \quad pq \leqslant (1 - \rho)(1 - \beta). \tag{6.18}$$

Then every solution of (6.17) exists globally for any initial data. If (6.18) does not hold, then all solutions of (6.17) exist globally for small initial data, and blow up in finite time for large initial data.

(2) *Suppose that either*

$$\rho > 1, \ p > 0, \ q = 0, \ \beta = 1, \ \lambda < 1, \ \rho \leqslant 1 + p(1 - \lambda)/\lambda, \tag{6.19}$$

or

$$\beta > 1, \ q > 0, \ p = 0, \ \rho = 1, \ \lambda < 1, \ \beta \leqslant 1 + q(1 - \lambda)/\lambda. \tag{6.20}$$

Furthermore, when $\rho = 1 + p(1 - \lambda)/\lambda$ in (6.19), or $\beta = 1 + q(1 - \lambda)/\lambda$ in (6.20), we also assume $\lambda < 2/3$. Then, for any initial data $u_0, v_0 \geqslant 0, \not\equiv 0$, the corresponding solution of (6.17) blows up in finite time.

(3) *If one of (6.19) and (6.20) is not valid, then every solution of (6.17) exists globally for small initial data.*

Proof. Step 1. Assume that (6.18) holds. When $pq = 0$, the conclusion is obvious. When $pq > 0$, we have $\dfrac{1-\rho}{p}\dfrac{1-\beta}{q} \geqslant 1$. There exist $0 < m, l < 1$ such that

$$(1 - \rho)/p \geqslant l/m, \quad (1 - \beta)/q \geqslant m/l. \tag{6.21}$$

Denote $k = 1/m + 1/l$, and let w be the unique global solution of the linear problem

$$\begin{cases} w_t - \Delta w = kw, & x \in \Omega, \quad t > 0, \\ w = 1, & x \in \partial\Omega, \ t > 0, \\ w = 1 + u_0^{1/m}(x) + v_0^{1/l}(x), & x \in \Omega, \quad t = 0. \end{cases} \tag{6.22}$$

As is well known, w exists globally and $w \geqslant 1$. Define $\bar{u} = w^m$ and $\bar{v} = w^l$. In view of (6.21) and (6.22), we can verify

$$\bar{u}_t \geqslant \Delta \bar{u} + \bar{u}^\rho \bar{v}^p, \quad \bar{v}_t \geqslant \Delta \bar{v} + \bar{u}^q \bar{v}^\beta \quad \text{for all } x \in \Omega, \ t > 0.$$

Obviously, $\bar{u}, \bar{v} > 0$ on $\partial\Omega \times [0, \infty)$ and $\bar{u}(x,0) > u_0(x), \bar{v}(x,0) > v_0(x)$ in $\overline{\Omega}$. The comparison principle gives $(u, v) \leqslant (\bar{u}, \bar{v})$, and so (u, v) exists globally. Here, $(u, v) \leqslant (\bar{u}, \bar{v})$ means that $u \leqslant \bar{u}$ and $v \leqslant \bar{v}$.

Step 2. Assume that (6.19) holds and $u_0, v_0 \geq 0, \not\equiv 0$. Then v satisfies

$$\begin{cases} v_t - \Delta v = v, & x \in \Omega, \quad t > 0, \\ v = 0, & x \in \partial\Omega, \ t > 0, \\ v(x, 0) = v_0(x), & x \in \Omega. \end{cases}$$

According to the maximum principle and Lemma 3.7, we may think that $v_0 \geq \varepsilon\phi$ in Ω for some $\varepsilon > 0$. Therefore, $v \geq \varepsilon e^{(1-\lambda)t}\phi$, and of course

$$\begin{cases} u_t - \Delta u \geq \varepsilon^p e^{p(1-\lambda)t}\phi^p u^\rho, & x \in \Omega, \quad t > 0, \\ u = 0, & x \in \partial\Omega, \ t > 0, \\ u(x, 0) = u_0(x) \geq 0, \not\equiv 0, & x \in \Omega. \end{cases} \tag{6.23}$$

When $\rho < 1 + p(1-\lambda)/\lambda$, we choose $\Omega_0 \Subset \Omega$ so that the first eigenvalue λ_0 of $-\Delta$ in Ω_0 with the homogeneous Dirichlet boundary condition satisfies $\rho < 1 + p(1-\lambda_0)/\lambda_0$. Denote $a = \min_{\overline{\Omega}_0} \phi(x)$. Then $a > 0$, and

$$u_t - \Delta u \geq \varepsilon^p a^p e^{\lambda_0(\rho-1)t} u^\rho, \quad x \in \Omega_0, \ t > 0. \tag{6.24}$$

Let ϕ_0 be the eigenfunction corresponding to λ_0 and satisfy $\int_{\Omega_0} \phi_0(x) = 1$. Multiplying (6.24) by ϕ_0 and integrating the result over Ω_0 firstly, and using the Hölder inequality secondly, we have

$$\frac{\mathrm{d}}{\mathrm{d}t}\int_{\Omega_0} u\phi_0 \geq -\lambda_0 \int_{\Omega_0} u\phi_0 + \varepsilon^p a^p e^{\lambda_0(\rho-1)t}\left(\int_{\Omega_0} u\phi_0\right)^\rho,$$

which implies

$$\frac{\mathrm{d}}{\mathrm{d}t}\left(e^{\lambda_0 t}\int_{\Omega_0} u\phi_0\right) \geq \varepsilon^p a^p\left(e^{\lambda_0 t}\int_{\Omega_0} u\phi_0\right)^\rho.$$

Due to the fact that $\int_{\Omega_0} u(x, 0)\phi_0(x) > 0$ and $\rho > 1$, it derives that $\int_{\Omega_0} u(x, t)\phi_0(x)$ blows up in finite time, and so does u.

If $\rho = 1 + p(1-\lambda)/\lambda$ and $\lambda < 2/3$, then $(\rho - 1 - p)/(\rho - 1) > -1$. Since $\partial_{\boldsymbol{n}}\phi < 0$ on $\partial\Omega$, it is clear that

$$b := \int_\Omega \phi^{(\rho-1-p)/(\rho-1)}(x) < \infty.$$

The Hölder inequality gives

$$\int_\Omega u\phi \leq \left(\int_\Omega u^\rho\phi^{1+p}\right)^{1/\rho}\left(\int_\Omega \phi^{(\rho-1-p)/(\rho-1)}\right)^{(\rho-1)/\rho}$$
$$= b^{(\rho-1)/\rho}\left(\int_\Omega u^\rho\phi^{1+p}\right)^{1/\rho}.$$

Multiplying (6.23) by ϕ and integrating the result over Ω, from above we have

$$\frac{\mathrm{d}}{\mathrm{d}t}\int_\Omega u\phi \geq -\lambda \int_\Omega u\phi + \varepsilon^p b^{1-\rho}e^{\lambda(\rho-1)t}\left(\int_\Omega u\phi\right)^\rho.$$

Similar to the above, it can be deduced that u blows up in finite time.

When (6.20) holds, the proof is similar.

Step 3. If one of (6.18), (6.19), and (6.20) does not hold, we first prove that each solution (u, v) of (6.17) exists globally for small initial data. To this end, let ψ be the unique solution of

$$-\Delta\psi = 1 \ \text{ in } \Omega, \quad \psi = 0 \ \text{ on } \partial\Omega.$$

Then $0 \leqslant \psi \leqslant C$ for some constant $C > 0$.

Case 1: $\rho > 1$ and $q > 0$. We can find positive constants a, b satisfying

$$a \geqslant a^\rho b^p (1 + C)^{\rho+p} \ \text{ and } \ b \geqslant a^q b^\beta (1 + C)^{q+\beta}. \tag{6.25}$$

Then the pair $(\bar{u}, \bar{v}) = (a(1 + \psi), b(1 + \psi))$ is an upper solution of (6.17) as long as $u_0 \leqslant a(1 + \psi)$, $v_0 \leqslant b(1 + \psi)$. Consequently, (u, v) exists globally.

When $\beta > 1$ and $p > 0$, the proof is similar.

Case 2: $q = 0$ and $\beta \neq 1$. Similarly to the case 1, there exists $b > 0$ such that $v \leqslant b(1 + \psi)$ whenever $v_0 \leqslant b(1 + \psi)$, and then

$$u_t - \Delta u \leqslant b^p (1 + \psi)^p u^\rho.$$

If $\rho \leqslant 1$, then u exists globally. If $\rho > 1$, then $\bar{u} = a(1 + \psi(x))$ satisfies

$$\bar{u}_t - \Delta\bar{u} \geqslant b^p (1 + \psi)^p \bar{u}^\rho$$

provided that $0 < a^{\rho-1} \leqslant b^{-p}(1 + C)^{-(p+\rho)}$. Of course $u \leqslant \bar{u}$ when $u_0 \leqslant a(1 + \psi)$, and thereby (u, v) exists globally when $u_0 \leqslant a(1 + \psi)$ and $v_0 \leqslant b(1 + \psi)$.

Case 3: $q = 0, \beta = 1$ and $\lambda \geqslant 1$. In such a case, we have $v \leqslant \phi$ provided $v_0(x) \leqslant \phi(x)$. Similar to the case 2, it can be proved that u exists globally if u_0 is properly small.

Case 4: $\rho > 1, p > 0, q = 0, \beta = 1, \lambda < 1$ and $\rho > 1 + p(1 - \lambda)/\lambda$. Similar to arguments in Step 2, it can be shown that $v \leqslant \phi e^{(1-\lambda)t}$ if $v_0 \leqslant \phi$. Therefore

$$u_t \leqslant \Delta u + \phi^p e^{p(1-\lambda)t} u^\rho \leqslant \Delta u + C e^{p(1-\lambda)t} u^\rho.$$

Since $(\rho - 1)\lambda > p(1 - \lambda)$, we can find $0 < \theta < \lambda$ for which $(\rho - 1)\theta > p(1 - \lambda)$. Choosing $a > 0$ small enough, by a direct computation, we know that $\bar{u} = a e^{-\theta t}\phi(x)$ satisfies

$$\bar{u}_t - \Delta\bar{u} \geqslant C e^{p(1-\lambda)t}\bar{u}^\rho.$$

The comparison principle yields $u \leqslant \bar{u}$ so long as $u_0 \leqslant a\phi$, and hence (u, v) exists globally when $u_0 \leqslant a\phi$ and $v_0 \leqslant \phi$.

Case 5: $\rho \leqslant 1, \beta \leqslant 1$ and $pq > (1 - \rho)(1 - \beta)$. In such a case, we can find two positive constants a and b satisfying (6.25). Similar to the proof of Case 1, (u, v) exists globally provided $u_0 \leqslant a(1 + \psi)$ and $v_0 \leqslant b(1 + \psi)$.

For other cases, their proofs can be argued similarly.

Step 4. It will be shown that if (6.18) does not hold, then each solution (u, v) of (6.17) blows up in finite time for large initial data.

<u>Case 1</u>: $\rho > 1$. It is well known that $v \geqslant \mathrm{e}^{-\lambda t}\phi$ whenever $v_0 \geqslant \phi$. Accordingly, u satisfies

$$u_t - \Delta u \geqslant \mathrm{e}^{-\lambda p t}\phi^p u^\rho. \tag{6.26}$$

If $p \geqslant \rho - 1$, then $\theta := p/(\rho - 1) \geqslant 1$. Multiplying (6.26) by ϕ^θ and integrating the result over Ω one has

$$\frac{\mathrm{d}}{\mathrm{d}t}\int_\Omega u\phi^\theta \geqslant -\lambda\theta\int_\Omega u\phi^\theta + \theta(\theta-1)\int_\Omega u\phi^{\theta-2}|\nabla\phi|^2 + \mathrm{e}^{-\lambda p t}\int_\Omega u^\rho\phi^{p+\theta}$$

$$\geqslant -\lambda\theta\int_\Omega u\phi^\theta + |\Omega|^{1-\rho}\mathrm{e}^{-\lambda p t}\left(\int_\Omega u\phi^\theta\right)^\rho,$$

and consequently,

$$\frac{\mathrm{d}}{\mathrm{d}t}\left(\mathrm{e}^{-\lambda\theta t}\int_\Omega u\phi^\theta\right) \geqslant -2\lambda\theta\mathrm{e}^{-\lambda\theta t}\int_\Omega u\phi^\theta + |\Omega|^{1-\rho}\left(\mathrm{e}^{-\lambda\theta t}\int_\Omega u\phi^\theta\right)^\rho.$$

Thereby, u blows up in finite time when $\displaystyle\int_\Omega u_0(x)\phi^\theta(x)$ is sufficiently large.

If $p < \rho - 1$, multiplying (6.26) by ϕ and integrating the result over Ω, we have

$$\frac{\mathrm{d}}{\mathrm{d}t}\int_\Omega u\phi \geqslant -\lambda\int_\Omega u\phi + \left(\int_\Omega \phi^{(\rho-1-p)/(\rho-1)}\right)^{1-\rho}\mathrm{e}^{-\lambda p t}\left(\int_\Omega u\phi\right)^\rho.$$

Thus, u blows up in finite time so long as $\displaystyle\int_\Omega u_0(x)\phi(x)$ is sufficiently large.

For the case when $\beta > 1$, the proof is similar to the case $\rho > 1$.

<u>Case 2</u>: $\rho, \beta \leqslant 1$ and $pq > (1-\rho)(1-\beta)$. We choose $m, l > 1$ satisfying $(p+1-\beta)/(q+1-\rho) = m/l$. Then $(1-\rho)/p < l/m$, $(1-\beta)/q < m/l$ and $\sigma := m(\rho-1)+lp = l(\beta-1)+mq > 0$. Take $k = \min\{1/m, 1/l\}$, and write w as the unique non-negative solution of

$$\begin{cases} w_t - \Delta w = kw^{1+\sigma}, & x \in \Omega_0, \quad t > 0, \\ w = 0, & x \in \partial\Omega_0, \ t > 0, \\ w(x,0) = w_0(x) \geqslant 0, & x \in \Omega_0, \end{cases}$$

where $\Omega_0 \Subset \Omega$. Then w blows up in finite time $T_{\max}$ provided that initial datum w_0 is suitable large. Set $\underline{u} = w^m$ and $\underline{v} = w^l$. After careful calculations we obtain

$$\Delta\underline{u} + \underline{u}^p\underline{v}^p = mw^{m-1}\Delta w + m(m-1)w^{m-2}|\nabla w|^2 + w^{mp+pl}$$

$$\geqslant mw^{m-1}(\Delta w + \frac{1}{m}w^{1+\sigma})$$

$$\geqslant mw^{m-1}w_t$$

$$= \underline{u}_t, \quad x \in \Omega_0, \ 0 < t < T_{\max},$$

$$\Delta\underline{v} + \underline{u}^q\underline{v}^\beta \geqslant \underline{v}_t, \quad x \in \Omega_0, \ 0 < t < T_{\max}.$$

Define $w_0 = 0$ in $\Omega \setminus \Omega_0$. If $u_0 \geqslant \phi + w_0^m$ and $v_0 \geqslant \phi + w_0^l$, then the maximum principle yields $u, v > 0$ in $\overline{\Omega}_0 \times (0, T_{\max})$, and then the comparison principle causes $u > \underline{u}$ and $v > \underline{v}$ in $\overline{\Omega}_0 \times (0, T_{\max})$. Therefore, (u, v) blows up in finite time. $\qquad\square$

6.2.2 Blowup rate estimates

We consider a special case of (6.17), namely a problem as follows:

$$\begin{cases} u_t - \Delta u = u^\rho v^p, & x \in B_R, \ t > 0, \\ v_t - \Delta v = u^q v^\beta, & x \in B_R, \ t > 0, \\ u = v = 0, & |x| = R, \ t > 0, \\ u = u_0(x), \ v = v_0(x), & x \in B_R, \ t = 0, \end{cases} \tag{6.27}$$

where $B_R = B_R(0)$ is a ball in $\mathbb{R}^n$ centred at the origin with radius $R > 0$. Initial data u_0 and v_0 are continuous and non-negative functions which vanish on the boundary $|x| = R$. The content of this part is due to [107].

Theorem 6.12 (Blowup rate estimates) *Let (u, v) be a smooth solution of (6.27), and let the following conditions hold:*

(i) *$\rho = 0$ or $\rho \geqslant 1$, $\beta = 0$ or $\beta \geqslant 1$, and $p, q \geqslant 1$ with $p + 1 > \beta$ and $q + 1 > \rho$;*

(ii) *u_0, v_0 are radially symmetric and decreasing;*

(iii) *u and v blow up simultaneously in finite time $T_{\max} < \infty$, and $u_t, v_t \geqslant 0$.*

Then there are positive constants c and C such that

$$\begin{cases} c(T_{\max} - t)^{-\theta} \leqslant \max\limits_{0 \leqslant |x| \leqslant R} u(x, t) = u(0, t) \leqslant C(T_{\max} - t)^{-\theta}, & 0 \leqslant t < T_{\max}, \\ c(T_{\max} - t)^{-\sigma} \leqslant \max\limits_{0 \leqslant |x| \leqslant R} v(x, t) = v(0, t) \leqslant C(T_{\max} - t)^{-\sigma}, & 0 \leqslant t < T_{\max}, \end{cases} \tag{6.28}$$

where

$$\theta = \frac{p + 1 - \beta}{pq - (\rho - 1)(\beta - 1)} > 0, \qquad \sigma = \frac{q + 1 - \rho}{pq - (\rho - 1)(\beta - 1)} > 0.$$

Proof. Firstly, as u_0 and v_0 are radially symmetric and decreasing, so do u and v, i.e., $u = u(r, t)$ and $v = v(r, t)$ with $r = |x|$ and $u_r \leqslant 0, v_r \leqslant 0$.

Step 1. Since u and v blow up simultaneously in finite time $T_{\max}$, and $u_t \geqslant 0$ and $v_t \geqslant 0$, it follows that, for any $0 < t_0 < T_{\max}$, either $u_t(x, t_0) \not\equiv 0$ or $v_t(x, t_0) \not\equiv 0$ in $\bar{B}_R$. Otherwise, $(u(x, t_0), v(x, t_0))$ is a positive equilibrium solution of (6.27), and thereby (u, v) cannot blow up in finite time. Let $\varphi = u_t$ and $\psi = v_t$. Then we have

$$\begin{cases} \varphi_t = \Delta \varphi + \rho u^{\rho - 1} v^p \varphi + p u^\rho v^{p-1} \psi, & x \in B_R, \ t_0 \leqslant t < T_{\max}, \\ \psi_t = \Delta \psi + q u^{q-1} v^\beta \varphi + \beta u^q v^{\beta - 1} \psi, & x \in B_R, \ t_0 \leqslant t < T_{\max}, \\ \varphi = \psi = 0, & |x| = R, \ t_0 \leqslant t < T_{\max}, \\ \varphi(x, t_0) \geqslant 0, \ \psi(x, t_0) \geqslant 0, & x \in \bar{B}_R, \end{cases}$$

and $\varphi(x, t_0) \not\equiv 0$ or $\psi(x, t_0) \not\equiv 0$ in $\bar{B}_R$. The maximum principle shows that

$$\varphi > 0, \ \psi > 0 \quad \text{in} \ B_R \times (t_0, T_{\max}),$$
$$\partial_{\boldsymbol{n}} \varphi < 0, \ \partial_{\boldsymbol{n}} \psi < 0 \quad \text{on} \ \partial B_R \times (t_0, T_{\max}).$$

Using Lemma 3.7 we can show that, for any $t_0 < t_1 < T_{\max}$, there exists $\varepsilon > 0$ such that $\varphi \geqslant \varepsilon u^\rho v^p$ and $\psi \geqslant \varepsilon u^q v^\beta$ for $t = t_1$ and $x \in B_R$. That is,

$$\Delta u + u^\rho v^p \geqslant \varepsilon u^\rho v^p \quad \text{and} \quad \Delta v + u^q v^\beta \geqslant \varepsilon u^q v^\beta \quad \text{for } t = t_1, \ |x| \leqslant R.$$

Without loss of generality, we may assume that $t_1 = 0$, i.e.,

$$\Delta u_0 + u_0^\rho v_0^p \geqslant \varepsilon u_0^\rho v_0^p \quad \text{and} \quad \Delta v_0 + u_0^q v_0^\beta \geqslant \varepsilon u_0^q v_0^\beta, \quad |x| \leqslant R. \tag{6.29}$$

Set

$$w = u_t - \varepsilon u^\rho v^p \quad \text{and} \quad z = v_t - \varepsilon u^q v^\beta.$$

In view of that $\nabla u \cdot \nabla v = u_r v_r \geqslant 0$, from the condition (i) and (6.29), we have

$$\begin{aligned}
w_t - \Delta w &= (u_t - \Delta u)_t - \varepsilon \rho u^{\rho-1} v^p (u_t - \Delta u) - \varepsilon p u^\rho v^{p-1}(v_t - \Delta v) \\
&\quad + \varepsilon u^{\rho-2} v^{p-2}[\rho(\rho-1)|v\nabla u|^2 + 2\rho p u v \nabla u \cdot \nabla v + p(p-1)|u\nabla v|^2] \\
&\geqslant (u_t - \Delta u)_t - \varepsilon \rho u^{\rho-1} v^p (u_t - \Delta u) - \varepsilon p u^\rho v^{p-1}(v_t - \Delta v) \\
&= \rho u^{\rho-1} v^p w + p u^\rho v^{p-1} z, \quad |x| < R, \ 0 < t < T_{\max}, \\
z_t - \Delta z &\geqslant q u^{q-1} v^\beta w + \beta u^q v^{\beta-1} z, \quad |x| < R, \ 0 < t < T_{\max}
\end{aligned}$$

and

$$w(x, 0) \geqslant 0, \quad z(x, 0) \geqslant 0, \quad |x| < R,$$
$$w(x, t) = z(x, t) = 0, \quad |x| = R, \ 0 < t < T_{\max}.$$

The maximum principle asserts that $w \geqslant 0$ and $z \geqslant 0$, i.e., $u_t \geqslant \varepsilon u^\rho v^p$ and $v_t \geqslant \varepsilon u^q v^\beta$. In particular,

$$u_t(0, t) \geqslant \varepsilon u^\rho(0, t) v^p(0, t) \quad \text{and} \quad v_t(0, t) \geqslant \varepsilon u^q(0, t) v^\beta(0, t), \ 0 < t < T_{\max}. \tag{6.30}$$

On the other hand, the condition (ii) implies $\Delta u(0, t) \leqslant 0$ and $\Delta v(0, t) \leqslant 0$, i.e.,

$$u_t(0, t) \leqslant u^\rho(0, t) v^p(0, t) \quad \text{and} \quad v_t(0, t) \leqslant u^q(0, t) v^\beta(0, t), \quad 0 < t < T_{\max}. \tag{6.31}$$

Step 2. It follows from (6.30) and (6.31) that

$$\varepsilon u^{\rho-q} v^{p-\beta} v_t(0, t) \leqslant u_t(0, t) \leqslant \frac{1}{\varepsilon} u^{\rho-q} v^{p-\beta} v_t(0, t). \tag{6.32}$$

Integrating the right-hand side of (6.32) from 0 to t, we obtain

$$\frac{1}{q+1-\rho} u^{q+1-\rho}(0, s)\big|_0^t \leqslant \frac{1}{\varepsilon(p+1-\beta)} v^{p+1-\beta}(0, s)\big|_0^t.$$

Since $v(0, t) \to \infty$ as $t \to T_{\max}$ and $p + 1 > \beta$, it confirms

$$u^{q+1-\rho}(0, t) \leqslant C_1 v^{p+1-\beta}(0, t) \quad \text{whenever } 0 < T_{\max} - t \ll 1 \tag{6.33}$$

for some $C_1 > 0$. Analogously, by the left-hand side of (6.32), we have

$$v^{p+1-\beta}(0,t) \leqslant C_2 u^{q+1-p}(0,t) \quad \text{whenever } 0 < T_{\max} - t \ll 1 \qquad (6.34)$$

for some $C_2 > 0$. It then follows from (6.33), (6.34) and (6.30) that

$$u_t(0,t) \geqslant \varepsilon_1 u^{(1+\theta)/\theta}(0,t) \quad \text{and} \quad v_t(0,t) \geqslant \varepsilon_2 v^{(1+\sigma)/\sigma}(0,t).$$

Thereby, for some positive constant C,

$$u(0,t) \leqslant C(T_{\max} - t)^{-\theta} \quad \text{and} \quad v(0,t) \leqslant C(T_{\max} - t)^{-\sigma} \quad \text{for } 0 < T_{\max} - t \ll 1.$$

Similarly, combining (6.33), (6.34) and (6.31) we get, for positive constant c,

$$u(0,t) \geqslant c(T_{\max} - t)^{-\theta} \quad \text{and} \quad v(0,t) \geqslant c(T_{\max} - t)^{-\sigma} \quad \text{for } 0 < T_{\max} - t \ll 1.$$

Estimates (6.28) are achieved, and hence the proof is complete. $\qquad\qquad\square$

6.3 A SYSTEM COUPLED IN EQUATION AND BOUNDARY

In this section, we study the following parabolic system coupled in an equation and a boundary condition

$$\begin{cases} u_t - \Delta u = v^p, \quad v_t - \Delta v = 0, & x \in \Omega, \quad t > 0, \\ \partial_{\boldsymbol{n}} u = 0, \quad \partial_{\boldsymbol{n}} v = u^q, & x \in \partial\Omega, \quad t > 0, \\ u = u_0(x), \quad v = v_0(x), & x \in \Omega, \quad t = 0, \end{cases} \qquad (6.35)$$

where p, q are positive constants; initial data $u_0, v_0 \in C^2(\overline{\Omega})$, and $u_0, v_0 \geqslant 0, \not\equiv 0$ in $\overline{\Omega}$ which satisfy $\partial_{\boldsymbol{n}} u_0 = 0$ and $\partial_{\boldsymbol{n}} v_0 = u_0^q$ on $\partial\Omega$.

Let $G(x,y,t-s)$ be the *Green function* for heat equation with the homogeneous Neumann boundary condition, i.e., for each fixed $x \in \overline{\Omega}$, $G(x,y,t)$ satisfies

$$\begin{cases} G_t - \Delta_y G = 0, & y \in \Omega, \quad t > 0, \\ \partial_{\boldsymbol{n}_y} G = 0, & y \in \partial\Omega, \quad t > 0 \end{cases}$$

and

$$\lim_{t \to 0}[G(x,y,t) - \Gamma(x-y,t)]\big|_{y \in \Omega} = 0 \quad \text{in the sense of measure,}$$

where

$$\Gamma(x-y,t) = \begin{cases} \dfrac{1}{(4\pi t)^{n/2}} \exp\left(-\dfrac{|x-y|^2}{4t}\right), & t > 0, \\ 0, \quad t \leqslant 0, \quad (x,t) \neq (y,0). \end{cases}$$

Then we have the following representation formulae

$$u(x,t) = \int_\Omega G(x,y,t)u_0(y)\mathrm{d}y + \int_0^t \int_\Omega G(x,y,t-s)v^p(y,s)\mathrm{d}y\mathrm{d}s, \qquad (6.36)$$

$$v(x,t) = \int_\Omega G(x,y,t)v_0(y)\mathrm{d}y + \int_0^t \int_{\partial\Omega} G(x,y,t-s)u^q(y,s)\mathrm{d}S_y\mathrm{d}s. \qquad (6.37)$$

Invoking above representation formulae and the contraction mapping principle, we can establish local existence for solutions of (6.35).

Contents of this section refer to [19, 49, 50, 108].

6.3.1 Properties of the Green function $G(x, y, t - s)$

It is clear that
$$G(x, y, t - s) = G(y, x, t - s).$$

We fix $\varepsilon > 0$. Thanks to the maximum principle, $G(x, y, \varepsilon) > 0$ for $x, y \in \overline{\Omega}$, and thereby
$$\min_{x, y \in \overline{\Omega}} G(x, y, \varepsilon) = c_0(\varepsilon) > 0.$$

By using of the maximum principle again, it deduces
$$G(x, y, t - s) \geqslant c_0(\varepsilon), \quad x, y \in \overline{\Omega}, \ t - s \geqslant \varepsilon. \tag{6.38}$$

Let w be the unique solution of
$$\begin{cases} w_t - \Delta w = 0, & x \in \Omega, \quad t > 0, \\ \partial_{\boldsymbol{n}} w = 0, & x \in \partial\Omega, \ t > 0, \\ w(x, 0) = 1, & x \in \Omega. \end{cases}$$

Then $w = 1$, and by the representation formula we have
$$1 = w(x, t) = \int_{\Omega} G(x, y, t) \mathrm{d}y, \quad x \in \overline{\Omega}, \ t \geqslant 0. \tag{6.39}$$

Now, we are going to demonstrate that there exists a constant $c_0 > 0$ such that
$$\int_{\partial\Omega} G(x, y, t - s) \mathrm{d}S_y \geqslant c_0, \quad x \in \overline{\Omega}, \ t > s \geqslant 0. \tag{6.40}$$

To this aim, we first prove that, for some small $\varepsilon_0 > 0$,
$$\frac{c_0(\varepsilon_0)}{\sqrt{t - s}} \leqslant \int_{\partial\Omega} G(x, y, t - s) \mathrm{d}S_y \leqslant \frac{C_0}{\sqrt{t - s}}, \quad x \in \partial\Omega, \ 0 < t - s \leqslant \varepsilon_0. \tag{6.41}$$

It is well-known that (see [38]) there exist constants $C > c > 0$ such that
$$G(x, y, t - s) \leqslant \frac{C}{(t - s)^{n/2}} \exp\left(-c\frac{|x - y|^2}{t - s}\right), \quad x, y \in \overline{\Omega}, \ 0 < t - s \leqslant 1. \tag{6.42}$$

The upper bound in (6.41) can be deduced from (6.42).

In the following, we will derive the lower bound in (6.41). It causes from the construction of $G(x, y, t - s)$ that
$$G(x, y, t - s) = \Gamma(x - y, t - s) + h(x, y, t - s), \tag{6.43}$$

where $h(x, y, \tau)$ solves
$$\begin{cases} h_\tau = \Delta_y h, & y \in \Omega, \quad \tau > 0, \\ \partial_{\boldsymbol{n}_y} h = -\partial_{\boldsymbol{n}_y} \Gamma, & y \in \partial\Omega, \ \tau > 0, \\ h(x, y, 0) = 0, & y \in \Omega. \end{cases}$$

Obviously,

$$-\partial_{\boldsymbol{n}_y}\Gamma(x-y,\tau) = -\frac{2(x-y)\cdot\boldsymbol{n}_y}{\pi^{n/2}(4\tau)^{1+n/2}}\exp\left(-\frac{|x-y|^2}{4\tau}\right).$$

The representation formula and the estimate (6.42) lead to, for $x \in \partial\Omega$,

$$\int_{\partial\Omega}|h(x,y,\tau)|\mathrm{d}S_y$$

$$\leqslant \int_{\partial\Omega}\int_0^\tau\int_{\partial\Omega}\frac{C|(x-z)\cdot\boldsymbol{n}_z|}{(\tau-s)^{n/2}s^{1+n/2}}\exp\left(-\frac{|x-z|^2}{4s}-c\frac{|y-z|^2}{\tau-s}\right)\mathrm{d}S_z\mathrm{d}s\mathrm{d}S_y$$

$$= \int_0^\tau\int_{\partial\Omega}\int_{\partial\Omega}\frac{C|(x-z)\cdot\boldsymbol{n}_z|}{(\tau-s)^{n/2}s^{1+n/2}}\exp\left(-\frac{|x-z|^2}{4s}-c\frac{|y-z|^2}{\tau-s}\right)\mathrm{d}S_y\mathrm{d}S_z\mathrm{d}s. \quad (6.44)$$

Clearly,

$$\int_{\partial\Omega}\exp\left(-c\frac{|y-z|^2}{\tau-s}\right)\mathrm{d}S_y \leqslant C'(\tau-s)^{(n-1)/2}. \quad (6.45)$$

As Ω is of class C^2, we can find $0 < s_0 \ll 1$ and $C_1 > 0$ such that for all $x \in \partial\Omega$

$$|(x-z)\cdot\boldsymbol{n}_z|\mathrm{d}S_z \leqslant C_1|x'-z'|^2\mathrm{d}z' \quad \text{for } z \in \partial\Omega\cap\{|z-x| < s_0\}, \quad (6.46)$$

where $z' = (z_1,\ldots,z_{n-1})$, $x' = (x_1,\ldots,x_{n-1})$. Consequently,

$$\int_{\partial\Omega\cap\{|z-x|<s_0\}}\frac{|(x-z)\cdot\boldsymbol{n}_z|}{s^{1+n/2}}\exp\left(-\frac{|x-z|^2}{4s}\right)\mathrm{d}S_z$$

$$\leqslant C_1\int_{|z'-x'|<s_0}\frac{|x'-z'|^2}{s^{1+n/2}}\exp\left(-\frac{|x'-z'|^2}{4s}\right)\mathrm{d}z'$$

$$\leqslant \frac{C_1'}{s^{1/2}}\int_{\mathbb{R}^{n-1}}|\eta|^2\mathrm{e}^{-|\eta|^2}\mathrm{d}\eta$$

$$\leqslant \frac{C_1''}{s^{1/2}}.$$

It is not hard to show that there exists a positive constant C_2 depending on s_0, or equivalently, depending on $\partial\Omega$, so that

$$\int_{\partial\Omega\cap\{|z-x|\geqslant s_0\}}\frac{|(x-z)\cdot\boldsymbol{n}_z|}{s^{1+n/2}}\exp\left(-\frac{|x-z|^2}{4s}\right)\mathrm{d}S_z \leqslant \frac{C_2}{s^{1/2}}.$$

Anyhow we have, for all $0 < s < \tau$,

$$\int_{\partial\Omega}\frac{|(x-z)\cdot\boldsymbol{n}_z|}{s^{1+n/2}}\exp\left(-\frac{|x-z|^2}{4s}\right)\mathrm{d}S_z \leqslant \frac{C_2+C_1''}{s^{1/2}}. \quad (6.47)$$

It follows from (6.44)–(6.47) that

$$\int_{\partial\Omega} |h(x,y,\tau)| dS_y \leqslant CC' \int_0^\tau \int_{\partial\Omega} \frac{|(x-z) \cdot \boldsymbol{n}_z|}{s^{1+n/2}\sqrt{\tau-s}} \exp\left(-\frac{|x-z|^2}{4s}\right) dS_z ds$$

$$\leqslant C_3 \int_0^\tau \frac{ds}{s^{1/2}\sqrt{\tau-s}}$$

$$\leqslant C_4 \quad \text{for } 0 < \tau \leqslant 1, \ x \in \partial\Omega. \tag{6.48}$$

On the other hand, it is clear that

$$\int_{\partial\Omega} \Gamma(x-y,\tau) dS_y \geqslant \frac{c}{\sqrt{\tau}} \quad \text{for } 0 < \tau \leqslant 1, \ x \in \partial\Omega \tag{6.49}$$

for some positive constant c. We choose $\varepsilon_0 > 0$ small enough so that

$$\frac{c}{\sqrt{\tau}} - C_4 \geqslant \frac{c}{2\sqrt{\tau}} \quad \text{for } 0 < \tau \leqslant \varepsilon_0. \tag{6.50}$$

The lower bound in (6.41) can be deduced by (6.43) and (6.48)–(6.50).

Combining (6.38) and (6.41), we conclude (6.40).

6.3.2 Critical exponents

In this subsection, we will show that every solution of (6.35) is global if and only if $pq \leqslant 1$.

Theorem 6.13 *Assume $pq \leqslant 1$. Then every solution of (6.35) exists globally.*

Proof. We hope to seek a global upper solution $(\bar{u}, \bar{v})$ of problem (6.35). Let $h(x)$ be a solution of

$$\begin{cases} \Delta h = k = \dfrac{|\partial\Omega|}{|\Omega|} & \text{in } \Omega, \\ \partial_n h = 1 & \text{on } \partial\Omega. \end{cases}$$

Without loss of generality, we may assume $h(x) > 0$ in $\overline{\Omega}$. Let

$$\ell = \max_{\overline{\Omega}}\{h(x) + |\nabla h(x)|\}, \quad m = \max\{\|u_0\|_\infty, \|v_0\|_\infty\},$$

$$\gamma = m^{q-1}, \quad \sigma = \max\{m^{p-1}e^{p\gamma\ell}, \gamma(k+\gamma\ell^2)/q\},$$

and define

$$\bar{u} = me^{\sigma t}, \quad \bar{v} = me^{q\sigma t + \gamma h(x)}.$$

By a series of computations we find that $(\bar{u}, \bar{v})$ satisfies

$$\begin{cases} \bar{u}_t \geqslant \Delta\bar{u} + \bar{v}^p, \quad \bar{v}_t \geqslant \Delta\bar{v}, & x \in \Omega, \quad t > 0, \\ \partial_n\bar{u} \geqslant 0, \quad \partial_n\bar{v} \geqslant \bar{u}^q, & x \in \partial\Omega, \quad t > 0, \\ \bar{u}(x,0) \geqslant u_0(x), \quad \bar{v}(x,0) \geqslant v_0(x), & x \in \Omega. \end{cases}$$

This indicates that $(\bar{u}, \bar{v})$ is a desired upper solution. $\qquad\square$

To establish blowup results, we need the following relationship between solution components u and v.

Lemma 6.14 *Let (u, v) be a non-negative solution of (6.35) with the maximal existence time $T_{\max} \leqslant \infty$. Then there exists a positive constant $c = c(p, q, \Omega)$ such that*

(1) *when $p, q \geqslant 1$, it holds*

$$\int_{\partial\Omega} u^q(x, t)\mathrm{d}S_x \geqslant c\left(\int_0^t \int_\Omega v^p(y, s)\mathrm{d}y\mathrm{d}s\right)^q \quad \textit{for all } 0 < t < T_{\max}; \quad (6.51)$$

(2) *when $p > 1 > q$ and $pq > 1$, we have*

$$\int_{\partial\Omega} u^q(x, t)\mathrm{d}S_x \geqslant ct^{q(1-p)}\left(\int_0^t \int_\Omega v(y, s)\mathrm{d}y\mathrm{d}s\right)^{pq} \quad \textit{for all } 0 < t < T_{\max}; \quad (6.52)$$

(3) *when $q > 1 > p$ and $pq > 1$, if $T_{\max} > 1$ then*

$$\int_\Omega v^p(x, t) \geqslant ct^{p(1-q)}\left(\int_0^t \int_{\partial\Omega} u(y, s)\mathrm{d}S_y\mathrm{d}s\right)^{pq} \quad \textit{for all } 1 \leqslant t < T_{\max}. \quad (6.53)$$

Proof. By using the maximum principle and Hopf boundary lemma, we know that $u, v > 0$ in $\overline{\Omega} \times (0, T_{\max})$.

(1) Proof of (6.51). By means of (6.36), (6.40) and the Hölder inequality, it can be deduced that

$$\int_{\partial\Omega} u^q(x, t)\mathrm{d}S_x \geqslant \int_{\partial\Omega}\left(\int_0^t \int_\Omega G(x, y, t - s)v^p(y, s)\mathrm{d}y\mathrm{d}s\right)^q \mathrm{d}S_x$$

$$\geqslant |\partial\Omega|^{1-q}\left(\int_{\partial\Omega}\int_0^t \int_\Omega G(x, y, t - s)v^p(y, s)\mathrm{d}y\mathrm{d}s\mathrm{d}S_x\right)^q$$

$$= |\partial\Omega|^{1-q}\left(\int_0^t \int_\Omega v^p(y, s)\int_{\partial\Omega} G(x, y, t - s)\mathrm{d}S_x\mathrm{d}y\mathrm{d}s\right)^q$$

$$\geqslant c_0^q|\partial\Omega|^{1-q}\left(\int_0^t \int_\Omega v^p(y, s)\mathrm{d}y\mathrm{d}s\right)^q \quad \text{for all } 0 < t < T_{\max}.$$

(2) Proof of (6.52). Making use of the Hölder inequality and (6.39), it concludes

$$\int_0^t \int_\Omega G(x, y, t - s)v^p(y, s)\mathrm{d}y\mathrm{d}s \geqslant \frac{\left(\int_0^t \int_\Omega G(x, y, t - s)v(y, s)\mathrm{d}y\mathrm{d}s\right)^p}{\left(\int_0^t \int_\Omega G(x, y, t - s)\mathrm{d}y\mathrm{d}s\right)^{p-1}}$$

$$= t^{1-p}\left(\int_0^t \int_\Omega G(x, y, t - s)v(y, s)\mathrm{d}y\mathrm{d}s\right)^p.$$

This together with (6.36) and (6.40) derives

$$\int_{\partial\Omega} u^q(x, t)\mathrm{d}S_x \geqslant c_1 t^{q(1-p)}\int_{\partial\Omega}\left(\int_0^t \int_\Omega G(x, y, t - s)v(y, s)\mathrm{d}y\mathrm{d}s\right)^{pq} \mathrm{d}S_x$$

$$\geqslant c_2 t^{q(1-p)}\left(\int_0^t \int_\Omega v(y, s)\int_{\partial\Omega} G(x, y, t - s)\mathrm{d}S_x\mathrm{d}y\mathrm{d}s\right)^{pq}$$

$$\geqslant c_2 c_0^{pq} t^{q(1-p)}\left(\int_0^t \int_\Omega v(y, s)\mathrm{d}y\mathrm{d}s\right)^{pq} \quad \text{for all } 0 < t < T_{\max}.$$

(3) Proof of (6.53). Note that $G(x, y, t - s)$ is bounded when $t - s > \varepsilon_0$, and (6.41) holds when $t - s \leqslant \varepsilon_0$. It follows that

$$\int_0^t \int_{\partial\Omega} G(x, y, t - s) \mathrm{d}S_y \mathrm{d}s \leqslant C_1 t, \quad x \in \overline{\Omega}, \; 1 \leqslant t < T_{\max}.$$

This together with (6.37), the Hölder inequality and (6.39) derives that

$$\int_\Omega v^p(x, t) \mathrm{d}x \geqslant \int_\Omega \left(\int_0^t \int_{\partial\Omega} G(x, y, t - s) u^q(y, s) \mathrm{d}S_y \mathrm{d}s \right)^p$$

$$\geqslant \int_\Omega \frac{\left(\int_0^t \int_{\partial\Omega} G(x, y, t - s) u(y, s) \mathrm{d}S_y \mathrm{d}s \right)^{pq}}{\left(\int_0^t \int_{\partial\Omega} G(x, y, t - s) \mathrm{d}S_y \mathrm{d}s \right)^{p(q-1)}}$$

$$\geqslant (C_1 t)^{p(1-q)} \left(\int_0^t \int_{\partial\Omega} u(y, s) \int_\Omega G(x, y, t - s) \mathrm{d}x \mathrm{d}S_y \mathrm{d}s \right)^{pq}$$

$$= (C_1 t)^{p(1-q)} \left(\int_0^t \int_{\partial\Omega} u(y, s) \mathrm{d}S_y \mathrm{d}s \right)^{pq} \quad \text{for all } 1 \leqslant t < T_{\max}.$$

Proofs are complete. □

Theorem 6.15 *Suppose $pq > 1$. Then all non-negative and non-trivial solutions of* (6.35) *blow up in finite time.*

Proof. By using the maximum principle and the Hopf boundary lemma, we see that u and v are positive. In view of the preceding lemma, we treat three cases respectively.

Case 1: $p, q \geqslant 1$. Introduce a function

$$F(t) = \int_0^t \int_\Omega v(x, s) \mathrm{d}x \mathrm{d}s.$$

Utilizing (6.37) and (6.39), we see that $F(t)$ satisfies

$$F(t) \geqslant \int_0^t \int_\Omega \int_\Omega G(x, y, s) v_0(y) \mathrm{d}y \mathrm{d}x \mathrm{d}s$$

$$= \int_0^t \int_\Omega v_0(y) \int_\Omega G(x, y, s) \mathrm{d}x \mathrm{d}y \mathrm{d}s$$

$$= \int_0^t \int_\Omega v_0(y) \mathrm{d}y \mathrm{d}s =: c_3 t. \tag{6.54}$$

Moreover, integrating by parts and recalling (6.51) we conclude

$$F''(t) = \int_\Omega v_t(x, t) \mathrm{d}x = \int_{\partial\Omega} u^q(x, t) \mathrm{d}S_x$$

$$\geqslant c \left(\int_0^t \int_\Omega v^p(y, s) \mathrm{d}y \mathrm{d}s \right)^q \geqslant c_4 t^{q(1-p)} F^{pq}(t).$$

Since $F'(t) > 0$, we may multiply the above inequality by $F'(t)$ and integrate by parts, and keep $p \geqslant 1$ in mind, to obtain

$$(F'(t))^2 \geqslant c_5^2 t^{q(1-p)} F^{pq+1}(t),$$

or, equivalently,

$$F'(t) \geqslant c_5 t^{q(1-p)/2} F^{(pq+1)/2}(t). \qquad (6.55)$$

Choose a constant δ satisfying

$$1 < \delta < \frac{1}{2} \min\{pq+1, q+3\}.$$

We then use (6.55) and the lower bound in (6.54) for $F^{(pq+1)/2-\delta}(t)$ to get

$$F'(t) \geqslant c_6 t^{(q+1)/2-\delta} F^{\delta}(t),$$

i.e., $F'(t)F^{-\delta}(t) \geqslant c_6 t^{(q+1)/2-\delta}$. It follows that

$$F^{1-\delta}(t) \leqslant F^{1-\delta}(\eta) + c_7 \left(\eta^{(q+3)/2-\delta} - t^{(q+3)/2-\delta} \right) \to -\infty$$

as $t \to \infty$. This indicates that $F(t)$ cannot be global, so (u,v) cannot be global either.

Case 2: $p > 1 > q$. Define $F(t)$ as above. Recalling (6.52), the proof is essentially the same as that in Case 1, and it is omitted here.

Case 3: $q > 1 > p$. Assume conversely that the solution (u,v) of (6.35) exists globally. Thanks to (6.40),

$$\int_{\partial\Omega} \int_{\Omega} G(x,y,t)u_0(y)\mathrm{d}y\mathrm{d}S_x = \int_{\Omega} \int_{\partial\Omega} G(x,y,t)u_0(y)\mathrm{d}S_x\mathrm{d}y$$

$$\geqslant c_0 \int_{\Omega} u_0(y)\mathrm{d}y, \quad t > 0. \qquad (6.56)$$

On the other hand, applying (6.40) and (6.53) successively we derive that, for $t \geqslant 1$,

$$\int_{\partial\Omega} \int_0^t \int_{\Omega} G(x,y,t-s)v^p(y,s)\mathrm{d}y\mathrm{d}s\mathrm{d}S_x \geqslant c_0 \int_0^t \int_{\Omega} v^p(y,s)\mathrm{d}y\mathrm{d}s$$

$$\geqslant c_1 \int_0^t s^{p(1-q)} \left[\int_0^s \int_{\partial\Omega} u(y,\tau)\mathrm{d}S_y\mathrm{d}\tau \right]^{pq} \mathrm{d}s,$$

which, combined with (6.36) and (6.56), leads to

$$\int_{\partial\Omega} u(x,t)\mathrm{d}S_x \geqslant a + b \int_0^t s^{p(1-q)} \left(\int_0^s \int_{\partial\Omega} u(y,\tau)\mathrm{d}S_y\mathrm{d}\tau \right)^{pq} \mathrm{d}s, \quad t \geqslant 1$$

for some positive constants a and b. Set

$$H(t) = \int_0^t \int_{\partial\Omega} u(y,s)\mathrm{d}S_y\mathrm{d}s, \quad t \geqslant 1.$$

Then

$$H'(t) \geqslant a + b \int_0^t s^{p(1-q)} H^{pq}(s)\mathrm{d}s, \quad t \geqslant 1.$$

Integrating above inequality from 1 to t, then we assert that, with $a_1 = a/2$

$$H(t) \geqslant a_1 t + b \int_1^t \int_1^\tau s^{p(1-q)} H^{pq}(s) ds d\tau$$

$$= a_1 t + b \int_1^t (t-s) s^{p(1-q)} H^{pq}(s) ds$$

$$\geqslant a_1 t + b t^{p(1-q)} \int_1^t (t-s) H^{pq}(s) ds, \quad t \geqslant 2.$$

Choosing $\tau \geqslant 2$, we have

$$H(t) \geqslant a_1 \tau + b_1 \tau^{p(1-q)} \int_\tau^t (t-s) H^{pq}(s) ds, \quad \tau \leqslant t \leqslant 2\tau$$

with $b_1 = 2^{p(1-q)} b$. Thus, by the comparison principle, $H(t) \geqslant h(t)$ in $[\tau, 2\tau]$, where $h(t)$ is a solution of

$$h(t) = a_1 \tau + b_1 \tau^{p(1-q)} \int_\tau^t (t-s) h^{pq}(s) ds, \quad \tau \leqslant t \leqslant 2\tau.$$

Obviously, $h(t)$ satisfies

$$h''(t) = b_1 \tau^{p(1-q)} h^{pq}(t), \quad \tau \leqslant t \leqslant 2\tau,$$

$$h(\tau) = a_1 \tau, \quad h'(\tau) = 0.$$

Multiplying the differential equation by $h'(t)$ and integrating the result from τ to t we ascertain

$$h'(t) = b_2 \tau^{p(1-q)/2} [h^{pq+1}(t) - h^{pq+1}(\tau)]^{1/2}$$

with $b_2 = \sqrt{2b_1/(pq+1)}$, and thereby,

$$b_2 \tau^{(p-pq+2)/2} = \int_{h(\tau)}^{h(2\tau)} \frac{dz}{\sqrt{z^{pq+1} - h^{pq+1}(\tau)}} = h^{(1-pq)/2}(\tau) \int_1^{\frac{h(2\tau)}{h(\tau)}} \frac{dy}{\sqrt{y^{pq+1} - 1}}.$$

It is easy to see that

$$\int_1^{\frac{h(2\tau)}{h(\tau)}} \frac{dy}{\sqrt{y^{pq+1} - 1}} \leqslant \int_1^2 \frac{dy}{\sqrt{y-1}} + \int_2^\infty \frac{dy}{\sqrt{y^{pq+1} - 1}} \leqslant 2 + \sqrt{2} \int_2^\infty \frac{dy}{y^{\frac{pq+1}{2}}} =: b_3.$$

Thus, we get, recalling $h(\tau) = a_1 \tau$,

$$b_2 \tau^{(p-pq+2)/2} \leqslant b_3 h^{(1-pq)/2}(\tau) = b_3 (a_1 \tau)^{(1-pq)/2} =: b_4 \tau^{(1-pq)/2},$$

which is equivalent to $\tau^{(p+1)/2} \leqslant b_4/b_2$. This is impossible for sufficiently large τ, and the proof is complete. $\square$

6.3.3   Blowup on the boundary

In this subsection, it is always assumed that (u, v) is a solution of (6.35) which blows up in finite time $T_{\max} < \infty$. We shall derive boundary blowup results.

Theorem 6.16   *When $p, q \geqslant 1$, blowup can occur only on the boundary.*

Proof. The proof will be divided into three steps.

Step 1. We first establish another relationship between the solution components u and v. It follows from (6.37) and (6.39) that

$$
\begin{aligned}
\int_\Omega v^p(x, t) \mathrm{d}x &\geqslant \int_\Omega \left(\int_0^t \int_{\partial\Omega} G(x, y, t - s) u^q(y, s) \mathrm{d}S_y \mathrm{d}s \right)^p \\
&\geqslant |\Omega|^{1-p} \left(\int_0^t \int_{\partial\Omega} u^q(y, s) \int_\Omega G(x, y, t - s) \mathrm{d}x \mathrm{d}S_y \mathrm{d}s \right)^p \\
&= |\Omega|^{1-p} \left(\int_0^t \int_{\partial\Omega} u^q(y, s) \mathrm{d}S_y \mathrm{d}s \right)^p.
\end{aligned}
\tag{6.57}
$$

Step 2. Set $A(t) = \|u(\cdot, t)\|_{q, \partial\Omega}$ and $B(t) = \|v(\cdot, t)\|_{p, \Omega}$. By virtue of (6.51) and (6.57), we have, with $c = |\Omega|^{1-p}$,

$$
A(t) \geqslant c^{1/q} \int_0^t B^p(s) \mathrm{d}s =: c^{1/q} f(t), \quad 0 \leqslant t < T_{\max},
$$

$$
B(t) \geqslant c^{1/p} \int_0^t A^q(s) \mathrm{d}s =: c^{1/p} g(t), \quad 0 \leqslant t < T_{\max}.
$$

As a consequence,

$$
f'(t) \geqslant c g^p(t), \quad g'(t) \geqslant c f^q(t), \quad 0 \leqslant t < T_{\max},
\tag{6.58}
$$

which implies

$$
(fg)'(t) \geqslant c g^{p+1}(t) + c f^{q+1}(t), \quad 0 \leqslant t < T_{\max}.
\tag{6.59}
$$

Define $\mu = (p+1)/(pq-1)$ and $\theta = (q+1)/(pq-1)$. We shall use the idea of [74] to prove

$$
f(t) \leqslant C(T_{\max} - t)^{-\mu}, \quad g(t) \leqslant C(T_{\max} - t)^{-\theta}, \quad 0 \leqslant t < T_{\max}.
\tag{6.60}
$$

In fact, set $k = (1 + p + q + pq)/(2 + p + q)$, then

$$
1/(k - 1) = \mu + \theta, \quad k/(p+1) + k/(q+1) = 1,
$$

and $k > 1$ since $pq > 1$. Application of the Young inequality in (6.59) confirms

$$
(fg)'(t) \geqslant C(c, p, q)(fg)^k(t), \quad 0 \leqslant t < T_{\max}.
$$

Integrating the above inequality from t to $T_{\max}$, we obtain

$$
f(t)g(t) \leqslant C(T_{\max} - t)^{-1/(k-1)} = C(T_{\max} - t)^{-\mu-\theta}, \quad 0 \leqslant t < T_{\max}.
\tag{6.61}
$$

Now, we prove (6.60) by contradiction, and assume that there exist sequences $\{t_i\}$ and $\{C_i\}$ with $t_i \to T^-_{\max}$ and $C_i \to \infty$ as $i \to \infty$ such that

$$f(t_i) \geqslant C_i(T_{\max} - t_i)^{-\mu}. \tag{6.62}$$

Integrating two inequalities in (6.58) from t_i to t, and using the monotonicity of $f(t)$ and (6.62), we conclude that, for $t_i \leqslant t < T_{\max}$,

$$g(t) \geqslant g(t_i) + \int_{t_i}^t cf^q(s)\mathrm{d}s \geqslant cf^q(t_i)(t - t_i) \geqslant cC_i^q(T_{\max} - t_i)^{-\mu q}(t - t_i),$$

$$f(t) \geqslant f(t_i) + \int_{t_i}^t cg^p(s)\mathrm{d}s \geqslant c^{p+1}C_i^{pq}(T_{\max} - t_i)^{-\mu pq}\int_{t_i}^t (s - t_i)^p\mathrm{d}s$$

$$= (p + 1)^{-1}c^{p+1}C_i^{pq}(T_{\max} - t_i)^{-\mu pq}(t - t_i)^{p+1}.$$

It then follows that

$$f(t)g(t) \geqslant (p + 1)^{-1}c^{p+2}C_i^{pq+q}(T_{\max} - t_i)^{-\mu q(p+1)}(t - t_i)^{p+2}.$$

Taking $t = t^* = (T_{\max} + t_i)/2$ in the above inequality, it yields

$$f(t^*)g(t^*) \geqslant (p + 1)^{-1}c^{p+2}C_i^{pq+q}(T_{\max} - t_i)^{-\mu q(p+1)}(1/2)^{p+2}(T_{\max} - t_i)^{p+2}$$

$$= (p + 1)^{-1}(c/2)^{p+2}C_i^{pq+q}(T_{\max} - t_i)^{-\mu - \theta}$$

$$= (p + 1)^{-1}(c/2)^{p+2}C_i^{pq+q}2^{-\mu - \theta}(T_{\max} - t^*)^{-\mu - \theta}.$$

This contradicts (6.61) since $C_i \to \infty$. As a result, the first estimate of (6.60) is valid. The second estimate of (6.60) can be achieved in a similar way.

Step 3. For any open subset $\Omega_1 \subset \Omega$ with $\partial\Omega_1 \in C^2$ and $\mathrm{dist}(\Omega_1, \partial\Omega) = \varepsilon > 0$, we can take $\Omega_2 \Subset \Omega$ so that $\Omega_1 \Subset \Omega_2$, $\mathrm{dist}(\Omega_1, \partial\Omega_2) \geqslant \varepsilon/3$ and $\mathrm{dist}(\Omega_2, \partial\Omega) \geqslant \varepsilon/3$. It is well known that for any $\varepsilon > 0$,

$$0 \leqslant G(x, y, t - s) \leqslant C_\varepsilon \quad \text{for } x, y \in \overline{\Omega}, \ |x - y| \geqslant \varepsilon/3, \ 0 \leqslant s < t < T_{\max}.$$

This together with (6.37) and the second estimate of (6.60) deduces

$$\max_{\overline{\Omega}_2} v(\cdot, t) \leqslant C_0 + C_\varepsilon \int_0^t A^q(s)\mathrm{d}s$$

$$= C_0 + C_\varepsilon g(t)$$

$$\leqslant C_1(T_{\max} - t)^{-\theta}, \quad 0 \leqslant t < T_{\max}. \tag{6.63}$$

We shall use the idea of [49] to construct a control function in order to show that v cannot blow up in the interior of Ω_1. Define $d(x) = d(x, \partial\Omega_1)$ and

$$\phi(x) = d^2(x), \quad x \in N_\sigma(\partial\Omega_1) := \{x \in \Omega_1 : d(x) < \sigma\}.$$

Owing to $\partial\Omega_1 \in C^2$, the function $\phi \in C^2(\overline{N_\sigma(\partial\Omega_1)})$ if σ is small enough. By the direct calculation we see that there exists a constant $C > 0$ such that

$$\Delta\phi - \frac{(\theta + 1)|\nabla\phi|^2}{\phi} \geqslant -C \quad \text{in } \overline{N_\sigma(\partial\Omega_1)}$$

provided that σ is small enough. We next extend ϕ to a function in $\overline{\Omega}_1$, still denoted by itself, such that $\phi \in C^2(\overline{\Omega}_1)$ and $\phi \geqslant c_0 > 0$ in $\Omega_1 \setminus N_\sigma(\partial\Omega_1)$. Naturally,

$$\Delta\phi - \frac{(\theta+1)|\nabla\phi|^2}{\phi} > -\rho \quad \text{in } \overline{\Omega}_1$$

for some $\rho > 0$. Set

$$w = \frac{C}{[\phi(x) + \rho(T_{\max} - t)]^\theta}, \quad x \in \Omega_1, \ 0 < t < T_{\max}.$$

Careful computations yield

$$w_t - \Delta w = \frac{C\theta}{[\phi + \rho(T_{\max} - t)]^{1+\theta}}\left(\rho + \Delta\phi - \frac{(\theta+1)|\nabla\phi|^2}{\phi + \rho(T_{\max} - t)}\right) > 0.$$

Take C to be large enough so that $w(x,0) \geqslant v(x,0)$ in $\overline{\Omega}_1$ and $C\rho^{-\theta} > C_1$. Then

$$w\big|_{\partial\Omega_1} = C\rho^{-\theta}(T_{\max} - t)^{-\theta} \geqslant v\big|_{\partial\Omega_1}$$

in accordance with (6.63). The comparison principle gives

$$v(x,t) \leqslant w(x,t) = \frac{C}{[\phi(x) + \rho(T_{\max} - t)]^\theta}, \quad x \in \Omega_1, \ 0 \leqslant t < T_{\max}.$$

This shows that v cannot blow up in the interior of Ω_1, and so v cannot blow up in the interior of Ω by the arbitrariness of Ω_1.

Analogously, u cannot blow up in the interior of Ω. □

6.3.4 Blowup rate estimates

In this part, we study a special case of (6.35) of the form

$$\begin{cases} u_t - \Delta u = v^p, \quad v_t - \Delta v = 0, & x \in B_R, \quad t > 0, \\ \partial_n u = 0, \quad \partial_n v = u^q, & x \in \partial B_R, \quad t > 0, \\ u(x,0) = u_0(x), \quad v(x,0) = v_0(x), & x \in B_R, \end{cases} \quad (6.64)$$

here B_R is a ball in $\mathbb{R}^n$ centred at the origin with radius $R > 0$, initial data $u_0(x) = u_0(r)$, $v_0(x) = v_0(r)$ with $r = |x|$ are positive, radially symmetric C^2 functions and satisfy

$$v_0'(r) \geqslant 0 \ \text{ for } 0 \leqslant r < R, \quad u_0'(R) = 0, \quad v_0'(R) = u_0^q(R).$$

Under above conditions, it is easy to show that any solution (u,v) of (6.64) is positive and radially symmetric, and satisfies $v_r(r,t) \geqslant 0$. By results of §6.3.2, all solutions of (6.64) exist globally if and only if $pq \leqslant 1$.

It is assumed that $pq > 1$, and components u and v of the solution (u,v) of (6.64) blow up simultaneously in the same time $T_{\max} < \infty$. For any $t > 0$, we denote

$$h(t) := \max_{0 \leqslant \tau \leqslant t} u(R,\tau), \quad f(t) := \max_{0 \leqslant \tau \leqslant t} \max_{\overline{B}_R} u(\cdot,\tau), \quad g(t) := \max_{0 \leqslant \tau \leqslant t} \max_{\overline{B}_R} v(\cdot,\tau).$$

Then $f(t) \geqslant h(t)$, and $h(t)$, $f(t)$ and $g(t)$ are monotonically increasing in t and

$$f(t), \ g(t) \to \infty \quad \text{as } t \to T_{\max}.$$

Moreover, the maximum principle gives $g(t) = \max_{0 \leqslant \tau \leqslant t} v(R, \tau)$. Denote $\gamma = \max_{\overline{B}_R} u_0'(r)$, and let $w(r, t) = u_r(r, t) + \gamma$. Then we have

$$w_t = w_{rr} + \frac{n-1}{r} w_r - \frac{n-1}{r^2} w + \frac{n-1}{r^2} \gamma + p v^{p-1} v_r$$

$$\geqslant w_{rr} + \frac{n-1}{r} w_r - \frac{n-1}{r^2} w, \quad 0 < r < R, \ 0 < t < T_{\max}.$$

Since $w(0, t) = w(R, t) = \gamma \geqslant 0$, and $w(r, 0) \geqslant 0$ for $0 \leqslant r \leqslant R$, the maximum principle implies $w \geqslant 0$, i.e., $u_r \geqslant -\gamma$. For any $t > 0$, we choose $r(t) \in [0, R]$ so that $u(r(t), t) = \max_{\overline{B}_R} u(\cdot, t)$. Integrating the inequality $u_r(r, t) \geqslant -\gamma$ from $r(t)$ to R with respect to r we conclude

$$u(R, t) \geqslant u(r(t), t) - \gamma(R - r(t)) \geqslant u(r(t), t) - \gamma R.$$

Consequently, when t closes to $T_{\max}$,

$$h(t) = \max_{0 \leqslant \tau \leqslant t} u(R, \tau) \geqslant \max_{0 \leqslant \tau \leqslant t} u(r(\tau), \tau) - \gamma R = f(t) - \gamma R \geqslant \frac{1}{2} f(t). \qquad (6.65)$$

Define

$$\rho = \frac{2+p}{2(pq-1)} \quad \text{and} \quad \sigma = \frac{1+2q}{2(pq-1)}. \qquad (6.66)$$

We shall use the idea of [17] (the proof of (2.2) there) to prove the following lemma.

Lemma 6.17 *There exists a positive constant ε such that*

$$\varepsilon g^{1/\sigma}(t) \leqslant f^{1/\rho}(t), \quad \varepsilon f^{1/\rho}(t) \leqslant g^{1/\sigma}(t) \quad \text{for all } t \in (T_{\max}/2, T_{\max}). \qquad (6.67)$$

Proof. On the contrary we assume that the first inequality of (6.67) was not valid. Then there would exist $t_i \to T_{\max}$ such that

$$g^{-1/\sigma}(t_i) f^{1/\rho}(t_i) \to 0. \qquad (6.68)$$

For each t_i, we can choose $(\hat{x}_i, \hat{t}_i) \in \partial B_R \times (0, t_i]$ for which $v(\hat{x}_i, \hat{t}_i) = g(t_i)$. On account that $g(t_i) \to \infty$, it can be inferred that $\hat{t}_i \to T_{\max}$ as $i \to \infty$. Define

$$\lambda_i = \lambda(t_i) = g^{-1/(2\sigma)}(t_i)$$

and

$$\begin{cases} \varphi_i(y, s) = \lambda_i^{2\rho} u(\lambda_i R_i y + \hat{x}_i, \lambda_i^2 s + \hat{t}_i), \quad (y, s) \in \overline{\Omega}_i \times I_i(T_{\max}), \\ \psi_i(y, s) = \lambda_i^{2\sigma} v(\lambda_i R_i y + \hat{x}_i, \lambda_i^2 s + \hat{t}_i), \quad (y, s) \in \overline{\Omega}_i \times I_i(T_{\max}), \end{cases} \qquad (6.69)$$

where

$$\Omega_i = \{y : \lambda_i R_i y + \hat{x}_i \in B_R\}, \quad I_i(T_{\max}) = (-\lambda_i^{-2}\hat{t}_i, \lambda_i^{-2}(T_{\max} - \hat{t}_i)),$$

and R_i is a rotation operator such that $(-1, 0, \ldots, 0)$ is the exterior normal vector of $\partial\Omega_i$ at $y = 0$. By use of expressions of ρ and σ, direct calculations tell us

$$\begin{cases} \varphi_{is} - \Delta\varphi_i = \psi_i^p, \quad \psi_{is} - \Delta\psi_i = 0, \quad y \in \Omega_i, \quad s \in I_i(T_{\max}), \\ \partial_{\boldsymbol{n}}\varphi_i = 0, \quad \partial_{\boldsymbol{n}}\psi_i = \varphi_i^q, \quad y \in \partial\Omega_i, \quad s \in I_i(T_{\max}) \end{cases} \tag{6.70}$$

and

$$\begin{cases} \psi_i(0,0) = 1, \quad 0 \leqslant \psi_i(y, s) \leqslant 1, \quad y \in \Omega_i, \ s \in (-\lambda_i^{-2}\hat{t}_i, 0], \\ 0 \leqslant \varphi_i(y, s) \leqslant f(t_i)g^{-\rho/\sigma}(t_i), \quad y \in \Omega_i, \ s \in (-\lambda_i^{-2}\hat{t}_i, 0]. \end{cases} \tag{6.71}$$

For any given $k > 0$, we can find $i_0(k) > 0$ such that

$$\Sigma_k := \{|y - y^k| < k\} \subset \Omega_i, \quad [-k, 0] \subset (-\lambda_i^{-2}\hat{t}_i, 0] \quad \text{for all } i \geqslant i_0(k),$$

where $y^k = (k, 0, \ldots, 0)$. Recalling (6.70) and (6.71), and using the interior L^p estimate and the interior Schauder estimate successively, we obtain

$$\|(\varphi_i, \psi_i)\|_{C^{2+\mu, 1+\mu/2}(\overline{\Sigma}_k \times [-k, 0])} \leqslant C_k \quad \text{for all } i \geqslant i_0(k).$$

Using the Cantor-Hilbert diagonal method, we can prove that there exist a subsequence of $\{(\varphi_i, \psi_i)\}$, which is still denoted by itself, and non-negative functions φ and ψ such that

$$(\varphi_i, \psi_i) \to (\varphi, \psi) \quad \text{in } C^{2,1}(\overline{\Sigma}_k \times [-k, 0])$$

for any $k > 0$, and (φ, ψ) satisfies

$$\varphi_s - \Delta\varphi = \psi^p, \quad y \in \mathbb{R}_+^n := \{y \in \mathbb{R}^n, \ y_1 > 0\}, \ s \in (-\infty, 0]. \tag{6.72}$$

Obviously, φ and ψ are continuous at $(0,0)$. Moreover, $\varphi \equiv 0$ and $\psi(0,0) = 1$ by (6.68) and (6.71). This contradicts (6.72).

Remembering (6.65), the second inequality of (6.67) can be proved similarly. $\square$

Lemma 6.18 ([34]) *Consider the following problem in half space*

$$\begin{cases} \varphi_s - \Delta\varphi = \psi^p, \quad \psi_s - \Delta\psi = 0, \quad y \in \mathbb{R}_+^n, \quad s > 0, \\ -\partial_{y_1}\varphi = 0, \quad -\partial_{y_1}\psi = \varphi^q, \quad y_1 = 0, \quad s > 0, \\ \varphi(y, s) \geqslant 0, \quad \psi(y, s) \geqslant 0. \end{cases} \tag{6.73}$$

Let ρ, σ be given by (6.66). If either $\max\{\rho, \sigma\} > n/2$, or $\max\{\rho, \sigma\} = n/2$ and $p, q \geqslant 1$, then all non-trivial solutions are not global.

Considering that the proof of Lemma 6.18 is too long, we omit details here.

Theorem 6.19 (Blowup rate estimates)

(1) *There exists a constant $c > 0$ such that*

$$f(t) \geqslant c(T_{\max} - t)^{-\rho}, \quad g(t) \geqslant c(T_{\max} - t)^{-\sigma}, \quad 0 < t < T_{\max}. \qquad (6.74)$$

(2) *Assume that one of the followings holds:*

(2a) $u_t, v_t \geqslant 0$; (2b) $\max\{\rho, \sigma\} > n/2$; (2c) $\max\{\rho, \sigma\} = n/2$ *and* $p, q \geqslant 1$.

Then there exists a constant $C > 0$ such that

$$f(t) \leqslant C(T_{\max} - t)^{-\rho}, \quad g(t) \leqslant C(T_{\max} - t)^{-\sigma}, \quad 0 < t < T_{\max}. \qquad (6.75)$$

Proof. Let $G(x, y, t-s)$ be the Green function for heat equation with the homogeneous Neumann boundary condition. Then we conclude that, for $0 < s < t < T_{\max}$,

$$u(x, t) = \int_{B_R} G(x, y, t - s)u(y, s)\mathrm{d}y + \int_s^t \int_{B_R} G(x, y, t - \tau)v^p(y, \tau)\mathrm{d}y\mathrm{d}\tau, \qquad (6.76)$$

$$v(x, t) = \int_{B_R} G(x, y, t - s)v(y, s)\mathrm{d}y + \int_s^t \int_{\partial B_R} G(x, y, t - \tau)u^q(R, \tau)\mathrm{d}S_y\mathrm{d}\tau. \qquad (6.77)$$

(1) Taking advantage of (6.67), it follows from (6.76) that

$$f(t) \leqslant f(s) + \int_s^t g^p(t)\mathrm{d}\tau \leqslant f(s) + C(t - s)f^{p\sigma/\rho}(t), \quad 0 < s < t < T_{\max}.$$

Owing to our assumption, $f(t) \to \infty$ as $t \to T_{\max}$. For any given $s \in (0, T_{\max})$, we can select a $t \in (s, T_{\max})$ such that $f(t) = 2f(s)$. Accordingly, we have

$$2f(s) \leqslant f(s) + C(t - s)f^{p\sigma/\rho}(s) \leqslant f(s) + C(T_{\max} - s)f^{p\sigma/\rho}(s), \quad 0 < s < T_{\max},$$

which implies $f(s) \geqslant c(T_{\max} - s)^{-\rho}$ for some constant $c > 0$. The first inequality of (6.74) is obtained, and then the second inequality of that is derived immediately by utilizing (6.67).

(2) When (2a) holds we have

$$h(t) = u(R, t), \quad f(t) = \max_{\overline{B}_R} u(\cdot, t), \quad g(t) = v(R, t).$$

Recalling (6.65) and (6.67), it follows from (6.77) that

$$g(t) \geqslant c \int_s^t \frac{f^q(\tau)}{\sqrt{t - \tau}}\mathrm{d}\tau \geqslant c \int_s^t \frac{g^{q\rho/\sigma}(\tau)}{\sqrt{t - \tau}}\mathrm{d}\tau, \quad 0 < s < t < T_{\max}. \qquad (6.78)$$

We shall use the idea of [49] to prove the second inequality in (6.75). Once this is done, using the second inequality of (6.67), we can get the first inequality in (6.75) immediately.

Denote $\eta = q\rho/\sigma$ and

$$J(t) = \int_s^t \frac{g^\eta(\tau)}{\sqrt{T_{\max} - \tau}}\mathrm{d}\tau.$$

Then $g(t) \geqslant cJ(t)$ by (6.78), and so

$$J'(t) = \frac{g^\eta(t)}{\sqrt{T_{\max} - t}} \geqslant \frac{c^\eta J^\eta(t)}{\sqrt{T_{\max} - t}},$$

which implies

$$\int_{J(t)}^{J(T_{\max})} \frac{\mathrm{d}J}{J^\eta} \geqslant c^\eta \int_t^{T_{\max}} \frac{\mathrm{d}t}{\sqrt{T_{\max} - t}} = 2c^\eta \sqrt{T_{\max} - t}.$$

Note that $\eta > 1$ due to $pq > 1$. The above inequality leads to

$$\frac{J^{1-\eta}(t)}{\eta - 1} \geqslant 2c^\eta \sqrt{T_{\max} - t} + \frac{J^{1-\eta}(T_{\max})}{\eta - 1} \geqslant 2c^\eta \sqrt{T_{\max} - t},$$

and thereafter,

$$J(t) \leqslant C(T_{\max} - t)^{-1/[2(\eta-1)]}. \tag{6.79}$$

On the other hand, for $t^* = 2t - T_{\max}$ (we assume that t is close to $T_{\max}$ here),

$$J(t) \geqslant \int_{t^*}^t \frac{g^\eta(\tau)}{\sqrt{T_{\max} - \tau}} \mathrm{d}\tau \geqslant g^\eta(t^*) \int_{t^*}^t \frac{\mathrm{d}\tau}{\sqrt{T_{\max} - \tau}} = C' g^\eta(t^*) \sqrt{T_{\max} - t}.$$

Combining this inequality with (6.79), we get

$$g(t^*) \leqslant C_1 (T_{\max} - t)^{-1/[2(\eta-1)]} = C_2 (T_{\max} - t^*)^{-1/[2(\eta-1)]},$$

and it is exactly the second inequality of (6.75) due to $1/[2(\eta - 1)] = \sigma$.

When (2b) or (2c) holds, we first use ideas of [35, 47] to prove the second inequality in (6.75), and then get the first inequality in (6.75) by using the second inequality of (6.67).

For any $t_0 \in (T_{\max}/2, T_{\max})$, define

$$t_0^+ := \max\{t \in (t_0, T_{\max}) : g(t) = 2g(t_0)\}$$

and $\lambda(t_0) = g^{-1/(2\sigma)}(t_0)$. We claim

$$\lambda^{-2}(t_0)(t_0^+ - t_0) \leqslant M, \quad T_{\max}/2 < t_0 < T_{\max} \tag{6.80}$$

for some constant $M > 0$ which is independent of t_0. If (6.80) was not valid, then there would exist $t_i \to T_{\max}^-$ satisfying $\lambda^{-2}(t_i)(t_i^+ - t_i) \to \infty$. Same as the proof of Lemma 6.17, we can choose $(\hat{x}_i, \hat{t}_i) \in \partial B_R \times (0, t_i]$ for which $v(\hat{x}_i, \hat{t}_i) = g(t_i)$. Define $\varphi_i(y, s)$ and $\psi_i(y, s)$ as in (6.69). Then (6.70) holds, and (6.71) becomes

$$\psi_i(0,0) = 1, \quad 0 \leqslant \psi_i(y,s) \leqslant \lambda^{2\sigma}(t_i)g(t_i^+) = 2, \quad y \in \Omega_i, \ s \in I_i(t_i^+),$$

$$0 \leqslant \varphi_i(y,s) \leqslant \lambda^{2\rho}(t_i)f(t_i^+) \leqslant \varepsilon^{-\rho} 2^{\rho/\sigma}, \quad y \in \Omega_i, \ s \in I_i(t_i^+),$$

where (6.67) was used in the last inequality. Same as the proof of Lemma 6.17, we can find two $C^{2,1}$ functions $\varphi(y, s)$ and $\psi(y, s)$ to solve (6.73) which satisfy $\psi(0,0) = 1$,

$\psi \geqslant 0$ and $\varphi \geqslant 0$. Since the assumption (2b) or (2c) holds, by Lemma 6.18, (φ, ψ) blows up in finite time. It is a contradiction, and hence (6.80) holds.

Now we use an idea from [47, Lemma 3.1] to deduce the second inequality of (6.75). Fix $t_0 \in (T_{\max}/2, T_{\max})$, and denote $t_1 = t_0^+$, $t_2 = t_1^+, \ldots, t_{i+1} = t_i^+, \ldots$. Then $g(t_{i+1}) = 2g(t_i)$, and $t_{i+1} - t_i \leqslant Mg^{-1/\sigma}(t_i)$ by (6.80). Clearly, $t_i \to T_{\max}$ and

$$
T_{\max} - t_0 = \sum_{i=0}^{\infty}(t_{i+1} - t_i) \leqslant M \sum_{i=0}^{\infty} g^{-1/\sigma}(t_i)
$$

$$
= Mg^{-1/\sigma}(t_0) \sum_{i=0}^{\infty} 2^{-i/\sigma} = Cg^{-1/\sigma}(t_0).
$$

This implies the second inequality of (6.75). □

In the study of blowup rates, blowup sets, blowup profiles and blowup boundary layers of blowup solutions for parabolic systems, a basic prerequisite is that all components of the solution must blow up simultaneously. This phenomenon is called simultaneous blowup. This book does not cover this aspect. Interested readers can refer to [100, 94, 67, 76] and the references therein.

EXERCISES

6.1 For problem (6.27), prove that if initial data $u_0, v_0 \in C^2(\overline{\Omega})$ satisfy

$$
-\Delta u_0 \leqslant u_0^\rho v_0^p, \quad -\Delta v_0 \leqslant u_0^q v_0^\sigma, \quad x \in B_R(0),
$$

then $u_t, v_t \geqslant 0$.

6.2 Prove (6.46).

6.3 Let $f(s)$ be a continuous function in $s \geqslant 0$ and $f(s) > 0$. Prove that if the integral $\displaystyle\int_0^\infty \frac{ds}{f(s)} < \infty$ then any solution u of

$$
\begin{cases}
u_t - \Delta u = f(u), & x \in \Omega, \ t > 0, \\
\partial_{\boldsymbol{n}} u = 0, & x \in \partial\Omega, \ t > 0, \\
u(x, 0) = u_0(x) \geqslant 0, \not\equiv 0, \ x \in \Omega
\end{cases}
$$

must blow up in finite time.

6.4 (Kalantarov-Ladyzhenskaya [53]) Suppose that

$$
f''(t)f(t) - (1 + \rho)(f'(t))^2 > af(t)f'(t) - bf^2(t), \quad t > 0
$$

with constants $\rho, a, b > 0$. Prove that if

$$
f(0) > 0, \quad f'(0) + \frac{a - \sqrt{a^2 + 4\rho b}}{2\rho} f(0) > 0,
$$

then $f(t)$ blows up in finite time.

6.5 Consider the following problem

$$\begin{cases} u_t - \Delta u = |u|^{p-1}u, & x \in \Omega, \ t > 0, \\ u = 0, & x \in \partial\Omega, \ t > 0, \\ u(x,0) = u_0(x), & x \in \Omega, \end{cases}$$

where $p > 1$. Investigate finite time blowup of solution by use of the eigenfunction method, the energy method and the concavity method, respectively, and compare their advantages and disadvantages.

6.6 Let $p > q > 1$ and $b > 0$. Consider following problem

$$\begin{cases} u_t - \Delta u = u^p - bu^q, & x \in \Omega, \quad t > 0, \\ u = 0, & x \in \partial\Omega, \ t > 0, \\ u(x,0) = u_0(x) \geqslant 0, & x \in \Omega, \end{cases}$$

where $u_0 \in C^2(\overline{\Omega})$ and $u_0 = 0$ on $\partial\Omega$. Please study global existence and finite time blowup of solution (using the eigenfunction method, the energy method and the concavity method, respectively).

6.7 Let $p, q \geqslant 1$ and $pq > 1$. Investigate global existence and finite time blowup of solution of

$$\begin{cases} u_t - \Delta u = v^p - u, & x \in \Omega, \quad t > 0, \\ v_t - \Delta v = u^q - v, & x \in \Omega, \quad t > 0, \\ u = v = 0, & x \in \partial\Omega, \quad t > 0, \\ u(x,0) \geqslant 0, \quad v(x,0) \geqslant 0, & x \in \Omega. \end{cases}$$

CHAPTER 7

Time-Periodic Parabolic Boundary Value Problems

"It seems at first that this fact [the existence of periodic solutions] could not be of any practical interest whatsoever [however] what renders these periodic solutions so precious is that they are, so to speak, the only breach through which we may try to penetrate a stronghold previously reputed to be impregnable."

—Henri Poincaré

The importance of periodic solutions in understanding behaviours of ordinary differential equations is well known. In recent decades, time-periodic solutions of partial differential equations have gained prominence, and have been studied by employing a variety of methods. This chapter focuses on the time-periodic parabolic boundary value problems. The upper and lower solutions method for time-periodic parabolic boundary value problems, and the principal eigenvalue of periodic parabolic eigenvalue problems form the bulk of the chapter. As applications, a diffusive logistic equation (including the unbounded domain case) and a diffusive competition model are systematically studied. For the diffusive logistic equation, the existence and uniqueness of positive solutions are established; in the bounded domain case, it is proved that the unique positive solution is globally asymptotically stable in the periodic sense. For the diffusive competition model, the existence of positive solutions is also established.

Throughout this chapter we always assume that coefficients of $\mathscr{L}$ are time-periodic with period T and satisfy the condition $(\mathbf{C})$ given in §1.1, and Ω is a bounded domain of class $C^{2+\alpha}$.

7.1   THE UPPER AND LOWER SOLUTIONS METHOD FOR SCALAR EQUATIONS

We first study the following *time-periodic parabolic boundary value problem*:

$$\begin{cases} \mathscr{L}u = f(x,t,u) & \text{in } Q_T, \\ \mathscr{B}u = 0 & \text{on } S_T, \\ u(x,0) = u(x,T) & \text{in } \Omega, \end{cases} \tag{7.1}$$

where $\mathscr{B}u = u$ or $\mathscr{B}u = \partial_n u + bu$ with $b \geqslant 0$ and $b \in C^{1+\alpha,\,(1+\alpha)/2}(\overline{S}_T)$, f and b are *time-periodic* with period T.

Nonlinear periodic diffusion equations of the form (7.1) arise naturally, such as in problems stemming from population ecology, where data depends periodically on time (seasonal or daily variations, for example).

The basis of comparison principle is the maximum principle. We have realized the importance and roles of comparison principle in the study of parabolic problems. For time-periodic parabolic boundary value problems, there is also the maximum principle which takes the following version.

Theorem 7.1 (Maximum principle) *Assume* $c \geqslant 0$, *and* $c \not\equiv 0$ *when* $\mathscr{B} = \partial_n$. *Let* $u \in C(\overline{Q}_T) \cap C^{2,1}(Q_T)$ *when* $\mathscr{B}u = u$ *and* $u \in C^{1,0}(\overline{Q}_T) \cap C^{2,1}(Q_T)$ *when* $\mathscr{B}u = \partial_n u + bu$. *If* u *satisfies*

$$\begin{cases} \mathscr{L}u \geqslant 0 & in \ \ Q_T, \\ \mathscr{B}u \geqslant 0 & on \ \ S_T, \\ u(x,0) \geqslant u(x,T) & in \ \ \Omega, \end{cases}$$

then $u \geqslant 0$ *in* Q_T. *Moreover, if one of the above three inequalities is not identical, then* $u > 0$ *in* Q_T.

Proof. Suppose, to the contrary, that $u(x_0, t_0) = \min_{\overline{Q}_T} u < 0$. We may assume $t_0 > 0$ on account of $u(x_0, T) \leqslant u(x_0, 0)$. Using the Hopf boundary lemma, we can conclude $x_0 \in \Omega$. Then the strong maximum principle leads to $u \equiv u(x_0, t_0)$ in Q_{t_0}. Certainly, $u(x,0) \equiv u(x_0, t_0)$ in Ω. By the assumption, $u(x, T) \leqslant u(x,0) \equiv u(x_0, t_0)$ in Ω. As a consequence, $t_0 = T$ and $u \equiv u(x_0, t_0) < 0$ in Q_T. Note that $c \geqslant 0$, and $c \not\equiv 0$ when $\mathscr{B} = \partial_n$. This is impossible. Hence $u \geqslant 0$ in Q_T.

If there is $(x_0, t_0) \in Q_T$ such that $u(x_0, t_0) = 0 = \min_{\overline{Q}_T} u$, then $u \equiv 0$ in Q_{t_0} by Corollary 1.25. Hence $u(x, T) \equiv u(x, 0) \equiv 0$ in Ω due to $u(x, 0) \geqslant u(x, T) \geqslant 0$ in Ω, and in turn $u \equiv 0$ in Q_T. This indicates that $\mathscr{L}u \equiv 0$ in Q_T and $\mathscr{B}u \equiv 0$ on S_T. $\square$

Existence and uniqueness of solutions of initial-boundary value problem for linear equation and the comparison principle are foundations of establishing the upper and lower solutions method of initial-boundary value problem for semilinear equations. However, existence and uniqueness of solutions are not completely clear for time-periodic parabolic boundary value problem of linear equations. To establish the upper and lower solutions method of (7.1), we must adopt a different approach from initial-boundary value problems. Based on the maximum principle (Theorem 7.1), we define upper and lower solutions of (7.1) as follows.

Definition 7.2 *Let* $u \in C(\overline{Q}_T) \cap C^{2,1}(Q_T)$ *when* $\mathscr{B}u = u$ *and* $u \in C^{1,0}(\overline{Q}_T) \cap C^{2,1}(Q_T)$ *when* $\mathscr{B}u = \partial_n u + bu$. *Such a function* u *is referred to as an* upper solution (*a* lower solution) *of* (7.1) *if*

$$\begin{cases} \mathscr{L}u \geqslant (\leqslant) f(x, t, u) & in \ \ Q_T, \\ \mathscr{B}u \geqslant (\leqslant) 0 & on \ \ S_T, \\ u(x, 0) \geqslant (\leqslant) u(x, T) & in \ \ \Omega. \end{cases}$$

In order to set up the upper and lower solutions method, it is generally sufficient for initial-boundary value problems to have upper and lower solutions, and for boundary value problems of elliptic equations to have ordered upper and lower solutions. However, these are not enough for time-periodic parabolic boundary value problem (7.1), and a new hypothesis is needed, that is, there must be a better function between $\underline{u}(x,0)$ and $\bar{u}(x,0)$.

We set

$$W_{p,\mathscr{B}}^2(\Omega) = \mathring{W}_p^1(\Omega) \cap W_p^2(\Omega)$$

when $\mathscr{B}u = u$, and

$$W_{p,\mathscr{B}}^2(\Omega) = \left\{ u \in W_p^2(\Omega) : \left[\partial_{\boldsymbol{n}} u + b(x,0)u\right]_{\partial\Omega} = 0 \right\}$$

when $\mathscr{B}u = \partial_{\boldsymbol{n}} u + bu$.

Theorem 7.3 (The upper and lower solutions method) *Suppose that* $\bar{u} \geqslant \underline{u}$ *are an upper solution and a lower solution of* (7.1), *respectively. Let* $f(\cdot,u) \in C^{\alpha,\alpha/2}(\overline{Q}_T)$ *be uniform with respect to* $u \in [\underline{c},\bar{c}]$ *and* $f_u \in C(\overline{Q}_T \times [\underline{c},\bar{c}])$, *where* $\underline{c} = \min_{\overline{Q}_T} \underline{u}$ *and* $\bar{c} = \max_{\overline{Q}_T} \bar{u}$. *When* $\mathscr{B}u = u$, *we further assume* $f(x,0,0) = 0$ *on* $\partial\Omega$. *Fix* $p > 1 + n/2$. *If there exists at least one* $u_0 \in W_{p,\mathscr{B}}^2(\Omega)$ *satisfying* $\underline{u}(x,0) \leqslant u_0(x) \leqslant \bar{u}(x,0)$, *then problem* (7.1) *has at least one solution* $u \in C^{2+\alpha,1+\alpha/2}(\overline{\Omega} \times [0,T])$ *which satisfies*

$$\underline{u} \leqslant u \leqslant \bar{u} \quad in \ \overline{Q}_T.$$

Proof. The idea here is inspired by [46, Section 21]. We only deal with the case $\mathscr{B}u = u$.

Step 1. Define a set

$$V = \left\{ u_0 \in W_{p,\mathscr{B}}^2(\Omega) : \underline{u}(x,0) \leqslant u_0(x) \leqslant \bar{u}(x,0) \ \text{in} \ \Omega, \ \|u_0\|_{W_p^2(\Omega)} \leqslant C(T,\underline{c},\bar{c}) \right\},$$

where $C(T,\underline{c},\bar{c})$ is a positive constant which will be given in the following (7.4). Then V is a bounded and closed convex set of $W_p^2(\Omega)$ and $V \neq \emptyset$. For any given $u_0 \in V$, let us investigate the problem

$$\begin{cases} \mathscr{L}u = f(x,t,u) & \text{in} \ Q_T, \\ u = 0 & \text{on} \ S_T, \\ u(x,0) = u_0(x) & \text{in} \ \Omega. \end{cases} \tag{7.2}$$

Obviously, $\bar{u}$ and $\underline{u}$ are an upper solution and a lower solution of (7.2), respectively. Thus (7.2) has a unique solution $u \in W_p^{2,1}(Q_T)$ which satisfies $\underline{u} \leqslant u \leqslant \bar{u}$. It follows that

$$u(x,T) \geqslant \underline{u}(x,T) \geqslant \underline{u}(x,0) \quad \text{and} \quad u(x,T) \leqslant \bar{u}(x,T) \leqslant \bar{u}(x,0).$$

Set $F(x,t) = f(x,t,u(x,t))$. Then $F \in C^{\alpha,\alpha/2}(\overline{Q}_T)$ as $u \in W_p^{2,1}(Q_T) \hookrightarrow C^{\alpha,\alpha/2}(\overline{Q}_T)$. Using Theorem 2.10, we conclude

$$u \in W_p^{2,1}(Q_T) \cap C^{2+\alpha,1+\alpha/2}(\overline{\Omega} \times (0,T]).$$

This leads to $u(x,T) \in \overset{\circ}{W}^1_p(\Omega) \cap W^2_p(\Omega) = W^2_{p,\mathscr{B}}(\Omega)$. Define an operator $\mathscr{F} : V \to W^2_{p,\mathscr{B}}(\Omega) \cap \langle \underline{u}(x,0), \bar{u}(x,0) \rangle$ by

$$\mathscr{F}(u_0) = u(x,T).$$

In the following, we will prove that $\mathscr{F} : V \to V$ is compact.

Step 2. Thanks to $\underline{u} \leqslant u \leqslant \bar{u}$, it follows

$$|F(x,t)| \leqslant C(\underline{c},\bar{c}) := \max_{\overline{Q}_T \times [\underline{c},\bar{c}]} |f(x,t,s)| \quad \text{for all} \ \ u_0 \in V.$$

In view of Theorem 1.8, for any $0 < \delta < T$, there exists $C(\delta) > 0$ so that

$$\|u\|_{W^{2,1}_p(\Omega \times (\delta,T))} \leqslant C(\delta) \left(\|F\|_{p,Q_T} + \|u\|_{p,Q_T}\right) \leqslant C(\delta,\underline{c},\bar{c}) \quad \text{for all} \ \ u_0 \in V.$$

The embedding theorem affirms

$$|u|_{\alpha,\overline{\Omega} \times [\delta,T]} \leqslant C(\delta,\underline{c},\bar{c}) \quad \text{for all} \ \ u_0 \in V,$$

which implies

$$|F|_{\alpha,\overline{\Omega} \times [\delta,T]} \leqslant C(\delta,\underline{c},\bar{c}) \quad \text{for all} \ \ u_0 \in V.$$

For any $0 < \delta < \tau < T$, taking advantage of Theorem 2.10 and Theorem 1.14 (2), we can find a constant $C(\delta,\tau)$ such that

$$|u|_{2+\alpha,\overline{\Omega} \times [\tau,T]} \leqslant C(\delta,\tau) \left(|F|_{\alpha,\overline{\Omega} \times [\delta,T]} + \|u\|_{\infty,\Omega \times (\delta,T)}\right)$$
$$\leqslant C(\delta,\tau,\underline{c},\bar{c}) \quad \text{for all} \ \ u_0 \in V.$$

Certainly,

$$\|\mathscr{F}(u_0)\|_{C^{2+\alpha}(\overline{\Omega})} = \|u(x,T)\|_{C^{2+\alpha}(\overline{\Omega})} \leqslant C(\delta,\tau,\underline{c},\bar{c}) \quad \text{for all} \ \ u_0 \in V. \tag{7.3}$$

Take $\delta = T/3$ and $\tau = T/2$. Then we have

$$\|u(x,T)\|_{W^2_p(\Omega)} \leqslant C'\|u(x,T)\|_{C^{2+\alpha}(\overline{\Omega})} \leqslant C(T,\underline{c},\bar{c}) \quad \text{for all} \ \ u_0 \in V. \tag{7.4}$$

This shows that $\mathscr{F} : V \to V$.

Step 3. Now, we show that $\mathscr{F} : V \to V$ is compact. Recalling the above estimate (7.3), it is sufficient to prove that $\mathscr{F} : V \to V$ is continuous.

For $u_{i0} \in V$ with $u_{i0} \to u_0$ in $W^2_p(\Omega)$ as $i \to \infty$, let u_i be the unique solution of (7.2) with $u_i(x,0) = u_{i0}(x)$ and $w_i = u_i - u$. Write

$$f(x,t,u_i) - f(x,t,u) = w_i \int_0^1 f_u(x,t,u+sw_i)\mathrm{d}s =: -c_i w_i.$$

Then $c_i \in L^\infty(Q_T)$ and w_i satisfies

$$\begin{cases} \mathscr{L}w_i + c_i w_i = 0 & \text{in} \ \ Q_T, \\ w_i = 0 & \text{on} \ \ S_T, \\ w_i(x,0) = u_{i0}(x) - u_0(x) & \text{in} \ \ \Omega, \end{cases}$$

and in accordance with Theorem 1.1,

$$\|w_i\|_{W_p^{2,1}(Q_T)} \leqslant C\|u_{i0} - u_0\|_{W_p^2(\Omega)},$$

which implies

$$|u_i - u|_{\alpha, \overline{Q}_T} \leqslant C\|u_{i0} - u_0\|_{W_p^2(\Omega)} \to 0 \quad \text{as} \quad i \to \infty.$$

It follows that $\left\{|u_i|_{\alpha, \overline{Q}_T}\right\}$ is bounded, and so is $\left\{|f(\cdot, u_i(\cdot))|_{\alpha, \overline{Q}_T}\right\}$. We have known that $f(\cdot, u_i(\cdot)) \to f(\cdot, u(\cdot))$ in $C(\overline{Q}_T)$, and as a result of compactness, $f(\cdot, u_i(\cdot)) \to f(\cdot, u(\cdot))$ in $C^{\gamma, \gamma/2}(\overline{Q}_T)$ for any $0 < \gamma < \alpha$. The application of Schauder theory (Theorem 1.14 (2)) deduces

$$|u_i - u|_{2+\gamma, \overline{\Omega} \times [T/3, T]} \leqslant C\left(|f(x, t, u_i) - f(x, t, u)|_{\gamma, \overline{Q}_T} + \|u_i - u\|_{\infty, Q_T}\right) \to 0.$$

Certainly, $|u_i(x, T) - u(x, T)|_{2+\gamma, \overline{\Omega}} \to 0$, and then $\mathscr{F} : V \to V$ is continuous.

Step 4. Applying the Schauder fixed point theorem (Corollary A.3), $\mathscr{F}$ has at least one fixed point $u_0 \in V$, i.e., solution u of (7.2) with such u_0 as its initial datum satisfies $u(x, T) = u_0(x) = u(x, 0)$. Therefore, u is a solution of (7.1). Moreover, since $u = \mathscr{F}(u_0) \in C^{2+\alpha, 1+\alpha/2}(\overline{\Omega} \times (0, T])$, we see that, with $g = 0$,

$$u(x, T) = g(x, T) \quad \text{for} \quad x \in \partial\Omega,$$

$$g_t - a_{ij}D_{ij}u + b_iD_iu + cu = f(x, t, u) \quad \text{for} \quad x \in \partial\Omega, \quad t = T.$$

These imply the following compatibility conditions with $\varphi(x) = u(x, 0)$:

$$\varphi(x) = g(x, 0) \quad \text{for} \quad x \in \partial\Omega,$$

$$g_t - a_{ij}D_{ij}\varphi + b_iD_i\varphi + c\varphi = f(x, t, u) \quad \text{for} \quad x \in \partial\Omega, \quad t = 0$$

because $u(x, 0) = u(x, T)$ and f is time-periodic with period T. Thus, we have $u \in C^{2+\alpha, 1+\alpha/2}(\overline{Q}_T)$ by the Schauder theory. □

We should mention that the uniqueness in Theorem 7.3 is not valid. This is the most difference from initial-boundary value problems of parabolic equations. For time-periodic parabolic boundary value problems, some good properties of initial-boundary value problems may not hold, and however, some properties of elliptic boundary value problems are inherited.

From Step 2 in the proof of the following Theorem 7.10 we shall see that if a lower solution $\underline{u}$ (an upper solution $\bar{u}$) satisfies the compatibility condition at $t = 0$ and $x \in \partial\Omega$, then we can take $\underline{u}(x, 0)$ $(\bar{u}(x, 0))$ as initial datum to construct an iteration sequence $z(x, t+iT)$ so that $z(x, t+iT) \to u(x, t)$ as $i \to \infty$ and $u(x, t)$ is the minimal (maximal) solution of (7.1) located between $\underline{u}$ and $\bar{u}$.

7.2 TIME-PERIODIC PARABOLIC EIGENVALUE PROBLEMS

In this section, we will briefly discuss the *time-periodic parabolic eigenvalue problems*

$$\begin{cases} \mathscr{L}\phi = \lambda\phi & \text{in} \quad Q_T, \\ \mathscr{B}\phi = 0 & \text{on} \quad S_T, \\ \phi(x, 0) = \phi(x, T) & \text{in} \quad \Omega, \end{cases} \tag{7.5}$$

where the operator $\mathscr{B}$ is defined as at the first paragraph of §7.1, and $b = b(x)$ is independent of t. Some contents of this part refer to [46]. The notion of principal eigenvalue for time-periodic parabolic boundary value problems was introduced by Lazer [61] (see also [14]).

Same as elliptic operators, a number λ (real or complex) is called an *eigenvalue* of (7.5) if problem (7.5) has a non-trivial solution for such λ, and the corresponding non-trivial solution is called the corresponding *eigenfunction* to λ. A real eigenvalue λ is called *principal eigenvalue* of (7.5) if its corresponding eigenfunction is positive.

Theorem 7.4 ([46, Proposition 14.4]) *The principal eigenvalue λ_1 of (7.5) exists uniquely. Furthermore, if $c \geqslant 0$, and $c \not\equiv 0$ when $\mathscr{B} = \partial_{\boldsymbol{n}}$, then $\lambda_1 > 0$. In the case $c = 0$ and $\mathscr{B} = \partial_{\boldsymbol{n}}$, we have $\lambda_1 = 0$.*

Define two spaces $\mathbb{X}_0$, $\mathbb{X}_2$:

$$\mathbb{X}_0 = \left\{ w \in C^{\alpha,\alpha/2}(\overline{Q}_T) : w \text{ is } T\text{-periodic in time } t \right\},$$

$$\mathbb{X}_2 = \left\{ w \in C^{2+\alpha,1+\alpha/2}(\overline{Q}_T) : \mathscr{B}w = 0 \text{ on } S_T, \text{ and } w \text{ is } T\text{-periodic in time } t \right\},$$

as well as an operator L:

$$D(L) = \mathbb{X}_2, \quad L = \mathscr{L} : D(L) \to \mathbb{X}_0.$$

The eigenvalue problem (7.5) can be formulated as an abstract eigenvalue problem

$$L\phi = \lambda\phi \quad \text{in } \mathbb{X}_0. \tag{7.6}$$

We next study variation of the principal eigenvalue with respect to the coefficient c, the boundary condition and the domain Ω. The following theorem will play an important role in the study of variation of principal eigenvalue.

Theorem 7.5 ([46, Theorem 16.6]) *For the inhomogeneous problem*

$$L\phi - \lambda\phi = h, \quad h \in \mathbb{X}_0, \ h \geqslant 0, \not\equiv 0, \tag{7.7}$$

let λ_1 be the principal eigenvalue of (7.6) and assume $\lambda_1 > 0$.

(1) *If $0 < \lambda < \lambda_1$, then problem (7.7) has a unique solution $\phi \in \mathbb{X}_0$ and $\phi > 0$.*

(2) *If $\lambda \geqslant \lambda_1$, then (7.7) has no positive solution, and no solution at all for $\lambda = \lambda_1$.*

We first explore the dependence of the principal eigenvalue on coefficient c.

Theorem 7.6 (Monotonicity of principal eigenvalue) *Let $\lambda_1 = \lambda_1(c)$ be the principal eigenvalue of (7.6). Then $\lambda_1(c)$ is strictly increasing in c.*

Proof. Suppose $c_1 \leqslant c_2$ and $c_1 \not\equiv c_2$. Let ϕ_i be the corresponding positive eigenfunction to $\lambda_1(c_i)$, $i = 1, 2$. We may think that $\lambda_1(c_1) > 0$ and $\lambda_1(c_2) > 0$. Assume conversely that $\lambda_1(c_1) \geqslant \lambda_1(c_2)$. Write $\mathscr{L} = \mathscr{L}(c) := L_0 + c$. Then we conclude

$$
\begin{aligned}
\mathscr{L}(c_2)(\phi_1 - \phi_2) &= L_0\phi_1 + c_2\phi_1 - \lambda_1(c_2)\phi_2 \\
&\geqslant, \not\equiv L_0\phi_1 + c_1\phi_1 - \lambda_1(c_2)\phi_2 \\
&= \lambda_1(c_1)\phi_1 - \lambda_1(c_2)\phi_2 \\
&\geqslant \lambda_1(c_2)(\phi_1 - \phi_2),
\end{aligned}
$$

namely

$$
\mathscr{L}(c_2)(\phi_1 - \phi_2) - \lambda_1(c_2)(\phi_1 - \phi_2) =: h \geqslant 0, \not\equiv 0.
$$

This is a contradiction with Theorem 7.5 (2). $\qquad\square$

In order to emphasize the dependence of principal eigenvalue on boundary conditions, when $\mathscr{B}u = u$ we write $L = L_D$ and $\lambda_1 = \lambda_1^D$, when $\mathscr{B}u = \partial_n u$ we write $L = L_N$ and $\lambda_1 = \lambda_1^N$, and when $\mathscr{B}u = \partial_n u + b(x)u$ with $b \in C^{1+\alpha}(\partial\Omega)$ and $b \geqslant 0, \not\equiv 0$ we write $L = L_R$ and $\lambda_1 = \lambda_1^R$.

Theorem 7.7 (Monotonicity of principal eigenvalue) *Relations $\lambda_1^N < \lambda_1^R < \lambda_1^D$ hold.*

Proof. Adding a multiple of ϕ on both sides of (7.6), we may assume $c > 0$ in $\overline{Q}_T$, and then $\lambda_1^N, \lambda_1^R, \lambda_1^D > 0$. Let ϕ_R be the corresponding positive eigenfunction to λ_1^R. Then

$$
L_R\phi_R = \lambda_1^R\phi_R \quad \text{and} \quad \lambda_1^R\phi_R > 0 \quad \text{in } Q_T,
$$

and $\phi_R > 0$ on S_T. Take $0 < \varepsilon < \min\{\lambda_1^R, \lambda_1^D, \min_{\overline{Q}_T} c\}$. Then $(\lambda_1^R - \varepsilon)\phi_R > 0$. Applying Theorem 7.5 (1), we see that inhomogeneous problem

$$
L_D v - \varepsilon v = (\lambda_1^R - \varepsilon)\phi_R
$$

has a unique positive solution v. Obviously $v = 0$ on S_T. Thus, the function $z = \phi_R - v$ satisfies

$$
\begin{cases}
\mathscr{L}z - \varepsilon z = 0 & \text{in } Q_T, \\
z > 0 & \text{on } S_T, \\
z(x, 0) = z(x, T) & \text{in } \Omega.
\end{cases}
$$

As $c(x, t) - \varepsilon > 0$, by Theorem 7.1, $z > 0$, i.e., $\phi_R > v$ in Q_T. Therefore, $L_D v - \varepsilon v = (\lambda_1^R - \varepsilon)\phi_R > (\lambda_1^R - \varepsilon)v$, which implies that v solves

$$
L_D v - \lambda_1^R v =: h > 0, \quad v > 0.
$$

By Theorem 7.5 (2), it holds $\lambda_1^R < \lambda_1^D$.

The proof of $\lambda_1^N < \lambda_1^R$ is left to readers as an exercise. $\qquad\square$

Finally, we discuss the dependence of the principal eigenvalue on domain Ω.

Theorem 7.8 (Monotonicity of principal eigenvalue) *Suppose that $\Omega_1 \subset \Omega_2$ and $\Omega_1 \neq \Omega_2$. Let $L_D(\Omega_i)$ be the corresponding operator in Ω_i with the homogeneous Dirichlet boundary condition $(\mathscr{B}u = u)$, and $\lambda_1^D(\Omega_i)$ be the principal eigenvalue of $L_D(\Omega_i)$, $i = 1, 2$. Then $\lambda_1^D(\Omega_1) > \lambda_1^D(\Omega_2)$.*

Proof. Let ϕ_i be the corresponding positive eigenfunction to $\lambda_1^D(\Omega_i)$, $i = 1, 2$. Similar to the proof of Theorem 7.7, we may assume that $c > 0$ in $\overline{\Omega}_2 \times [0, T]$ and $\lambda_1^D(\Omega_i) > 0$, $i = 1, 2$. Then $\phi_2 \geq 0$, $\not\equiv 0$ on $\partial\Omega_1 \times [0, T]$, and

$$L_D(\Omega_2)\phi_2 = \lambda_1^D(\Omega_2)\phi_2 \quad \text{and} \quad \lambda_1^D(\Omega_2)\phi_2 > 0 \quad \text{in } \Omega_2 \times [0, T].$$

As above, we take $0 < \varepsilon < \min\{\lambda_1^D(\Omega_1), \lambda_1^D(\Omega_2), \min_{\overline{\Omega}_2 \times [0,T]} c\}$. Then $(\lambda_1^D(\Omega_2) - \varepsilon)\phi_2 > 0$. In view of Theorem 7.5 (1), the inhomogeneous problem

$$L_D(\Omega_1)v - \varepsilon v = (\lambda_1^D(\Omega_2) - \varepsilon)\phi_2$$

has a unique positive solution v. Certainly, $v = 0$ on $\partial\Omega_1 \times (0, T]$. As a result, the function $z = \phi_2 - v$ satisfies

$$\begin{cases} \mathscr{L}z - \varepsilon z = 0 & \text{in } \Omega_1 \times (0, T], \\ z \geq 0, \not\equiv 0 & \text{on } \partial\Omega_1 \times (0, T], \\ z(x, 0) = z(x, T) & \text{in } \Omega_1. \end{cases}$$

As $c(x, t) - \varepsilon > 0$, Theorem 7.1 gives $z > 0$, i.e., $\phi_2 > v$ in $\Omega_1 \times (0, T]$. Therefore, $L_D(\Omega_1)v - \varepsilon v = (\lambda_1^D(\Omega_2) - \varepsilon)\phi_2 > (\lambda_1^D(\Omega_2) - \varepsilon)v$. This shows that v is a positive solution of

$$L_D(\Omega_1)v - \lambda_1^D(\Omega_2)v =: h > 0.$$

According to Theorem 7.5 (2), $\lambda_1^D(\Omega_2) < \lambda_1^D(\Omega_1)$. □

In the following we consider a special case

$$\begin{cases} \phi_t - d\phi_{xx} - a(x, t)\phi = \lambda\phi, & 0 < x < \ell, \ 0 < t \leq T, \\ \mathscr{B}\phi(0, t) = \phi(\ell, t) = 0, & 0 < t \leq T, \\ \phi(x, 0) = \phi(x, T), & 0 < x < \ell, \end{cases} \tag{7.8}$$

where $\ell > 0$, $\mathscr{B}u = \beta u - (1 - \beta)u_x$ and β is a constant satisfying $0 \leq \beta \leq 1$. The function $a \in C^{\alpha,\alpha/2}(\overline{Q}_T)$ is time-periodic with period T.

Theorem 7.9 *The principal eigenvalue of (7.8) exists uniquely, denoted by $\lambda(\ell; d, a)$. And the following statements hold:*

(1) $\lambda(\ell; d, a)$ *is strictly monotone decreasing in a and ℓ;*

(2) *when $\beta = 1$ we denote $\lambda(\ell; d, a)$ by $\lambda^D(\ell; d, a)$. Then $\lambda(\ell; d, a) \leq \lambda^D(\ell; d, a)$ for all $0 \leq \beta \leq 1$;*

(3) $\lim_{d\to\infty} \lambda(\ell; d, a) = \lim_{\ell\to 0^+} \lambda(\ell; d, a) = \infty$.

Proof. The proof of existence and uniqueness of $\lambda(\ell; d, a)$ is similar to Theorem 7.4. Theorem 7.5 holds for problem (7.8). Conclusions (1) and (2) can be proved by the same ways as those of Theorems 7.6–7.8.

In the following we prove the conclusion (3). Define $\hat{a} = \max_{[0,\ell] \times [0,T]} a(x, t)$. Then $\lambda(\ell; d, a) \geqslant \lambda(\ell; d, \hat{a})$ since $\lambda(\ell; d, a)$ is monotone decreasing in a. Because $\hat{a}$ is a constant, we know that $\lambda(\ell; d, \hat{a})$ is the principal eigenvalue of

$$\begin{cases} -d\phi'' - \hat{a}\phi = \lambda\phi, & 0 < x < \ell, \\ \mathscr{B}\phi(0) = \phi(\ell) = 0, \end{cases}$$

and $\lim_{d \to \infty} \lambda(\ell; d, \hat{a}) = \infty$. Thereafter, $\lim_{d \to \infty} \lambda(\ell; d, a) = \infty$.

Define $\tilde{a} = \max_{[0,1] \times [0,T]} a(x, t)$. Similarly to above, $\lambda(\ell; d, a) \geqslant \lambda(\ell; d, \tilde{a})$ for all $0 < \ell \leqslant 1$, and $\lambda(\ell; d, \tilde{a})$ is the principal eigenvalue of elliptic problem

$$\begin{cases} -d\phi'' - \tilde{a}\phi = \lambda\phi, & 0 < x < \ell, \\ \mathscr{B}\phi(0) = \phi(\ell) = 0. \end{cases}$$

As $\lim_{\ell \to 0^+} \lambda(\ell; d, \tilde{a}) = \infty$, we have $\lim_{\ell \to 0^+} \lambda(\ell; d, a) = \infty$. $\qquad\square$

7.3 THE LOGISTIC EQUATION

In this section, we will use the principle eigenvalue and the upper and lower solutions method of time-periodic parabolic boundary value problem, together with the comparison principle of initial-boundary value problem to study the time-periodic problem of logistic equation. Some contents of this part take from [113].

7.3.1 The case of bounded domain

In this subsection we study the *time-periodic parabolic boundary value problem*

$$\begin{cases} u_t - d\Delta u = a(x, t)u - b(x, t)u^2 & \text{in } Q_T, \\ \mathscr{B}u = 0 & \text{on } S_T, \\ u(x, 0) = u(x, T) & \text{in } \Omega, \end{cases} \tag{7.9}$$

where $\mathscr{B}u = \beta u + (1 - \beta)\partial_{\boldsymbol{n}} u$ with constant $0 \leqslant \beta \leqslant 1$. We assume that functions $a, b \in C^{\alpha, \alpha/2}(\overline{Q}_T)$ are time-periodic with period T, $b_1 \leqslant b \leqslant b_2$ in $\overline{Q}_T$ for some positive constants b_1 and b_2.

Let $\lambda(a)$ be the principal eigenvalue of the following *time-periodic parabolic eigenvalue problem*

$$\begin{cases} \phi_t - d\Delta\phi - a(x, t)\phi = \lambda\phi & \text{in } Q_T, \\ \mathscr{B}\phi = 0 & \text{on } S_T, \\ \phi(x, 0) = \phi(x, T) & \text{in } \Omega. \end{cases} \tag{7.10}$$

Theorem 7.10 *If $\lambda(a) < 0$, then problem (7.9) has a unique positive solution u.*

Moreover, for any $v_0 \in C^1(\overline{\Omega}) \cap W_p^2(\Omega)$ satisfying $\mathscr{B}v_0 = 0$ on $\partial\Omega$ and $v_0 > 0$ in Ω, the unique solution v of the initial-boundary value problem

$$\begin{cases} v_t - d\Delta v = a(x,t)v - b(x,t)v^2 & in \ \ Q_\infty, \\ \mathscr{B}v = 0 & on \ \ S_\infty, \\ v(x,0) = v_0(x) & in \ \ \Omega \end{cases} \tag{7.11}$$

satisfies

$$\lim_{i\to\infty} v(x,t+iT) = u(x,t) \ \ uniformly \ in \ \overline{Q}_T. \tag{7.12}$$

This indicates that the unique positive solution u of (7.9) is globally asymptotically stable in the time-periodic sense.

Proof. The proof will be divided into three steps:

Step 1. The existence. Let ϕ be a positive eigenfunction corresponding to $\lambda(a)$. Take $\varepsilon > 0$, and set $\bar{u} = M$ and $\underline{u} = \varepsilon\phi$. Then $\bar{u}$ and $\underline{u}$ are ordered upper and lower solutions of (7.9) so long as $\varepsilon \ll 1$ and $M \gg 1$. Obviously, $\varepsilon\phi(x,0) \in W_p^2(\Omega)$, $\mathscr{B}(\varepsilon\phi(x,0)) = 0$ on $\partial\Omega$ and $\underline{u}(x,0) \leqslant \varepsilon\phi(x,0) \leqslant \bar{u}(x,0)$ in Ω. According to Theorem 7.3, problem (7.9) has at least one solution u which satisfies $\varepsilon\phi \leqslant u \leqslant M$.

Step 2. We now prove uniqueness. The strategy of this proof is to construct a monotone sequence to get a minimal positive solution and then use a standard comparison method. Let u be a positive solution of (7.9). Invoking Lemma 3.7, there exists $0 < \varepsilon \ll 1$ such that $u(x,0) \geqslant \varepsilon\phi(x,0)$ in $\overline{\Omega}$. Let z be the unique solution of (7.11) with initial datum $\varepsilon\phi(x,0)$. Note that the function $\xi := \varepsilon\phi$ satisfies

$$\xi_t - d\Delta\xi < a(x,t)\xi - b(x,t)\xi^2 \ \ in \ \ Q_T.$$

We have $\varepsilon\phi = \xi \leqslant z \leqslant u$ by the comparison principle. For any integer $i \geqslant 0$, let us define $z^i(x,t) = z(x,t+iT)$ for $(x,t) \in \overline{Q}_T$. Since a and b are time-periodic with period T, we see that z^i satisfies

$$\begin{cases} z_t^i - d\Delta z^i = a(x,t)z^i - b(x,t)(z^i)^2 & in \ \ Q_T, \\ \mathscr{B}z^i = 0 & on \ \ S_T, \\ z^i(x,0) = z(x,iT) & in \ \ \Omega. \end{cases}$$

Note that

$$z(x,0) = \varepsilon\phi(x,0) = \varepsilon\phi(x,T) \leqslant z(x,T) = z^1(x,0) \leqslant u(x,T) = u(x,0).$$

We can adopt the comparison principle to produce $\varepsilon\phi \leqslant z \leqslant z^1 \leqslant u$ in $\overline{Q}_T$ which in turn asserts

$$z^1(x,0) = z(x,T) \leqslant z^1(x,T) = z^2(x,0) \leqslant u(x,T) = u(x,0).$$

As above, $z^1 \leqslant z^2 \leqslant u$ in $\overline{Q}_T$. Applying the inductive method, we can prove that z^i

is monotonically increasing in i and $\varepsilon\phi \leqslant z^i \leqslant u$ in $\overline{Q}_T$ for all i. Consequently, there exists a positive function u_ε such that $z^i \to u_\varepsilon$ pointwisely in $\overline{Q}_T$ as $i \to \infty$. Clearly, $u_\varepsilon(x,0) = u_\varepsilon(x,T)$ since $z^{i+1}(x,0) = z^i(x,T)$. Based on the regularity theory and compactness argument, it can be shown that $z^i \to u_\varepsilon$ in $C^{2,1}(\overline{Q}_T)$, $\varepsilon\phi \leqslant u_\varepsilon \leqslant u$ in $\overline{Q}_T$ and u_ε satisfies the first two equations of (7.9). This shows that u_ε is a positive solution of (7.9).

To prove uniqueness, it suffices to illustrate that $u_\varepsilon = u$. Firstly, we can find a constant $\delta > 0$ so that $u_\varepsilon \geqslant \delta u$ in $\overline{Q}_T$ (refer to the proof of Lemma 3.7). Then the infimum

$$\sigma = \inf_{\substack{x \in \Omega \\ 0 \leqslant t \leqslant T}} \frac{u_\varepsilon(x,t)}{u(x,t)}$$

exists and is positive. Clearly, $\delta \leqslant \sigma \leqslant 1$ and $u_\varepsilon \geqslant \sigma u$ in $\overline{Q}_T$ as we have known $u_\varepsilon \leqslant u$. If we can show $\sigma = 1$, then $u_\varepsilon = u$. Assume on the contrary that $\sigma < 1$. Define $\varphi = u_\varepsilon - \sigma u$. Then $\varphi \geqslant 0$ and

$$\mathscr{B}\varphi = 0 \quad \text{on} \quad S_T, \quad \varphi(x,0) = \varphi(x,T) \quad \text{in} \quad \overline{\Omega}.$$

A straightforward calculation alleges

$$\begin{aligned}
\varphi_t - d\Delta\varphi &= a(x,t)\varphi - b(x,t)\left(u_\varepsilon^2 - \sigma u^2\right) \\
&> a(x,t)\varphi - b(x,t)\left(u_\varepsilon^2 - \sigma^2 u^2\right) \\
&\geqslant a(x,t)\varphi - 2b(x,t)u_\varepsilon\varphi.
\end{aligned}$$

By the maximum principle (Theorem 7.1), $\varphi > 0$ in $\Omega \times [0,T]$. As above, there exists $\rho > 0$ for which $\varphi \geqslant \rho u$, and thus $u_\varepsilon \geqslant (\sigma + \rho)u$ in $\overline{Q}_T$. This is a contradiction with the definition of σ, and thereupon $u_\varepsilon = u$.

Step 3. Proof of (7.12). Let v be the unique positive solution of (7.11). Using $v(x,\tau)$ instead of $v_0(x)$ for the fixed $\tau > 0$, we may assume that $\partial_n v_0 < 0$ on $\partial\Omega$ when $\beta = 1$, and $v_0(x) > 0$ in $\overline{\Omega}$ when $0 \leqslant \beta < 1$. Using Lemma 3.7, we conclude that there exist $0 < \varepsilon \ll 1$ and $k \gg 1$ such that

$$\varepsilon\phi(x,0) \leqslant v_0(x) \leqslant ku(x,0) \quad \text{for all} \ x \in \overline{\Omega}.$$

Clearly, ku is an upper solution of (7.9) as $k > 1$ and $b > 0$. Let z and Z be solutions of (7.11) with initial data $\varepsilon\phi(x,0)$ and $ku(x,0)$, respectively. Then $z \leqslant v \leqslant Z$ in Q_∞ by the comparison principle.

On the other hand, from the discussion of Step 2, we see that

$$\lim_{i \to \infty} z(x,t+iT) = u(x,t) \quad \text{uniformly in} \ \overline{Q}_T.$$

Analogously,

$$\lim_{i \to \infty} Z(x,t+iT) = u(x,t) \quad \text{uniformly in} \ \overline{Q}_T.$$

Recalling $z \leqslant v \leqslant Z$, we get (7.12), and the proof is complete. $\qquad\square$

7.3.2 The case of half line

In this subsection, we study the existence and uniqueness of positive solutions to the following time-periodic parabolic boundary value problem in half line

$$
\begin{cases}
u_t - du_{xx} = a(x,t)u - b(x,t)u^2, & 0 < x < \infty,\ 0 < t \leqslant T, \\
\mathscr{B}u(0,t) = 0, & 0 < t \leqslant T, \\
u(x,0) = u(x,T), & 0 < x < \infty,
\end{cases}
\tag{7.13}
$$

where $\mathscr{B}u = \beta u - (1-\beta)u_x$ and $0 \leqslant \beta \leqslant 1$ is a constant. Define

$$
\Pi_\ell = [0,\ell] \times [0,T] \quad \text{and} \quad \Pi_\infty = [0,\infty) \times [0,T]
$$

for simplicity. We assume that functions a and b satisfy

(I) $a,b \in C^{\alpha,\alpha/2}(\Pi_\infty) \cap L^\infty(\Pi_\infty)$ are time-periodic with period T, a is positive somewhere, and $b_1 \leqslant b \leqslant b_2$ for some positive constants b_1 and b_2.

The strategy in the following arguments is to approximate (7.13) with a time-periodic parabolic boundary value problem in a bounded interval $(0,\ell)$.

We first consider the following problem with $\ell > 0$

$$
\begin{cases}
u_t - du_{xx} = a(x,t)u - b(x,t)u^2, & 0 < x < \ell,\ 0 < t \leqslant T, \\
\mathscr{B}u(0,t) = u(\ell,t) = 0, & 0 < t \leqslant T, \\
u(x,0) = u(x,T), & 0 < x < \ell.
\end{cases}
\tag{7.14}
$$

Let $\lambda(\ell;d,a)$ be the principal eigenvalue of (7.8). Same as Theorem 7.10, we have

Theorem 7.11 *If $\lambda(\ell;d,a) < 0$, then (7.14) has a unique positive solution u_ℓ. Moreover, for any given $v_0 \in C^1([0,\ell]) \cap W_p^2((0,\ell))$ satisfying $\beta v_0(0) - (1-\beta)v_0'(0) = v_0(\ell) = 0$ and $v_0 > 0$ in $(0,\ell)$, the unique solution v of the initial-boundary value problem*

$$
\begin{cases}
v_t - dv_{xx} = a(x,t)v - b(x,t)v^2, & 0 < x < \ell,\ t > 0, \\
\mathscr{B}v(0,t) = v(\ell,t) = 0, & t > 0, \\
v(x,0) = v_0(x), & 0 < x < \ell
\end{cases}
$$

satisfies

$$
\lim_{i \to \infty} v(x,t+iT) = u_\ell(x,t) \quad \text{uniformly in } \Pi_\ell.
$$

Now, we give a sufficient condition to ensure $\lambda(\infty;d,a) := \lim_{\ell \to \infty} \lambda(\ell;d,a) < 0$.

Proposition 7.12 *Assume that the function $a(x,t)$ verifies*

(I1) *There exist constants $\varsigma > 0$, $-2 < \rho \leqslant 0$, $k > 1$ and x_i satisfying $x_i \to \infty$ as $i \to \infty$, so that $a(x,t) \geqslant \varsigma x^\rho$ in $[x_i, kx_i] \times [0,T]$.*

Then $\lambda(\infty;d,a) < 0$ for any $d > 0$.

Proof. Let $\lambda^D(a)$ and $\gamma(a)$, respectively, be the principal eigenvalues of

$$\begin{cases} \phi_t - d\phi_{xx} - a(x,t)\phi = \lambda\phi, & 0 < x < kx_i, \ 0 < t \leqslant T, \\ \phi(0,t) = \phi(kx_i,t) = 0, & 0 < t \leqslant T, \\ \phi(x,0) = \phi(x,T), & 0 < x < kx_i \end{cases}$$

and

$$\begin{cases} \psi_t - d\psi_{xx} - a(x,t)\psi = \gamma\psi, & x_i < x < kx_i, \ 0 < t \leqslant T, \\ \psi(x_i,t) = \psi(kx_i,t) = 0, & 0 < t \leqslant T, \\ \psi(x,0) = \psi(x,T), & x_i < x < kx_i. \end{cases}$$

Applying Theorem 7.9, we conclude

$$\lambda(kx_i; d, a) \leqslant \lambda^D(a) < \gamma(a).$$

Owing to $\rho \leqslant 0$ and $k > 1$, one has

$$a(x,t) \geqslant \varsigma x^\rho \geqslant \varsigma k^\rho x_i^\rho \quad \text{in} \ [x_i, kx_i] \times [0,T].$$

Since $\gamma(a)$ is monotonically decreasing in a, it follows $\gamma(a) \leqslant \gamma(\varsigma k^\rho x_i^\rho)$, and then

$$\lambda(kx_i; d, a) < \gamma(\varsigma k^\rho x_i^\rho) \quad \text{for all} \ i \geqslant 1. \tag{7.15}$$

Let $\psi(x)$ be the positive eigenfunction corresponding to $\gamma(\varsigma k^\rho x_i^\rho)$, and set $y = x/x_i$ and $\Psi(y) = \psi(x)$. Then Ψ satisfies

$$\begin{cases} -dx_i^{-2}\Psi_{yy} - \varsigma k^\rho x_i^\rho \Psi = \gamma(\varsigma k^\rho x_i^\rho)\Psi, & 1 < y < k, \\ \Psi(1) = \Psi(k) = 0. \end{cases}$$

Denote λ^* as the principal eigenvalue of

$$\begin{cases} -u'' = \lambda u, & 1 < y < k, \\ u(1) = u(k) = 0. \end{cases}$$

Then we conclude

$$\gamma(\varsigma k^\rho x_i^\rho) = d\lambda^* x_i^{-2} - \varsigma k^\rho x_i^\rho = x_i^{-2}(d\lambda^* - \varsigma k^\rho x_i^{2+\rho}) < 0 \quad \text{when} \ i \gg 1,$$

since $2 + \rho > 0$ and $x_i \to \infty$ as $i \to \infty$. This combined with (7.15) derives that $\lambda(\infty; d, a) < 0$. □

Theorem 7.13 *Assume that there exist constants $a_0, b_0, a^0, b^0 > 0$ and $-2 < \rho \leqslant 0$ such that*

$$\begin{cases} a_0 = \liminf_{x \to \infty} \dfrac{\min_{[0,T]} a(x,t)}{x^\rho}, & a^0 = \limsup_{x \to \infty} \dfrac{\max_{[0,T]} a(x,t)}{x^\rho}, \\ b_0 = \liminf_{x \to \infty} \min_{[0,T]} b(x,t), & b^0 = \limsup_{x \to \infty} \max_{[0,T]} b(x,t). \end{cases} \tag{7.16}$$

Then (7.13) has a unique positive solution $u \in C^{2+\alpha, 1+\alpha/2}(\Pi_\infty)$. Moreover,

$$\frac{a_0}{b^0} \leqslant \liminf_{x \to \infty} \frac{u(x,t)}{x^\rho} \leqslant \limsup_{x \to \infty} \frac{u(x,t)}{x^\rho} \leqslant \frac{a^0}{b_0} \tag{7.17}$$

uniformly in $[0, T]$.

Proof. Here we only handle the case $0 < \beta < 1$. This proof will be divided into three steps. In the first one, we construct the minimal positive solution of (7.13). The estimate (7.17) will be given in the second step. Finally, we show the uniqueness of positive solutions.

It is worth mentioning that (7.16) implies the assumption **(I1)**.

Step 1. The existence. In this step, we shall construct a positive solution $\underline{u}$ and prove that it is the minimal one.

Let $\ell > 0$ and $\lambda(\ell; d, a)$ be the principal eigenvalue of (7.8). Since the assumption **(I1)** holds, by use of Proposition 7.12, there exists $\ell_0 \gg 1$ so that $\lambda(\ell; d, a) < 0$ when $\ell \geqslant \ell_0$. For each fixed $\ell \geqslant \ell_0$, utilizing Theorem 7.11, problem (7.14) has a unique positive solution u_ℓ. Obviously, $u_\ell \leqslant \|a\|_\infty / b_1$ by the maximum principle, where b_1 is given by **(I)**. When $\ell^* > \ell$, it is easy to see that u_{ℓ^*} is an upper solution of (7.14). Let ϕ be the positive eigenfunction of (7.8) corresponding to $\lambda(\ell; d, a)$ and $\varepsilon > 0$ be a constant. Then $\varepsilon\phi$ is a positive lower solution of (7.14), $\varepsilon\phi \leqslant u_{\ell^*}$ provided $\varepsilon \ll 1$, and $\varepsilon\phi(x, 0) \in W_p^2((0, \ell))$, $\mathscr{B}(\varepsilon\phi(0, 0)) = \varepsilon\phi(\ell, 0) = 0$. Thus $u_{\ell^*} \geqslant u_\ell$ since u_ℓ is the unique positive solution of (7.14). This shows that u_ℓ is monotonically increasing in ℓ. Making use of the regularity theory and compactness argument, it can be proved that there exists a positive function $\underline{u} \in C^{2,1}(\Pi_\infty)$ such that $\lim_{\ell \to \infty} u_\ell = \underline{u}$ in $C^{2,1}(\Pi_L)$ for any $L > 0$ and $\underline{u}$ solves (7.13).

Let u be a positive solution of (7.13). Then $u \leqslant \|a\|_\infty / b_1$ by the maximum principle. Obviously, u is an upper solution of (7.14) for any given $\ell > 0$. As above, $u_\ell \leqslant u$ in Π_ℓ for all $\ell \geqslant \ell_0$, and hence $\underline{u} \leqslant u$ in Π_∞. This shows that $\underline{u}$ is the minimal positive solution of (7.13).

Step 2. Proof of (7.17). For any positive solution u of (7.13), we have known that $\underline{u} \leqslant u \leqslant \|a\|_\infty / b_1$ in Π_∞. Set

$$\underline{a}(x) = \min_{[0,T]} a(x, t), \quad \bar{a}(x) = \max_{[0,T]} a(x, t),$$

$$\underline{b}(x) = \min_{[0,T]} b(x, t), \quad \bar{b}(x) = \max_{[0,T]} b(x, t).$$

Remembering $\lambda(\infty; d, \bar{a}) \leqslant \lambda(\infty; d, a) < 0$ and invoking [22, Theorem 7.12], we have that the problem

$$\begin{cases} -d\bar{w}'' = \bar{a}(x)\bar{w} - \underline{b}(x)\bar{w}^2, & 0 < x < \infty, \\ \bar{w}'(0) = 0 \end{cases}$$

has a unique positive solution $\bar{w}$, and

$$\limsup_{x \to \infty} \frac{\bar{w}(x)}{x^\rho} \leqslant \frac{a^0}{b_0}. \tag{7.18}$$

For $\ell > \ell_0$ and $\theta = \|a\|_\infty / b_1$, where b_1 is given by **(I)**, the problem

$$\begin{cases} -d\bar{w}'' = \bar{a}(x)\bar{w} - \underline{b}(x)\bar{w}^2, & 0 < x < \ell, \\ \bar{w}'(0) = 0, \quad \bar{w}(\ell) = \theta \end{cases} \tag{7.19}$$

has a unique positive solution $\bar{w}_\ell$, and $\bar{w}_\ell$ is globally asymptotically stable. Therefore, the solution v of

$$\begin{cases} v_t - dv_{xx} = \bar{a}(x)v - \underline{b}(x)v^2, & 0 < x < \ell, \, t > 0, \\ v_x(0,t) = 0, \quad v(\ell, t) = \theta, & t \geqslant 0, \\ v(x,0) = \theta, & 0 < x < \ell \end{cases}$$

satisfies

$$\lim_{t \to \infty} v(x,t) = \bar{w}_\ell(x) \text{ uniformly in } [0, \ell].$$

On the other hand, as $u \leqslant \theta$, the comparison principle gives $u \leqslant v$ in $[0, \ell] \times [0, \infty)$. Hence,

$$u(x,t) = \lim_{i \to \infty} u(x, t + iT) \leqslant \lim_{i \to \infty} v(x, t + iT) = \bar{w}_\ell(x) \text{ for all } (x,t) \in \Pi_\ell.$$

Using the upper and lower solutions method for elliptic equations, uniqueness of positive solutions of (7.19), regularity theory for elliptic equations and compactness argument, we can show that $\bar{w}_\ell$ is monotonically decreasing in ℓ and $\lim_{\ell \to \infty} \bar{w}_\ell = \bar{w}$ uniformly in $[0, L]$ for any $L > 0$. Therefore, $u \leqslant \bar{w}$ in Π_∞, and by use of (7.18),

$$\limsup_{x \to \infty} \frac{u(x,t)}{x^\rho} \leqslant \limsup_{x \to \infty} \frac{\bar{w}(x)}{x^\rho} \leqslant \frac{a^0}{b_0} \text{ uniformly in } [0, T]. \tag{7.20}$$

Let us now consider the following problem

$$\begin{cases} -d\underline{w}'' = \underline{a}(x)\underline{w} - \bar{b}(x)\underline{w}^2, & 0 < x < \infty, \\ \underline{w}(0) = 0. \end{cases} \tag{7.21}$$

By virtue of Proposition 7.12, the first eigenvalue of

$$\begin{cases} -d\phi'' - \underline{a}(x)\phi = \lambda\phi, & 0 < x < \ell, \\ \phi(0) = \phi(\ell) = 0 \end{cases}$$

is negative when $\ell \gg 1$. And then, the problem

$$\begin{cases} -d\underline{w}'' = \underline{a}(x)\underline{w} - \bar{b}(x)\underline{w}^2, & 0 < x < \ell, \\ \underline{w}(0) = \underline{w}(\ell) = 0 \end{cases}$$

has a unique positive solution $\underline{w}_\ell$. Same as above, we can show that $\lim_{l \to \infty} \underline{w}_\ell = \underline{w}$ in $C^2([0, L])$ for any $L > 0$, where $\underline{w}$ is a positive solution of (7.21) and $\underline{w}'(0) > 0$. Obviously, $\underline{w}_\ell$ is a lower solution of (7.14). Hence $\underline{w}_\ell \leqslant u_\ell$ since u_ℓ is the unique positive solution of (7.14). This implies $\underline{w} \leqslant \underline{u} \leqslant u$ in Π_∞ on account of $\underline{w}_\ell \to \underline{w}$ and $u_\ell \to \underline{u}$ as $\ell \to \infty$.

If $\underline{w}' > 0$ in $[0, \infty)$, then $\underline{w}(x) \to \underline{w}_*$ as $x \to \infty$ for some positive constant $\underline{w}_*$. As $\rho \leqslant 0$, it is immediate to get

$$\liminf_{x\to\infty} \frac{u(x,t)}{x^\rho} \geqslant \liminf_{x\to\infty} \frac{\underline{w}(x)}{x^\rho} \geqslant \underline{w}_* \tag{7.22}$$

uniformly in $[0, T]$. In this case we claim that $\rho = 0$ and

$$\underline{w}_* \geqslant \frac{\displaystyle\liminf_{x\to\infty} \underline{a}(x)}{\displaystyle\limsup_{x\to\infty} \bar{b}(x)} = \frac{a_0}{b^0}. \tag{7.23}$$

In fact, if $\rho < 0$, then $\lim_{x\to\infty} a(x,t) = 0$ uniformly in $[0,T]$ by (7.16), and consequently $\lim_{x\to\infty} \underline{a}(x) = 0$. This is impossible by noting that function $\bar{b}$ has a positive lower bound, $\underline{w}' > 0$ in $[0, \infty)$ and $\underline{w}$ is bounded from above. Therefore, $\rho = 0$. Now, we are going to prove (7.23). If it was not valid, then there would exist $\varepsilon > 0$ and $x^* \gg 1$ so that $\underline{w}(x) < a_0/b^0 - \varepsilon$ for all $x \geqslant x^*$. For such an $\varepsilon > 0$, we can find $\delta > 0$ and $x_0 > x^*$ such that

$$\frac{a_0 - \delta}{b^0 + \delta} - \frac{a_0}{b^0} + \varepsilon =: \sigma > 0,$$

and $\underline{a}(x) > a_0 - \delta$ and $\bar{b}(x) < b^0 + \delta$ for all $x > x_0$. It then follows

$$-d\underline{w}'' = \underline{w}[\underline{a}(x) - \bar{b}(x)\underline{w}]$$
$$> \underline{w}\bar{b}(x)\left(\frac{\underline{a}(x)}{\bar{b}(x)} - \frac{a_0}{b^0} + \varepsilon\right)$$
$$> \sigma\bar{b}(x)\underline{w} \quad \text{for all } x > x_0,$$

which is impossible since $\sigma > 0$ and $\bar{b}(x)$ has a positive lower bound. Hence, (7.23) holds.

If $\underline{w}'(x_0) = 0$ for some $x_0 > 0$, we set

$$\underline{a}^*(x) = \underline{a}(x + x_0), \ \bar{b}^*(x) = \bar{b}(x + x_0), \ v(x) = \underline{w}(x + x_0) \ \text{ for } x \geqslant 0,$$
$$\underline{a}^*(x) = \underline{a}(x_0 - x), \ \bar{b}^*(x) = \bar{b}(x_0 - x), \ v(x) = \underline{w}(x_0 - x) \ \text{ for } x < 0.$$

Then the function $v(x)$ satisfies

$$-dv'' = \underline{a}^*(x)v - \bar{b}^*(x)v^2 \quad \text{in } \mathbb{R}. \tag{7.24}$$

Applying [22, Theorem 7.12], we conclude that $v(x)$ is the unique positive solution of (7.24) and satisfies

$$\liminf_{x\to\infty} \frac{v(x)}{x^\rho} \geqslant \frac{\displaystyle\liminf_{x\to\infty} \frac{\underline{a}^*(x)}{x^\rho}}{\displaystyle\limsup_{x\to\infty} \bar{b}^*(x)} = \frac{\displaystyle\liminf_{x\to\infty} \frac{\underline{a}(x)}{x^\rho}}{\displaystyle\limsup_{x\to\infty} \bar{b}(x)} = \frac{a_0}{b^0}.$$

Accordingly,

$$\liminf_{x \to \infty} \frac{u(x,t)}{x^\rho} \geqslant \liminf_{x \to \infty} \frac{w(x)}{x^\rho} = \liminf_{x \to \infty} \frac{v(x)}{x^\rho} \geqslant \frac{a_0}{b^0} \tag{7.25}$$

uniformly in $[0, T]$.

It follows from (7.20), (7.22), (7.23) and (7.25) that u satisfies (7.17).

Step 3. The uniqueness. Let u be a positive solution of (7.13). Then $u \geqslant \underline{u}$, here $\underline{u}$ is the minimal positive solution of (7.13) obtained in Step 1. In order to prove uniqueness, it suffices to show that $u \equiv \underline{u}$. If not, then $u \geqslant, \not\equiv \underline{u}$. It follows from (7.17) that there exists $k > 1$ such that $u \leqslant k\underline{u}$ in Π_∞. To arrive at a contradiction, we turn to a technique introduced by Marcus and Véron [81]. Define $z = u - (2k)^{-1}(u - \underline{u})$. Then

$$\underline{u} \geqslant, \not\equiv z \geqslant \frac{k+1}{2k}\underline{u} \quad \text{and} \quad \frac{2k}{2k+1}z + \frac{1}{2k+1}u = \underline{u}. \tag{7.26}$$

Noticing that function y^2 is convex for $y \in \mathbb{R}_+$, we conclude

$$\underline{u}^2 \leqslant \frac{2k}{2k+1}z^2 + \frac{1}{2k+1}u^2$$

from the second formula of (7.26). Since b is positive, a direct computation gives

$$z_t - dz_{xx} \geqslant a(x,t)z - b(x,t)z^2.$$

Obviously, for the large ℓ,

$$\mathscr{B}z(0,t) = 0, \ z(\ell, t) > 0 \text{ in } [0, T] \text{ and } z(x, 0) = z(x, T) \text{ in } [0, \ell].$$

Hence, z is an upper solution of (7.14). Let u_ℓ be the unique positive solution of (7.14). As above, $u_\ell \leqslant z$ in Π_ℓ. As a consequence, $\underline{u} \leqslant z$ in Π_∞ due to the fact that $\lim_{l \to \infty} u_\ell = \underline{u}$ in $C^{2,1}(\Pi_L)$ for any $L > 0$. This contradicts the first inequality of (7.26). Therefore, $u \equiv \underline{u}$, and the uniqueness is derived. □

7.4 THE UPPER AND LOWER SOLUTIONS METHOD FOR SYSTEMS

In this section, we study the upper and lower solutions method for the time-periodic parabolic boundary value problems of weakly coupled parabolic systems.

Let i_k, d_k and $[u]_{i_k}$, $[u]_{d_k}$ be defined as in §4.2.2. Now, we discuss the following *time-periodic parabolic boundary value problem* of *weakly coupled parabolic system*

$$\begin{cases} \mathscr{L}_k u_k = f_k\left(x, t, u_k, [u]_{i_k}, [u]_{d_k}\right) & \text{in } Q_T, \\ \mathscr{B}_k u_k = 0 & \text{on } S_T, \\ u_k(x, 0) = u_k(x, T) & \text{in } \Omega, \\ 1 \leqslant k \leqslant m, \end{cases} \tag{7.27}$$

where $\mathscr{L}_k$ is strongly parabolic operator, coefficients of $\mathscr{L}_k$ are time-periodic with period T and satisfy the condition (**C**) which is given in §1.1, $\mathscr{B}_k = a^k\partial_n + b^k$, either $a^k = 0$ and $b^k = 1$, or $a^k = 1$ and $b^k \geqslant 0$ with $b^k \in C^{1+\alpha, (1+\alpha)/2}(\overline{S}_T)$, f_k and b^k are as well time-periodic with period T.

Definition 7.14 *Given $\bar{u}$ and $\underline{u}$ with $\bar{u}_k, \underline{u}_k \in C(\overline{Q}_T) \cap C^{2,1}(Q_T)$ when $a^k = 0$, and $\bar{u}_k, \underline{u}_k \in C^{1,0}(\overline{Q}_T) \cap C^{2,1}(Q_T)$ when $a^k = 1$. Assume that $\underline{u} \leqslant \bar{u}$ and f is mixed quasi-monotonous in $\langle \underline{u}, \bar{u} \rangle$. We say that $(\bar{u}, \underline{u})$ is a pair of coupled ordered upper and lower solutions of (7.27) if*

$$\begin{cases} \mathscr{L}_k \bar{u}_k \geqslant f_k\left(x, t, \bar{u}_k, [\bar{u}]_{i_k}, [\underline{u}]_{d_k}\right) & in \ Q_T, \\ \mathscr{L}_k \underline{u}_k \leqslant f_k\left(x, t, \underline{u}_k, [\underline{u}]_{i_k}, [\bar{u}]_{d_k}\right) & in \ Q_T, \\ \mathscr{B}_k \bar{u}_k \geqslant 0 \geqslant \mathscr{B}_k \underline{u}_k & on \ S_T, \\ \bar{u}_k(x, 0) \geqslant \bar{u}_k(x, T), \ \underline{u}_k(x, 0) \leqslant \underline{u}_k(x, T) & in \ \Omega, \\ 1 \leqslant k \leqslant m. \end{cases}$$

Sometimes we call that $\bar{u}$ and $\underline{u}$ are coupled ordered upper and lower solutions *of* (7.27).

Theorem 7.15 (The upper and lower solutions method) *Let $\bar{u}$ and $\underline{u}$ be coupled ordered upper and lower solutions of (7.27) and f be mixed quasi-monotonous in $\langle \underline{u}, \bar{u} \rangle$. Assume that $f_k(\cdot, u) \in C^{\alpha, \alpha/2}(\overline{Q}_T)$ are uniform in $u \in \langle \underline{c}, \bar{c} \rangle$ and $\frac{\partial f_k}{\partial u_j} \in C(\overline{Q}_T \times \langle \underline{c}, \bar{c} \rangle)$ for all $1 \leqslant k, j \leqslant m$, where $\underline{c}$ and $\bar{c}$ are defined as in §4.2 (Theorem 4.4). Fix a number $p > 1 + n/2$. If there exists a function $u_0 \in W^2_{p,\mathscr{B}}(\Omega)$ satisfying*

$$\underline{u}(x, 0) \leqslant u_0(x) \leqslant \bar{u}(x, 0) \ \ in \ \overline{\Omega},$$

then (7.27) has at least one solution $u \in [C^{2+\alpha, 1+\alpha/2}(\overline{\Omega} \times [0, T])]^m$, and $\underline{u} \leqslant u \leqslant \bar{u}$, where

$$W^2_{p,\mathscr{B}}(\Omega) = W^2_{p,\mathscr{B}_1}(\Omega) \times \cdots \times W^2_{p,\mathscr{B}_m}(\Omega),$$

and $W^2_{p,\mathscr{B}_k}(\Omega)$ is defined as in §7.1 for $1 \leqslant k \leqslant m$.

Proof. This proof is a combination of proofs of Theorem 7.3 and Theorem 4.9.

Step 1. Define a set

$$V = \left\{ u_0 \in W^2_{p,\mathscr{B}}(\Omega) : \begin{array}{l} \underline{u}(x, 0) \leqslant u_0(x) \leqslant \bar{u}(x, 0) \ \ in \ \overline{\Omega}, \\ \|u_{k0}\|_{W^2_p(\Omega)} \leqslant C(T, \underline{c}, \bar{c}), \ 1 \leqslant k \leqslant m \end{array} \right\},$$

where $C(T, \underline{c}, \bar{c})$ is a positive constant which will be given in (7.29) below. Then V is a bounded and closed convex set of $[W^2_p(\Omega)]^m$ and $V \neq \emptyset$. For any given $u_0 \in V$, let us consider initial-boundary value problem of a weakly coupled system

$$\begin{cases} \mathscr{L}_k u_k = f_k\left(x, t, u_k, [u]_{i_k}, [u]_{d_k}\right) & in \ Q_T, \\ \mathscr{B}_k u_k = 0 & on \ S_T, \\ u_k(x, 0) = u_{k0}(x) & in \ \Omega, \\ 1 \leqslant k \leqslant m. \end{cases} \tag{7.28}$$

It is easily checked that $(\bar{u}, \underline{u})$ is a pair of coupled upper and lower solutions of (7.28). Hence, problem (7.28) has a unique solution u in $\langle \underline{u}, \bar{u} \rangle$ by Theorem 4.9. Certainly

$$\underline{u}(x, 0) \leqslant \underline{u}(x, T) \leqslant u(x, T) \leqslant \bar{u}(x, T) \leqslant \bar{u}(x, 0).$$

The remaining proof is similar to Theorem 7.3, and we shall give details for the completeness.

Step 2. Set $F_k(x,t) = f_k(x,t,u(x,t))$ and $\alpha_0 = \min\{1, 2-(n+2)/p\}$. Since

$$u_l \in W_p^{2,1}(Q_T) \hookrightarrow C^{\alpha,\alpha/2}(\overline{Q}_T) \text{ for all } 1 \leqslant l \leqslant m, \ 0 < \alpha < \alpha_0,$$

it asserts $F_k \in C^{\alpha,\alpha/2}(\overline{Q}_T)$, and hence $u_k \in W_p^{2,1}(Q_T) \cap C^{2+\alpha, 1+\alpha/2}(\overline{\Omega} \times (0,T])$ by Theorem 2.10. Certainly, $u_k(x,T) \in W_{p,\mathscr{B}_k}^2(\Omega)$. Define an operator $\mathscr{F} : V \to W_{p,\mathscr{B}}^2(\Omega) \cap \langle \underline{u}(x,0), \bar{u}(x,0) \rangle$ by

$$\mathscr{F}(u_0) = u(x,T).$$

As $\underline{u} \leqslant u \leqslant \bar{u}$, it deduces

$$|F_k(x,t)| \leqslant C(\underline{c}, \bar{c}) := \max_{\overline{Q}_T \times [\underline{c}, \bar{c}]} |f_k(x,t,u)| \text{ for all } u_0 \in V.$$

In view of Theorem 1.8, for any $0 < \delta < T$, there exists a positive constant $C_k(\delta)$ such that

$$\|u_k\|_{W_p^{2,1}(\Omega \times (\delta,T))} \leqslant C_k(\delta) \left(\|F_k\|_{p,Q_T} + \|u_k\|_{p,Q_T} \right) \leqslant C_k(\delta, \underline{c}, \bar{c}) \text{ for all } u_0 \in V.$$

Then, by the embedding theorem (Theorem A.7),

$$|u_k|_{\alpha, \overline{\Omega} \times [\delta,T]} \leqslant C_k(\delta, \underline{c}, \bar{c}) \text{ for all } u_0 \in V,$$

which implies

$$|F_k|_{\alpha, \overline{\Omega} \times [\delta,T]} \leqslant C_k(\delta, \underline{c}, \bar{c}) \text{ for all } u_0 \in V.$$

Set $\delta = T/3$ and $\tau = T/2$. Taking advantage of Theorem 1.14 (2), there is a constant C_k for which (2.15) holds, i.e.,

$$|u_k|_{2+\alpha, \overline{\Omega} \times [T/2, T]} \leqslant C_k \left(|F_k|_{\alpha, \overline{\Omega} \times [T/3, T]} + \|u_k\|_{\infty, \Omega \times (T/3, T)} \right)$$
$$\leqslant C_k(T, \underline{c}, \bar{c}) \text{ for all } u_0 \in V.$$

In particular,

$$\|u_k(\cdot, T)\|_{C^{2+\alpha}(\overline{\Omega})} \leqslant C_k(T, \underline{c}, \bar{c}) \text{ for all } u_0 \in V, \ 1 \leqslant k \leqslant m,$$

and consequently

$$\|\mathscr{F}(u_0)\|_{C^{2+\alpha}(\overline{\Omega})} = \|u(x,T)\|_{C^{2+\alpha}(\overline{\Omega})} \leqslant C(T, \underline{c}, \bar{c}) \text{ for all } u_0 \in V. \tag{7.29}$$

It follows that $\mathscr{F} : V \to V$.

Step 3. We shall show that $\mathscr{F} : V \to V$ is a compact operator. By the above estimate, it suffices to prove that $\mathscr{F} : V \to V$ is continuous.

Let $u_0^i \in V$ with $u_0^i \to u_0$ in $[W_p^2(\Omega)]^m$ as $i \to \infty$. Denote $u^i := u^i(x,t)$ as the unique solution of (7.28) with $u^i(x,0) = u_0^i(x)$. Set $w^i = u^i - u$ and write

$$f_k(x,t,u^i) - f_k(x,t,u) = \sum_{j=1}^m \frac{\partial f_k}{\partial u_j}(x,t,v^i(x,t))w_j^i =: -\sum_{j=1}^m c_{kj}^i(x,t)w_j^i,$$

where $v^i(x,t)$ is between $u^i(x,t)$ and $u(x,t)$. Clearly, $c^i_{kj} \in L^\infty(Q_T)$. It then follows that

$$\begin{cases} \mathscr{L}_k w^i_k + \displaystyle\sum_{j=1}^m c^i_{kj}(x,t) w^i_j = 0 & \text{in } Q_T, \\ \mathscr{B}_k w^i_k = 0 & \text{on } S_T, \\ w^i_k(x,0) = u^i_{k0}(x) - u_{k0}(x) & \text{in } \Omega, \end{cases}$$

and Theorem 1.20 gives

$$\|w^i_k\|_{W^{2,1}_p(Q_T)} \leqslant C\|u^i_{k0} - u_{k0}\|_{W^2_p(\Omega)}.$$

This implies

$$|u^i - u|_{\alpha, \overline{Q}_T} = |w^i|_{\alpha, \overline{Q}_T} \leqslant C\|u^i_0 - u_0\|_{W^2_p(\Omega)} \to 0,$$

and thereafter $f_k(\cdot, u^i(\cdot)) \to f_k(\cdot, u(\cdot))$ in $C(\overline{Q}_T)$ as $i \to \infty$ for all $1 \leqslant k \leqslant m$.

In almost exactly the same way as the proof of Theorem 7.3, we can verify that $f_k(\cdot, u^i(\cdot)) \to f_k(\cdot, u(\cdot))$ in $C^{\gamma, \gamma/2}(\overline{Q}_T)$ and $|u^i_k - u_k|_{2+\gamma, \overline{\Omega} \times [T/3, T]} \to 0$ as $i \to \infty$ for any $0 < \gamma < \alpha$, which illustrates that $\|u^i_k(x,T) - u_k(x,T)\|_{C^{2+\gamma}(\overline{\Omega})} \to 0$ for all $1 \leqslant k \leqslant m$. Thus $\mathscr{F}: V \to V$ is continuous.

Step 4. Repeating the last part in proving Theorem 7.3, we can show that (7.27) has at least one solution u and u possesses the desired properties. $\qquad\square$

7.5   A DIFFUSIVE COMPETITION MODEL

As an application we study the following time-periodic parabolic boundary value problem of competition model

$$\begin{cases} u_t - d_1\Delta u = a(x,t)u - u^2 - kuv & \text{in } Q_T, \\ v_t - d_2\Delta v = b(x,t)v - v^2 - huv & \text{in } Q_T, \\ \mathscr{B}_1 u = \mathscr{B}_2 v = 0 & \text{on } S_T, \\ u(x,0) = u(x,T), \quad v(x,0) = v(x,T) & \text{in } \Omega, \end{cases} \tag{7.30}$$

where d_i, k and h are positive constants, and $\mathscr{B}_i w = \beta_i w + (1-\beta_i)\partial_{\boldsymbol{n}} w$ with constants $0 \leqslant \beta_i \leqslant 1$, $i = 1, 2$. We assume that functions $a, b \in C^{\alpha, \alpha/2}(\overline{Q}_T)$ are time-periodic with period T. The content of this part refers to [118].

Let $\lambda^i(a)$ denote the principal eigenvalue of (7.10) with $d = d_i$ and $\mathscr{B} = \mathscr{B}_i$, $i = 1, 2$. Define $a_0 = \max_{\overline{Q}_T} a(x,t)$ and $b_0 = \max_{\overline{Q}_T} b(x,t)$.

From discussions of §7.3 we see that the strategy (key point) is to make sure that the principal eigenvalue is negative. The same is reasonable for the model (7.30).

Theorem 7.16   *Assume $a_0, b_0 > 0$ and $\lambda^1(a - kb_0) < 0$, $\lambda^2(b - ha_0) < 0$. Then there exist four positive time-periodic functions $u^*, u_*, v^*, v_* \in C^{2+\alpha, 1+\alpha/2}(\overline{Q}_T)$, for which both (u^*, v_*) and (u_*, v^*) are positive solutions of (7.30). Moreover, any positive solution (u,v) of (7.30) satisfies*

$$u_* \leqslant u \leqslant u^* \quad \text{and} \quad v_* \leqslant v \leqslant v^* \quad \text{in } \overline{Q}_T. \tag{7.31}$$

Proof. The proof will be divided into four steps.

Step 1. Construction of a pair of coupled ordered upper and lower solutions $((\bar{u}, \bar{v}), (\underline{u}, \underline{v}))$ of (7.30). First of all, $\lambda^1(a) < \lambda^1(a - kb_0) < 0$ since $b_0 > 0$. Based on Theorem 7.10, the following problem

$$\begin{cases} u_t - d_1 \Delta u = a(x, t)u - u^2 & \text{in } Q_T, \\ \mathscr{B}_1 u = 0 & \text{on } S_T, \\ u(x, 0) = u(x, T) & \text{in } \Omega \end{cases} \tag{7.32}$$

has a unique positive solution, denoted by $\bar{u}$. Moreover, the maximum principle gives $\bar{u} \leqslant a_0$ in $\overline{Q}_T$. Noticing $\lambda^2(b - ha_0) < 0$, the following problem

$$\begin{cases} v_t - d_2 \Delta v = [b(x, t) - ha_0]v - v^2 & \text{in } Q_T, \\ \mathscr{B}_2 v = 0 & \text{on } S_T, \\ v(x, 0) = v(x, T) & \text{in } \Omega \end{cases} \tag{7.33}$$

admits a unique positive solution, denoted by $\underline{v}$. Similarly, the following problems

$$\begin{cases} v_t - d_2 \Delta v = b(x, t)v - v^2 & \text{in } Q_T, \\ \mathscr{B}_2 v = 0 & \text{on } S_T, \\ v(x, 0) = v(x, T) & \text{in } \Omega \end{cases} \tag{7.34}$$

and

$$\begin{cases} u_t - d_1 \Delta u = [a(x, t) - kb_0]u - u^2 & \text{in } Q_T, \\ \mathscr{B}_1 u = 0 & \text{on } S_T, \\ u(x, 0) = u(x, T) & \text{in } \Omega \end{cases} \tag{7.35}$$

have unique positive solutions $\bar{v}$ and $\underline{u}$, respectively, and $\bar{v} \leqslant b_0$. Using the upper and lower solutions method and uniqueness of positive solutions for the logistic equation, we can prove $\underline{u} \leqslant \bar{u}$ and $\underline{v} \leqslant \bar{v}$. Taking into account of (7.32)–(7.35), it is easy to verify that $((\bar{u}, \bar{v}), (\underline{u}, \underline{v}))$ is a pair of coupled ordered upper and lower solutions of (7.30). Theorem 7.15 avers that problem (7.30) has at least one positive solution (u, v) satisfying $\underline{u} \leqslant u \leqslant \bar{u}$ and $\underline{v} \leqslant v \leqslant \bar{v}$.

Step 2. Construction of (u^*, v_*). Our intention is to use functions $\bar{u}$ and $\underline{v}$ obtained above to construct a monotone sequence $\{(w^i, z^i)\}$ and then determine its limit.

Let (w, z) be the unique solution of the initial-boundary value problem

$$\begin{cases} w_t - d_1 \Delta w = w\,(a(x, t) - w - kz) & \text{in } Q_\infty, \\ z_t - d_2 \Delta z = z\,(b(x, t) - z - hw) & \text{in } Q_\infty, \\ \mathscr{B}_1 w = \mathscr{B}_2 z = 0 & \text{on } S_\infty, \\ w(x, 0) = \bar{u}(x, 0), \quad z(x, 0) = \underline{v}(x, 0) & \text{in } \Omega. \end{cases} \tag{7.36}$$

Then w and z are positive, and the comparison principle gives

$$\underline{u} \leqslant w \leqslant \bar{u} \quad \text{and} \quad \underline{v} \leqslant z \leqslant \bar{v} \quad \text{in } Q_\infty.$$

For any chosen non-negative integer i, we define

$$w^i(x,t) = w(x,t+iT) \quad \text{and} \quad z^i(x,t) = z(x,t+iT).$$

Noting that a and b are time-periodic functions with period T, it is easy to see that (w^i, z^i) satisfies

$$\begin{cases} w_t^i - d_1\Delta w^i = w^i\left(a(x,t) - w^i - kz^i\right) & \text{in } Q_T, \\ z_t^i - d_2\Delta z^i = z^i\left(b(x,t) - z^i - hw^i\right) & \text{in } Q_T, \\ \mathscr{B}_1 w^i = \mathscr{B}_2 z^i = 0 & \text{on } S_T, \\ w^i(x,0) = w(x,iT), \quad z^i(x,0) = z(x,iT) & \text{in } \Omega. \end{cases}$$

Remembering

$$w^1(x,0) = w(x,T) \leqslant \bar{u}(x,T) = \bar{u}(x,0) = w(x,0),$$
$$z^1(x,0) = z(x,T) \geqslant \underline{v}(x,T) = \underline{v}(x,0) = z(x,0),$$

the comparison principle declares $w^1 \leqslant w$ and $z^1 \geqslant z$ in $\overline{Q}_T$. And then

$$w^2(x,0) = w^1(x,T) \leqslant w(x,T) = w^1(x,0),$$
$$z^2(x,0) = z^1(x,T) \geqslant z(x,T) = z^1(x,0).$$

As above, $w^2 \leqslant w^1$ and $z^2 \geqslant z^1$ in $\overline{Q}_T$. Utilizing the inductive method, we can show that w^i and z^i are, respectively, monotonically decreasing and monotonically increasing in i. As a consequence, limits $\lim_{i\to\infty} w^i(x,t) = u^*(x,t)$ and $\lim_{i\to\infty} z^i(x,t) = v_*(x,t)$ exist. In addition, $u^*, v_* > 0$ in Q_T. Obviously,

$$u^*(x,0) = u^*(x,T) \quad \text{and} \quad v_*(x,0) = v_*(x,T) \quad \text{for } x \in \Omega,$$

since $w^{i+1}(x,0) = w^i(x,T)$ and $z^{i+1}(x,0) = z^i(x,T)$. By use of the regularity theory and compactness argument, it can be proved that $w^i \to u^*$ and $z^i \to v_*$ in $C^{2,1}(\overline{Q}_T)$ as $i \to \infty$, and (u^*, v_*) satisfies the first three equations of (7.30). Therefore, (u^*, v_*) is a solution of (7.30).

Step 3. Construction of (u_*, v^*). The idea is similar to Step 2.

Let (φ, ψ) be the unique solution of the initial-boundary value problem

$$\begin{cases} \varphi_t - d_1\Delta\varphi = \varphi\left(a(x,t) - \varphi - k\psi\right) & \text{in } Q_\infty, \\ \psi_t - d_2\Delta\psi = \psi\left(b(x,t) - \psi - h\varphi\right) & \text{in } Q_\infty, \\ \mathscr{B}_1\varphi = \mathscr{B}_2\psi = 0, & \text{on } S_\infty, \\ \varphi(x,0) = \underline{u}(x,0), \quad \psi(x,0) = \bar{v}(x,0) & \text{in } \Omega. \end{cases}$$

Then φ and ψ are positive. Similar to above,

$$\lim_{i\to\infty} \varphi(x,t+iT) = u_*(x,t) \quad \text{and} \quad \lim_{i\to\infty} \psi(x,t+iT) = v^*(x,t) \quad \text{in } C^{2,1}(\overline{Q}_T),$$

$u_*, v^* > 0$ in Q_T, and (u_*, v^*) solves (7.30).

Obviously, $u^* \geqslant u_*$ and $v^* \geqslant v_*$.

Step 4. Proof of (7.31). Let (u, v) be a positive solution of (7.30). We only prove $u \leqslant u^*$ and $v \geqslant v_*$, as the proofs of $u \geqslant u_*$ and $v \leqslant v^*$ are similar.

Recall that u satisfies

$$
\begin{cases}
u_t - d_1 \Delta u < a(x,t)u - u^2 & \text{in } Q_T, \\
\mathscr{B}_1 u = 0 & \text{on } S_T, \\
u(x,0) = u(x,T) & \text{in } \Omega,
\end{cases}
$$

and $\bar{u}$ is the unique positive solution of (7.32). We can deduce $u \leqslant \bar{u}$ by the upper and lower solutions method and the uniqueness of $\bar{u}$. Moreover, $u \leqslant a_0$ in $\overline{Q}_T$ by the maximum principle. Consequently, v satisfies

$$
\begin{cases}
v_t - d_2 \Delta v \geqslant [b(x,t) - ha_0]v - v^2 & \text{in } Q_T, \\
\mathscr{B}_2 v = 0 & \text{on } S_T, \\
v(x,0) = v(x,T) & \text{in } \Omega.
\end{cases}
$$

It follows that $v \geqslant \underline{v}$ since $\underline{v}$ is the unique positive solution of (7.33). Noting that (w, z) is the unique positive solution of (7.36), $w(x,0) = \bar{u}(x,0) \geqslant u(x,0)$ and $z(x,0) = \underline{v}(x,0) \leqslant v(x,0)$, we conclude $u \leqslant w$ and $v \geqslant z$ by the comparison principle. Certainly

$$
u(x,t) = u(x, t+iT) \leqslant w(x, t+iT) \to u^*(x,t),
$$
$$
v(x,t) = v(x, t+iT) \geqslant z(x, t+iT) \to v_*(x,t).
$$

The proof is finished. $\qquad\qquad\qquad\qquad\qquad\qquad\qquad\qquad\qquad\qquad\qquad$ □

EXERCISES

7.1 Consider the time-periodic parabolic boundary value problem

$$
\begin{cases}
\mathscr{L}u = f(x,t,u) & \text{in } Q_T, \\
\mathscr{B}u = g(x,t) & \text{on } S_T, \\
u(x,0) = u(x,T) & \text{in } \Omega,
\end{cases}
$$

where $\mathscr{L}$, $\mathscr{B}$ and f are as in (7.1), and g is time-periodic with period T and $g \in W_p^{2,1}(Q_T)$. Establish the upper and lower solutions method.

7.2 Prove $\lambda_1^N < \lambda_1^R$ in Theorem 7.7.

7.3 Consider the time-periodic parabolic boundary value problem

$$
\begin{cases}
u_t - d\Delta u = a(x,t)u - b(x,t)u^k & \text{in } Q_T, \\
u = 0 & \text{on } S_T, \qquad\qquad \text{(P7.1)} \\
u(x,0) = u(x,T) & \text{in } \Omega,
\end{cases}
$$

where $k \geqslant 2$ is an integer, functions $a, b \in C^{\alpha, \alpha/2}(\overline{Q}_T)$ are time-periodic with period T, and $b_1 \leqslant b \leqslant b_2$ in $\overline{Q}_T$ for some positive constants b_1, b_2.

Let $\lambda(a)$ be the principal eigenvalue of (7.10) with boundary condition $\phi = 0$ on S_T. Prove the following conclusions.

(a) Suppose that $k \geqslant 3$ is odd. If $\lambda(a) < 0$, then problem (P7.1) admits a unique positive solution and a unique negative solution. If $\lambda(a) \geqslant 0$, then problem (P7.1) admits only the trivial solution $u = 0$.

(b) Suppose that $k \geqslant 2$ is even. If $\lambda(a) < 0$, then problem (P7.1) has a unique positive solution, and does not have any negative solution. If $\lambda(a) \geqslant 0$, then problem (P7.1) has no positive solution.

(c) Let $\lambda(a) < 0$ and u be the unique positive solution of problem (P7.1). Discuss stability of u in the time-periodic sense.

7.4 Prove that problem (7.19) has a unique positive solution $\bar{w}_\ell$ and $\bar{w}_\ell$ is globally asymptotically stable. Hint: use the upper and lower solutions method to prove that (7.19) has a minimal positive solution, and then use the integration method or the method of Step 3 in proving Theorem 7.13 to prove the uniqueness.

7.5 Complete the proof of Step 4 of Theorem 7.15.

7.6 Use the upper and lower solutions method to discuss the existence of positive solutions to a time-periodic parabolic boundary value problem

$$
\begin{cases}
u_t - \Delta u = u \left(a(x,t) - u - b(x,t) \dfrac{v}{v+\rho} \right) & \text{in } Q_T, \\[2mm]
v_t - \Delta v = v \left(c(x,t) \dfrac{u}{v+\rho} - f(x,t) - \dfrac{w}{w+\sigma} \right) & \text{in } Q_T, \\[2mm]
w_t - \Delta w = w \left(g(x,t) \dfrac{v}{w+\sigma} - h(x,t) \right) & \text{in } Q_T, \\[2mm]
u = v = w = 0 & \text{on } S_T, \\[2mm]
(u, v, w)|_{t=0} = (u, v, w)|_{t=T} & \text{in } \Omega,
\end{cases}
$$

where functions $a, b, c, f, g, h \in C^{\alpha, \alpha/2}(\overline{Q}_T)$ are time-periodic with period T, ρ and σ are positive constants.

CHAPTER 8

Free Boundary Problems from Ecology

Free boundary problems deal with systems of partial differential equations, where a part of the domain boundary is unknown. They arise in many contexts in the real world. Some examples include heat transfer problems with phase-changes such as from liquid to solid, oxidation or damage of surfaces of materials, infiltration of groundwater, distribution of oil in the process of oil exploitation, wound healing, tumor growth, option pricing, spreads of (invasive) species and epidemic. The earliest free boundary model was put forward by Joseph Stefan when he studied the interface between ice and water in the process of icing. Sometimes, this kind of problem is called the *Stefan problem*, and the corresponding *free boundary condition* is called the *Stefan condition*.

This chapter only focuses on free boundary problems from ecological models. The main content includes the existence and uniqueness, regularity and uniform estimates of solutions, longtime behaviour of solutions, criteria and dichotomy for spreading and vanishing, and the asymptotic speed of the free boundary. Examples of ecological models are studied in detail.

Throughout this chapter, for convenience, we use function $u(t,x)$ and the Banach spaces $W_p^{1,2}(Q)$, $C^{(i+\alpha)/2,\,i+\alpha}(Q)$ $(i=0,1,2)$ instead of $u(x,t)$ and $W_p^{2,1}(Q)$, $C^{i+\alpha,\,(i+\alpha)/2}(Q)$, respectively.

Most theoretical approaches are based on or started with single-species models. In consideration of the heterogeneous environment, the following problem

$$
\begin{cases}
u_t - d\Delta u = a(t,x)u - b(t,x)u^2, & t > 0, \ x \in \Omega, \\
\mathscr{B}u = 0, & t \geqslant 0, \ x \in \partial\Omega, \\
u(0,x) = u_0(x), & x \in \Omega
\end{cases}
$$

is a typical model to describe spread, persistence and extinction of new or invasive species and has received an astonishing amount of attention. In this model, $u(t,x)$ represents population density, constant $d > 0$ denotes diffusion (dispersal) rate, $a(t,x)$ and $b(t,x)$ respectively represent intrinsic growth rate and self-limitation coefficient of the species, Ω is a bounded domain of $\mathbb{R}^n$, boundary operator $\mathscr{B} = \beta + (1 - \beta)\partial_{\boldsymbol{n}}$ with constant $0 \leqslant \beta \leqslant 1$, $\boldsymbol{n}$ is the outward unit normal vector of boundary $\partial\Omega$.

In most spreading processes in the natural world, a spreading front can be observed. When a new or invasive species initially occupies a region Ω_0 with density $u_0(x)$, it is natural to expect that as time t increases Ω_0 will evolve into an expanding region $\Omega(t)$ with an expanding front $\partial\Omega(t)$, and initial function $u_0(x)$ will evolve into a positive function $u(t, x)$ vanishing on moving boundary $\partial\Omega(t)$.

8.1 DEDUCTION OF FREE BOUNDARY CONDITIONS

The content of this section refers to the reference [12].

In the process of population range expansion, population density near the propagating front is assumed to be close to zero. According to Fick's first law, diffusion flux of u is $J = -d\nabla u$. Outflow per unit area per unit time through boundary $\partial\Omega(t)$ is $-d\nabla u \cdot \boldsymbol{n}$, where $\boldsymbol{n}$ is the outward unit normal vector of $\partial\Omega(t)$. For a small increment Δt, during the period from t to $t + \Delta t$, individual numbers entering region $\Omega(t + \Delta t) \setminus \Omega(t)$ through boundary $\partial\Omega(t)$ are approximate to $-d\nabla u \cdot \boldsymbol{n} \times \Delta t$ (the contribution of reaction term can be ignored as it is close to zero near the boundary).

As the front enters a new unpopulated environment, pioneering members at the front, with very low population density, are particularly vulnerable (note that since only one species is considered here, some existing interacting species are regarded as part of environment). Thus, it is plausible to assume that as expanding front propagates the population suffers a loss of k units per unit volume at the front.

For simplicity, we assume that k is a constant for a given species in a given homogeneous environment. In order to simplify mathematics, here we only consider one dimensional case, i.e., $n = 1$. Let $x = g(t)$ and $x = h(t)$ be left and right free boundaries, respectively. Set $\Omega(t) = (g(t), h(t))$, then $\partial\Omega(t) = \{g(t), h(t)\}$ and

$$\Omega(t + \Delta t) \setminus \Omega(t) = (g(t + \Delta t), g(t)) \cup (h(t), h(t + \Delta t)).$$

Average densities of u in regions $(g(t + \Delta t), g(t))$ and $(h(t), h(t + \Delta t))$ are

$$\frac{du_x(t, g(t))\Delta t}{g(t) - g(t + \Delta t)} \quad \text{and} \quad \frac{-du_x(t, h(t))\Delta t}{h(t + \Delta t) - h(t)},$$

respectively. Evidently, they tend to

$$-du_x(t, g(t))/g'(t) \quad \text{and} \quad -du_x(t, h(t))/h'(t)$$

as $\Delta t \to 0$, respectively. These limits are known as *diffusion pressures* on free boundaries, which are *average population densities* there. If we assume that such densities near fronts are kept at the "preferred density level", then the above two quantities should be equal to k. We thus have the following *free boundary conditions*

$$g'(t) = -\mu u_x(t, g(t)) \quad \text{and} \quad h'(t) = -\mu u_x(t, h(t)) \quad \text{with } \mu = dk^{-1}.$$

If environments on the left and right are different, then the above free boundary conditions can be written as

$$g'(t) = -\mu_1 u_x(t, g(t)) \quad \text{and} \quad h'(t) = -\mu_2 u_x(t, h(t)) \quad \text{with } \mu_i = dk_i^{-1}.$$

If the preferred density level k is not a constant, then the above free boundary conditions become

$$g'(t) = -\mu_1(t, g(t))u_x(t, g(t)) \quad \text{and} \quad h'(t) = -\mu_2(t, h(t))u_x(t, h(t)).$$

8.2 EXISTENCE AND UNIQUENESS OF SOLUTIONS

In this section, we will establish a general result on existence and uniqueness of solutions. Most materials of this section are taken from [115]. Another method for the existence and uniqueness was given in [25].

We consider the following *free boundary problem*

$$\begin{cases} u_t - du_{xx} = f(t, x, u), & t > 0, \ 0 < x < h(t), \\ \mathscr{B}u(t, 0) = u(t, h(t)) = 0, & t > 0, \\ h'(t) = -\mu u_x(t, h(t)), & t > 0, \\ h(0) = h_0, \ u(0, x) = u_0(x), & 0 < x < h_0, \end{cases} \tag{8.1}$$

where h_0 denotes the size of *initial habitat*, $\mathscr{B}u = \beta u - (1 - \beta)u_x$ with constant $0 \leqslant \beta \leqslant 1$, and initial function $u_0(x)$ satisfies

$$u_0 \in W_p^2((0, h_0)) \ \text{with} \ p > 3, \ u_0 \geqslant 0, \ \not\equiv 0 \ \text{in} \ (0, h_0), \ \text{and} \ \mathscr{B}u_0(0) = u_0(h_0) = 0.$$

Function f satisfies

(J) $f(t, x, 0) \geqslant 0$. For any given $\tau, l, k > 0$, $f \in L^\infty((0, \tau) \times (0, l) \times (0, k))$ and there exists a constant $L_1(\tau, l, k)$ such that, for all $t \in [0, \tau]$, $x \in [0, l]$ and $u, v \in [0, k]$,

$$|f(t, x, u) - f(t, x, v)| \leqslant L_1(\tau, l, k)|u - v|.$$

Denote $h^* = -\mu u_0'(h_0)$ and

$$\mathcal{K} = \left\{ h_0, h^*, \|u_0\|_{W_p^2((0, h_0])} \right\}.$$

In order to save space, for any given interval $I \subset \overline{\mathbb{R}}_+ = [0, \infty)$, we write simply

$$I \times [0, h(t)] = \bigcup_{t \in I} \{t\} \times [0, h(t)], \quad I \times (0, h(t)) = \bigcup_{t \in I} \{t\} \times (0, h(t)).$$

For any given $0 < T < \infty$, we set

$$D_T = (0, T] \times (0, h(t)).$$

And for any given bounded set $Q \subset \mathbb{R}^2$ and constant $0 < \alpha < 1$, we write

$$|\cdot|_{i+\alpha, \overline{Q}} := \|\cdot\|_{C^{(i+\alpha)/2, \, i+\alpha}(\overline{Q})}, \quad i = 0, 1, 2.$$

Theorem 8.1 *Assume that the condition* **(J)** *holds and fix* $0 < \alpha < 1 - 3/p$.

(1) *There is a constant $0 < T \ll 1$ depending only on $\mathcal{K}$ such that (8.1) has at least one solution $(u, h) \in W_p^{1,2}(D_T) \times C^{1+\alpha/2}([0, T])$ and $u > 0$ in D_T, $h'(t) > 0$ in $(0, T]$. Moreover, there exists a constant $C(\mathcal{K}) > 0$ such that*

$$\|u\|_{W_p^{1,2}(D_T)} + |u|_{1+\alpha, \overline{D}_T} + \|h\|_{C^{1+\alpha/2}([0,T])} \leqslant C(\mathcal{K}). \tag{8.2}$$

(2) *If, in addition, f is also locally* Lipschitz *continuous in x, i.e., for any given τ, l, $k > 0$, there exists a constant $L_2(\tau, l, k)$ such that, for all $t \in [0, \tau]$, $x, y \in [0, l]$ and $u \in [0, k]$,*

$$|f(t, x, u) - f(t, y, u)| \leqslant L_2(\tau, l, k)|x - y|, \tag{8.3}$$

then the solution (u, h) of (8.1) is unique.

(3) *Under above assumptions, if we further assume that, for any given $\tau > 0$, $f(\cdot, x, u) \in C^{\alpha/2}([0, \tau])$ uniformly in (x, u) when (x, u) varies in a bounded set of $\mathbb{R}_+ \times \overline{\mathbb{R}}_+$, i.e., for any given τ, l, $k > 0$, there exists a constant $L_3(\tau, l, k)$ such that*

$$\|f(\cdot, x, u)\|_{C^{\alpha/2}([0,\tau])} \leqslant L_3(\tau, l, k) \quad \text{for all } x \in [0, l], \ u \in [0, k], \tag{8.4}$$

then the unique solution $u \in C^{1+\alpha/2, 2+\alpha}((0, T] \times [0, h(t)])$.

What we need to emphasize here is that condition (8.3) is not needed for the uniqueness of solutions of the initial value problem and initial-boundary value problem with known boundary. Condition (8.3) reveals the important difference between free boundary problem and initial value problem and initial-boundary value problem with known boundary.

Proof of Theorem 8.1. (1) We first straighten the free boundary. Let

$$y = x/h(t), \quad v(t, y) = u(t, h(t)y).$$

Then (v, h) satisfies

$$\begin{cases} v_t - \rho(t)v_{yy} - \xi(t)yv_y = f(t, h(t)y, v), & 0 < t \leqslant T, \ 0 < y < 1, \\ [\beta h(t)v - (1 - \beta)v_y](t, 0) = v(t, 1) = 0, & 0 < t \leqslant T, \\ v(0, y) = u_0(h_0 y), & 0 < y < 1 \end{cases} \tag{8.5}$$

and

$$h'(t) = -\mu \frac{1}{h(t)} v_y(t, 1), \quad 0 < t \leqslant T; \quad h(0) = h_0, \tag{8.6}$$

where $\rho(t) = dh^{-2}(t)$ and $\xi(t) = h'(t)/h(t)$. Problem (8.5) is an initial-boundary value problem with fixed boundary. We shall use the fixed point theorem to prove the existence of solution (v, h) of (8.5) and (8.6).

Let $T_1 = \min\left\{1, \frac{h_0}{2(1+h^*)}\right\}$. For $0 < T \leqslant T_1$, we set

$$\Theta_T = \{h \in C^1([0,T]) : h(0) = h_0,\ h'(0) = h^*,\ \|h' - h^*\|_{C([0,T])} \leqslant 1\}.$$

Clearly, Θ_T is a bounded and closed convex set of $C^1([0,T])$. For any given $h \in \Theta_T$, we can extend h along the tangential direction of h at point $t = T$ to a new function, denoted by itself, such that $h \in \Theta_{T_1}$. Therefore, when $h \in \Theta_T$ we have $h \in \Theta_{T_1}$ and

$$|h(t) - h_0| \leqslant T_1 \|h'\|_{C([0,T_1])} \leqslant T_1(1 + h^*) \leqslant h_0/2 \quad \text{for all } t \in [0, T_1],$$

which implies

$$h_0/2 \leqslant h(t) \leqslant 3h_0/2 \quad \text{for all } t \in [0, T_1].$$

For any given $h \in \Theta_T$, using the upper and lower solutions method we can show that there exists $0 < T_* \leqslant T_1$, depending only on $\mathcal{K}$ and the bound of f in $[0, T_1] \times [0, \frac{3h_0}{2}] \times [0, \|u_0\|_\infty + 1]$, such that problem (8.5) has a unique solution $v(t, y) = v(t, y; h)$, $v \in W_p^{1,2}(\Delta_{T_*}) \hookrightarrow C^{(1+\alpha)/2, 1+\alpha}(\Delta_{T_*})$ and

$$\|v\|_{W_p^{1,2}(\Delta_{T_*})} + |v|_{1+\alpha, \Delta_{T_*}} \leqslant C(\mathcal{K}, T_*, T_*^{-1}),$$

where

$$\Delta_{T_*} = [0, T_*] \times [0, 1].$$

Note that the function f is known and the bound of f in $[0, T_1] \times [0, \frac{3h_0}{2}] \times [0, \|u_0\|_\infty + 1]$ depends only on $\mathcal{K}$. We may think that T_* depends only on $\mathcal{K}$ and rewrite $C(\mathcal{K}, T_*, T_*^{-1})$ as $C_1(\mathcal{K})$. Therefore, when $0 < T \leqslant T_*$, the unique solution $v(t, y)$ of (8.5) satisfies

$$\|v\|_{W_p^{1,2}(\Delta_T)} + |v|_{1+\alpha, \Delta_T} \leqslant \|v\|_{W_p^{1,2}(\Delta_{T_*})} + |v|_{1+\alpha, \Delta_{T_*}} \leqslant C_1(\mathcal{K}). \tag{8.7}$$

Noticing that $v(0, y) \geqslant 0$, $\not\equiv 0$, $f(t, h(t)y, 0) \geqslant 0$ and $f(t, h(t)y, v)$ is Lipschitz continuous in v, we conclude $v > 0$ in $(0, T] \times (0, 1)$ and $v_y(t, 1) < 0$ in $(0, T]$ by Theorem 1.40 (2) and Lemma 1.39.

On the other hand, by continuous dependence on the given data, v continuously depends on $h \in \Theta_T$ in the space $C^{(1+\alpha)/2, 1+\alpha}(\Delta_T)$. For such a definite function v, the initial value problem (8.6) has a unique solution, denoted by $\tilde{h}(t) = \tilde{h}(t; h)$. Then $\tilde{h}(0) = h_0$, $\tilde{h}'(0) = h^*$ and

$$\tilde{h}'(t) > 0, \quad \tilde{h}' \in C^{\alpha/2}([0,T]), \quad \|\tilde{h}'\|_{C^{\frac{\alpha}{2}}([0,T])} \leqslant C_2(\mathcal{K}) \quad \text{for all } h \in \Theta_T. \tag{8.8}$$

Clearly, $\tilde{h}$ continuously depends on $v \in C^{(1+\alpha)/2, 1+\alpha}(\Delta_T)$ in the space $C^1([0,T])$, and so does on $h \in \Theta_T$. Now we define an operator $\mathcal{F} : \Theta_T \to C^1([0,T])$ by

$$\mathcal{F}(h) = \tilde{h}.$$

Obviously, $\mathcal{F}$ is continuous in Θ_T, and $h \in \Theta_T$ is a fixed point of $\mathcal{F}$ if and only if (v, h) solves (8.5).

According to (8.8), we know that $\mathcal{F}$ is compact, and

$$\|\tilde{h}' - h^*\|_{C([0,T])} \leqslant \|\tilde{h}'\|_{C^{\frac{\alpha}{2}}([0,T])} T^{\alpha/2} \leqslant C_2(\mathcal{K}) T^{\alpha/2}.$$

As a result, $\mathcal{F}$ maps Θ_T into itself if

$$T \leqslant \min\left\{1, \frac{h_0}{2(1 + h^*)}, C_2^{-2/\alpha}(\mathcal{K})\right\}.$$

Consequently, $\mathcal{F}$ has at least one fixed point $h \in \Theta_T$ by the Schauder fixed point theorem (Corollary A.3), and then problems (8.5) and (8.6) have at least one solution (v, h) defined in $[0, T]$. Clearly,

$$h \in C^{1+\alpha/2}([0, T]), \quad h'(t) > 0 \ \text{ for } t > 0,$$

and function $u(t, x) = v(t, h^{-1}(t)x)$ satisfies

$$u \in W_p^{1,2}(D_T) \cap C^{(1+\alpha)/2, 1+\alpha}(\overline{D}_T), \quad u > 0 \ \text{ in } \ D_T.$$

Thus (u, h) solves (8.1).

From the above arguments we see that the estimate (8.2) holds.

(2) Let $(u_i, h_i) \in W_p^{1,2}(D_T) \times C^{1+\alpha/2}([0, T])$ $(i = 1, 2)$ be two solutions of (8.1) and $0 < T \ll 1$. Then $u_i > 0$ in D_T, $h_i'(t) > 0$ in $(0, T]$, and (u_i, h_i) satisfies the estimate (8.2). We may assume that

$$h_0 \leqslant h_i \leqslant h_0 + 1 \ \text{ in } [0, T], \quad u_i \leqslant \|u_0\|_\infty + 1 \ \text{ in } [0, T] \times [0, h_i(t)], \quad i = 1, 2.$$

Define

$$v_i(t, y) = u_i\left(t, h_i(t)y\right), \quad 0 \leqslant t \leqslant T, \ 0 \leqslant y \leqslant 1$$

for $i = 1, 2$, and set $v = v_1 - v_2$, $h = h_1 - h_2$. Then

$$\begin{cases} v_t - \rho_1(t)v_{yy} - \xi_1(t)yv_y - a(t, y)v = [\rho_1(t) - \rho_2(t)] v_{2yy} \\ \quad + [\xi_1(t) - \xi_2(t)] yv_{2y} + b(t, y)yh, \quad 0 < t \leqslant T, \ 0 < y < 1, \\ \mathcal{B}_1v(t, 0) = -\beta h v_2(t, 0), \quad v(t, 1) = 0, \quad 0 < t \leqslant T, \\ v(0, y) = 0, \quad 0 < y < 1 \end{cases} \tag{8.9}$$

and

$$\begin{cases} h'(t) = \mu\left(\frac{1}{h_2}v_{2y}(t, 1) - \frac{1}{h_1}v_{1y}(t, 1)\right), \quad 0 < t \leqslant T, \\ h(0) = 0, \end{cases} \tag{8.10}$$

where $\mathcal{B}_1v = \beta h_1(t)v - (1 - \beta)v_y$ and

$$a(t, y) = \frac{f(t, h_1(t)y, v_1(t, y)) - f(t, h_1(t)y, v_2(t, y))}{v_1(t, y) - v_2(t, y)},$$

$$b(t, y) = \frac{f(t, h_1(t)y, v_2(t, y)) - f(t, h_2(t)y, v_2(t, y))}{(h_1(t) - h_2(t))y}.$$

By assumptions on f we see that $a, b \in L^\infty(\Delta_T)$, and $L^\infty(\Delta_T)$ norms of a and b depend only on h_0 and $\|u_0\|_\infty$. Remember $\|v_2\|_{W_p^{1,2}(\Delta_T)} \leqslant C_1(\mathcal{K})$ and $0 \leqslant h_i'(t) \leqslant C(\mathcal{K})$, $h_0 \leqslant h_i(t) \leqslant h_0 + 1$, $i = 1, 2$. Applying Theorem 1.6 to problem (8.9), we achieve

$$\|v\|_{W_p^{1,2}(\Delta_T)} \leqslant C_3(\mathcal{K}) \left(\|(h_1^{-2} - h_2^{-2})v_{2yy}\|_{p,\Delta_T} + \|(h_1'/h_1 - h_2'/h_2)y v_{2y}\|_{p,\Delta_T} \right.$$
$$+ \|byh\|_{p,\Delta_T} + \|\beta h v_2\|_{W_p^{1,2}(\Delta_T)} \right)$$
$$\leqslant C_4(\mathcal{K}) \|h\|_{C^1([0,T])}.$$

Noting the dimension of space is 1, and following discussions of [111, §5.4 and §5.5] for v and v_y without extension of v to a large domain, we can show that

$$[v]_{C^{\alpha/2, \alpha}(\Delta_T)}, \ [v_y]_{C^{\alpha/2, \alpha}(\Delta_T)} \leqslant C \|v\|_{W_p^{1,2}(\Delta_T)}$$

for some positive constant C independent of T^{-1}, where $[\cdot]_{C^{\alpha/2, \alpha}(\Delta_T)}$ is the Hölder semi-norm. As a result,

$$[v_y]_{C^{\alpha/2, \alpha}(\Delta_T)} \leqslant C C_4(\mathcal{K}) \|h\|_{C^1([0,T])}.$$

This combined with (8.10) yields

$$[h']_{C^{\alpha/2}([0,T])} \leqslant \mu [h_1^{-1} v_y(t,1)]_{C^{\alpha/2}([0,T])} + \mu [(h_1^{-1} - h_2^{-1}) v_{2y}(t,1)]_{C^{\alpha/2}([0,T])}$$
$$\leqslant C_5(\mathcal{K}) \|h\|_{C^1([0,T])}.$$

Because of $h(0) = h'(0) = 0$, it is deduced that

$$\|h\|_{C^1([0,T])} \leqslant 2T^{\alpha/2} \|h'\|_{C^{\alpha/2}([0,T])} \leqslant 2C_5(\mathcal{K}) T^{\alpha/2} \|h\|_{C^1([0,T])}.$$

We can choose $0 < \hat{T}(\mathcal{K}) \ll 1$ such that when $0 < T \leqslant \hat{T}(\mathcal{K})$, $h = 0$ and hence $v = 0$.

(3) Recall that f satisfies conditions **(J)**, (8.3) and (8.4). Using (8.7) and (8.8) we see that function $F(t,y) := f(t, h(t)y, v(t,y)) \in C^{\alpha/2, \alpha}(\Delta_T)$. Application of Theorem 2.10 to (8.5) deduces that $v \in C^{1+\alpha/2, 2+\alpha}((0,T] \times [0,1])$. Consequently, $u \in C^{1+\alpha/2, 2+\alpha}((0,T] \times [0, h(t)])$. □

Assume that conditions **(J)** and (8.3) hold. According to Theorem 8.1 (2), we can extend the solution to the right and define

$$T_{\max} = \sup\{T > 0 : (u, h) \in W_p^{1,2}(D_T) \times C^{1+\alpha/2}([0,T]) \text{ solves } (8.1)\}. \quad (8.11)$$

Such a time $T_{\max}$ is called the *maximal existence time* of solution (u, h) of (8.1).

To determine whether $T_{\max}$ is infinity, we first give an estimate of $h'(t)$.

Lemma 8.2 *Assume that conditions* **(J)** *and* (8.3) *hold. Let* $T > 0$, *and* $(u, h) \in W_p^{1,2}(D_T) \times C^{1+\alpha/2}([0,T))$ *be the unique solution of* (8.1) *verifying* $u \leqslant M$ *for some* $M > 0$. *If* f *is bounded in* $\mathbb{R}_+ \times \mathbb{R}_+ \times (0, M)$, *then there exists a positive constant* K, *which depends on* M *but not on* T, *such that*

$$h'(t) \leqslant 2\mu M K =: C(M) \quad in \ [0, T).$$

Proof. The idea of this proof comes from [25, Lemma 2.2]. Set

$$A = \sup_{\mathbb{R}_+ \times \mathbb{R}_+ \times (0,M)} f(t,x,u),$$

and define a comparison function by

$$w(t,x) = M \left[2K(h(t)-x) - K^2(h(t)-x)^2 \right]$$

for some appropriate positive constant $K > 1/h_0$ over the region

$$\Lambda = \{(t,x): \ 0 < t < T, \ h(t) - K^{-1} < x < h(t)\}.$$

First of all, one can easily compute that, for any $(t,x) \in \Lambda$,

$$w_t = 2MK[1 - K(h(t)-x)]h'(t) \geqslant 0, \quad -w_{xx} = 2MK^2.$$

It follows that, when $K^2 \geqslant \frac{A}{2dM}$,

$$w_t - dw_{xx} \geqslant 2dMK^2 \geqslant f(t,x,u) = u_t - du_{xx} \quad \text{in } \Lambda.$$

It is clear that

$$w(t,h(t)-K^{-1}) = M \geqslant u(t,h(t)-K^{-1}), \quad w(t,h(t)) = 0 = u(t,h(t))$$

for all $0 < t < T$. Taking advantage of

$$u_0(x) = -\int_x^{h_0} u_0'(y)dy \leqslant -\min_{[0,h_0]} u_0'(x)(h_0 - x), \ x \in [h_0 - K^{-1}, h_0],$$
$$w(0,x) \geqslant MK(h_0 - x), \ x \in [h_0 - K^{-1}, h_0],$$

we see that if $MK \geqslant -\min_{[0,h_0]} u_0'(x)$, then

$$u_0(x) \leqslant w(0,x) \ \text{in } [h_0 - K^{-1}, h_0].$$

Applying the comparison principle we deduce $w \geqslant u$ in Λ. It then leads to $u_x(t,h(t)) \geqslant w_x(t,h(t)) = -2MK$, and hence $h'(t) \leqslant 2\mu MK$. □

Theorem 8.3 *Assume that conditions* **(J)** *and* (8.3) *hold and f is bounded in* $\mathbb{R}_+ \times \mathbb{R}_+ \times (0,k)$ *for any given $0 < k < \infty$. Let $T_{\max}$ be given by* (8.11). *Then $T_{\max} < \infty$ implies that either*

$$\limsup_{t \to T_{\max}^-} \max_{[0,h(t)]} u(t,\cdot) = \infty,$$

or

$$\lim_{t \to T_{\max}^-} \|u(t,\cdot)\|_{W_p^2((0,h(t)))} = \infty.$$

Proof. Assume that $T_{\max} < \infty$ and there exists a positive constant M such that

$$u \leqslant M \quad \text{in} \quad [0, T_{\max}) \times [0, h(t)].$$

By Lemma 8.2, $h'(t) \leqslant C$ in $(0, T_{\max})$, and then $h_0 \leqslant h(t) \leqslant h_0 + C T_{\max}$. Let $v(t, y) = u(t, h(t)y)$. For any given $T < T_{\max}$, applying the L^p theory (Theorem 1.7) to (8.5) we can get $\|v\|_{W_p^{1,2}(\Delta_T)} \leqslant C_1(T_{\max})$ which is independent of T. It follows that $v \in W_p^{1,2}(\Delta_{T_{\max}})$ and $\|v\|_{W_p^{1,2}(\Delta_{T_{\max}})} \leqslant C_1(T_{\max})$. Therefore, $h \in C^{1+\alpha/2}([0, T_{\max}])$ and $\|h\|_{C^{1+\alpha/2}([0, T_{\max}])} \leqslant C_2(T_{\max})$ by (8.6).

By the Fubini theorem, $v(t, \cdot) \in W_p^2((0, 1))$ for almost $t \in [0, T_{\max}]$. Thus, $u(t, \cdot) \in W_p^2((0, h(t)))$ for almost $t \in [0, T_{\max}]$. We claim that

$$\lim_{t \to T_{\max}^-} \|u(t, \cdot)\|_{W_p^2((0, h(t)))} = \infty.$$

If this was not valid, then there would exist $t_i \in (0, T_{\max})$ with $t_i \nearrow T_{\max}$ and a positive constant C such that

$$\|u(t_i, \cdot)\|_{W_p^2((0, h(t_i)))} \leqslant C \quad \text{for all } i.$$

Regarding t_i and $(u(t_i, x), h(t_i))$ as an initial time and an initial datum, similar to the proof of Theorem 8.1, we can find a constant $0 < T \ll 1$, which depends only on $h(t_i)$, $h'(t_i)$, $\|u(t_i, \cdot)\|_{W_p^2((0, h(t_i)))}$ and $\|f\|_{\infty, \mathbb{R}_+ \times \mathbb{R}_+ \times (0, M+1)}$, such that (8.1) has a unique solution (u_i, h_i) in $[t_i, t_i + T]$. By the uniqueness, $(u, h) = (u_i, h_i)$ for $t_i \leqslant t \leqslant \min\{t_i + T, T_{\max}\}$. This shows that the solution (u, h) of (8.1) can be extended uniquely to $[0, t_i + T]$. Noting that

$$h(t_i), \; h'(t_i) \leqslant C_2(T_{\max}), \quad \|u(t_i, \cdot)\|_{W_p^2((0, h(t_i)))} \leqslant C,$$

and $C_2(T_{\max})$, C do not depend on i, we can choose T independent of i. Clearly, $t_i + T > T_{\max}$ when i is large. This is a contradiction with the definition of $T_{\max}$. □

Theorem 8.4 *Assume that conditions* **(J)**, *(8.3) and (8.4) hold, and f is bounded in $\mathbb{R}_+ \times \mathbb{R}_+ \times (0, k)$ for any given $k > 0$. Let $T_{\max}$ be given by (8.11). If $T_{\max} < \infty$, then*

$$\limsup_{t \to T_{\max}^-} \max_{[0, h(t)]} u(t, \cdot) = \infty.$$

Proof. Assume, on the contrary, that $T_{\max} < \infty$ and there exists M such that

$$u \leqslant M \quad \text{in} \quad [0, T_{\max}) \times [0, h(t)].$$

Let $v(t, y) = u(t, h(t)y)$. Similar to the above, we have $v \in W_p^{1,2}(\Delta_{T_{\max}})$, $h \in C^{1+\alpha/2}([0, T_{\max}])$ and

$$\|v\|_{W_p^{1,2}(\Delta_{T_{\max}})} \leqslant C_1(T_{\max}), \quad \|h\|_{C^{1+\alpha/2}([0, T_{\max}])} \leqslant C_1(T_{\max}).$$

Moreover, we can apply Theorem 2.10 to (8.5) to get $v \in C^{1+\alpha/2, 2+\alpha}((0, T_{\max}] \times$

$[0,1])$, and for any given $0 < \varepsilon < T_{\max}$, there holds $|v|_{2+\alpha, [\varepsilon, T_{\max}] \times [0,1]} \leqslant C_2(\varepsilon, T_{\max})$. Therefore, $u \in C^{1+\alpha/2, 2+\alpha}((0, T_{\max}] \times [0, h(t)])$ and

$$|u|_{2+\alpha, [\varepsilon, T_{\max}] \times [0, h(t)]} \leqslant C_3(\varepsilon, T_{\max}).$$

This shows that solution (u, h) exists in $[0, T_{\max}]$. Choose $t_i \in (0, T_{\max})$ with $t_i \nearrow T_{\max}$, and regard t_i and $(u(t_i, x), h(t_i))$ as an initial time and an initial datum. Similar to the second half of the proof of Theorem 8.3 we can get a contradiction with the definition of $T_{\max}$, and the proof is complete. □

Now, we consider a special case $f(t, x, u) = a(t, x)u - b(t, x)u^2$.

Theorem 8.5 *Let $f(t, x, u) = a(t, x)u - b(t, x)u^2$, and $a, b \in (C^{\alpha/2, 1} \cap L^\infty)(\overline{\mathbb{R}}_+ \times \overline{\mathbb{R}}_+)$ satisfy $b(t, x) \geqslant b_1 > 0$. Then the solution (u, h) of (8.1) exists globally and uniquely. Moreover, $u \in C^{1+\alpha/2, 2+\alpha}((0, T] \times [0, h(t)])$ and*

$$\begin{cases} 0 < u \leqslant \max \{\|u_0\|_\infty, \|a\|_\infty/b_1\} =: M & \text{in } D_\infty, \\ 0 < h'(t) \leqslant 2\mu M K & \text{in } \mathbb{R}_+, \end{cases} \tag{8.12}$$

where $D_\infty = \mathbb{R}_+ \times (0, h(t))$ and

$$K = \max \left\{ \frac{1}{h_0}, \sqrt{\frac{\|a\|_\infty}{2d}}, -\frac{\min_{[0, h_0]} u_0'}{M} \right\}.$$

Proof. Let $T_{\max}$ be the maximal existence time of (u, h). Using the maximum principle directly we have $0 < u \leqslant M$ in $D_{T_{\max}}$, and hence $T_{\max} = \infty$ by Theorem 8.4. Thanks to Lemma 8.2, $h'(t) \leqslant 2\mu M K$ in $[0, \infty)$. □

Assume that problem (8.1) has a unique global solution (u, h). Then the limit $\lim_{t \to \infty} h(t) := h_\infty \leqslant \infty$ exists since $h'(t) > 0$. When $h_\infty = \infty$, we call that u *spreads successfully* or that *spreading* happens. When $h_\infty < \infty$ and $\lim_{t \to \infty} \max_{[0, h(t)]} u(t, \cdot) = 0$, we call that u *vanishes* or that *vanishing* happens.

8.3   REGULARITY AND UNIFORM ESTIMATES

Regularity and uniform estimates are important elements in the study of parabolic partial differential equations. They will be used to study longtime behaviour of solutions and to determine dichotomy for spreading and vanishing. Materials of this section are taken from references [113, 120].

Theorem 8.6 (Regularity and uniform estimate) *Take $f(t, x, u) = a(t, x)u - b(t, x)u^2$. Assume $a, b \in (C^{0,1} \cap L^\infty)(\mathbb{R}_+ \times \mathbb{R}_+)$ and $b \geqslant b_1 > 0$ for some constant b_1. Then the unique global solution (u, h) of (8.1) satisfies*

$$|u|_{1+\alpha, \overline{\mathbb{R}}_+ \times [0, h(t)]} \leqslant C \tag{8.13}$$

for some constant $C > 0$, which implies

$$\|h'\|_{C^{\alpha/2}(\overline{\mathbb{R}}_+)} \leqslant C. \tag{8.14}$$

If, in addition, $a, b \in (C^{\alpha/2,1} \cap L^\infty)(\overline{\mathbb{R}}_+ \times \overline{\mathbb{R}}_+)$, then we have

$$u \in C^{1+\alpha/2,\,2+\alpha}(\mathbb{R}_+ \times [0, h(t)]), \quad h \in C^{1+(1+\alpha)/2}(\mathbb{R}_+), \tag{8.15}$$

and there exists a constant $C > 0$ such that

$$|u|_{2+\alpha,\,[1,\infty)\times[0,h(t)]} \leqslant C. \tag{8.16}$$

Proof. The idea of this proof is the same as that in Theorem 2.11, and we will divide it into several steps.

Step 1. Noticing Theorem 8.5, from the proof of Theorem 8.1 we can see that $u \in C^{(1+\alpha)/2,\,1+\alpha}(\overline{\mathbb{R}}_+ \times [0, h(t)])$ and $h \in C^{1+\alpha/2}(\overline{\mathbb{R}}_+)$.

Step 2. Proof of (8.15). Same as the proof of Theorem 8.1, we define $y = x/h(t)$ and $v(t, y) = u(t, h(t)y)$. Then (v, h) satisfies (8.5) for all $T > 0$. It is easily checked that function $F(t, y) = f(t, h(t)y, v(t, y))$ is in $C^{\alpha/2,\,\alpha}(\overline{\mathbb{R}}_+ \times \overline{\mathbb{R}}_+)$. Recall $h \in C^{1+\alpha/2}(\overline{\mathbb{R}}_+)$. For any given $\tau > 0$ and $0 < \varepsilon \ll 1$, applying Theorem 2.10 to problem (8.5) in $[\varepsilon, \tau] \times [0, 1]$ we have $v \in C^{1+\alpha/2,\,2+\alpha}([\varepsilon, \tau] \times [0, 1])$. Thus $u \in C^{1+\alpha/2,\,2+\alpha}([\varepsilon, \tau] \times [0, h(t)])$. Due to the arbitrariness of ε, one achieves

$$u \in C^{1+\alpha/2,\,2+\alpha}(D'_\tau) \implies u_x \in C^{(1+\alpha)/2,\,1+\alpha}(D'_\tau),$$

where

$$D'_\tau = (0, \tau] \times [0, h(t)].$$

Hence, by condition that $h'(t) = -\mu u_x(t, h(t))$, we have $h' \in C^{(1+\alpha)/2}((0, \tau])$, and thus (8.15) is proved.

Step 3. Proof of (8.13). Applying the L^p estimate and embedding theorem to (8.5) directly, we can easily get an estimate

$$|v|_{1+\alpha,\,[0,2]\times[0,1]} \leqslant C. \tag{8.17}$$

In order to introduce more methods to readers, for two cases that $h_\infty < \infty$ and $h_\infty = \infty$, we shall use different processes.

Case 1: $h_\infty < \infty$. For any chosen integer $i \geqslant 0$, we define $v^i(t, y) = v(t + i, y)$ and $h^i(t) = h(t + i)$. Then v^i satisfies

$$\begin{cases} v^i_t - \rho^i v^i_{yy} - \xi^i y v^i_y = f^i(t, y), & 0 < t \leqslant 3,\ 0 < y < 1, \\ [\beta h^i v^i - (1-\beta)v^i_y](t, 0) = v^i(t, 1) = 0, & 0 < t \leqslant 3, \\ v^i(0, y) = u(i, h(i)y), & 0 < y < 1, \end{cases} \tag{8.18}$$

where $\rho^i = \rho(t + i)$, $\xi^i = \xi(t + i)$ and $f^i(t, y) = f(t + i, h^i(t)y, v^i(t, y))$. Making use of the estimate (8.12) we know that functions v^i, ρ^i, ξ^i and f^i are uniformly bounded, and

$$\omega^i(r) := \max_{0 \leqslant s,t \leqslant 3,\,|s-t| \leqslant r} |\rho^i(s) - \rho^i(t)| \leqslant \frac{4d\mu}{h_0^3} MKr \to 0 \quad \text{as} \quad r \to 0$$

uniformly with respect to i. On the other hand, $\rho^i(t) \geqslant dh_\infty^{-2}$ for all $i \geqslant 0$ according to $h(t) \leqslant h_\infty < \infty$.

Remember boundary conditions of $v^i(t, y)$ at $y = 0, 1$ and $0 < (h^i(t))' \leqslant 2\mu MK$. Choosing $p > 3$, we can apply Theorem 1.8 to problem (8.18), and then use the embedding theorem to get

$$|v^i|_{1+\alpha,\, [1,3] \times [0,1]} \leqslant C.$$

This implies $|v|_{1+\alpha,\, A_i} \leqslant C$, where $A_i = [i+1, i+3] \times [0, 1]$. As these A_i overlap each other and C is independent of i, we have $|v|_{1+\alpha,\, [1,\infty) \times [0,1]} \leqslant C$. This together with (8.17) yields

$$|v|_{1+\alpha,\, [0,\infty) \times [0,1]} \leqslant C.$$

In view of

$$u_x(t, x) = h^{-1}(t) v_y(t, x/h(t)), \quad 0 \leqslant h'(t) \leqslant 2\mu MK,$$

the estimate (8.13) is obtained.

Case 2: $h_\infty = \infty$. The process is to find two integers i_0 and k, and make estimates in $[0, i_0] \times [0, h(t)]$, $[i_0, \infty) \times [0, h(t) - k]$ and $[i_0, \infty) \times [h(t) - k, h(t)]$, respectively.

(a) The estimate in $[0, i_0] \times [0, h(t)]$. For any fixed $i_0 < \infty$, in the same way as Case 1, we can show

$$|u|_{1+\alpha,\, [0,i_0] \times [0,h(t)]} \leqslant C(i_0). \tag{8.19}$$

(b) The estimate in $[i_0, \infty) \times [0, h(t) - k]$. Since $\rho^i(t) = \rho(t + i) \to 0$ as $i \to \infty$, it is useless to study problem (8.18) as above in order to get an estimate of $|u|_{1+\alpha,\, [i_0, \infty) \times [0, h(t) - k]}$. We shall handle u directly.

Same as above, for integer $i \geqslant 0$, we let $u^i(t, x) = u(t + i, x)$ and $h^i(t) = h(t + i)$. Then u^i satisfies

$$\begin{cases} u_t^i - d u_{xx}^i = g^i(t, x), & 0 < t \leqslant 3,\ 0 < x < h^i(t), \\ \mathscr{B}u^i(t, 0) = u^i(t, h^i(t)) = 0, & 0 < t \leqslant 3, \\ u^i(0, x) = u(i, x), & 0 < x < h(i), \end{cases} \tag{8.20}$$

where $g^i(t, x) = f(t + i, x, u^i(t, x))$. According to Theorem 8.5, we know that u^i and g^i are bounded uniformly on i and

$$h^i(t) \leqslant h(i) + 2\mu MKt \leqslant h(i) + 6\mu MK \quad \text{for all } 0 \leqslant t \leqslant 3.$$

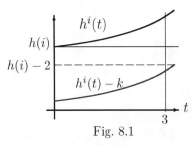

Fig. 8.1

Denote

$$k = 6\mu MK + 2.$$

Then $h(i) - 2 \geqslant h^i(t) - k$ for all $i \geqslant 0$ and $0 \leqslant t \leqslant 3$. As $h_\infty = \infty$, there exists an $i' \geqslant 0$ so that $h(i') > k + 1$ (see Fig. 8.1). In view of (8.19) we only need deal with the u^i for $i \geqslant i'$. Evidently,

$$h^i(t) - 3 \geqslant h(i) - 3 > k - 2 = 6\mu MK \quad \text{for all } 0 \leqslant t \leqslant 3,\ i \geqslant i'.$$

Choose $p > 3$. For any integer $0 \leqslant l \leqslant h(i) - 3$, we can adopt the local L^p estimate (Theorems 1.3 and 1.9) to problem (8.20) and derive that there exists a positive constant C, independent of l and i, such that

$$\|u^i\|_{W_p^{1,2}([1,3]\times[l,l+2])} \leqslant C \quad \text{for all } i \geqslant i'.$$

And then the embedding theorem avers

$$|u^i|_{1+\alpha,\,[1,3]\times[l,l+2]} \leqslant C \quad \text{for all } i \geqslant i', \ 0 \leqslant l \leqslant h(i) - 3.$$

Since these intervals $[l, l+2]$ overlap each other and C is independent of l, we declare

$$|u^i|_{1+\alpha,\,[1,3]\times[0,\,[h(i)]-1]} \leqslant C \quad \text{for all } i \geqslant i',$$

where $[h(i)]$ is an integral part of $h(i)$. The above estimate asserts

$$|u^i|_{1+\alpha,\,[1,3]\times[0,\,h(i)-2]} \leqslant C \quad \text{for all } i \geqslant i'.$$

Note that $h(i) - 2 \geqslant h^i(t) - k$ for $i \geqslant i'$ and $0 \leqslant t \leqslant 3$. The above estimate implies

$$|u^i|_{1+\alpha,\,[1,3]\times[0,\,h^i(t)-k]} \leqslant C \quad \text{for all } i \geqslant i',$$

which implies

$$|u|_{1+\alpha,\,B_i} \leqslant C \quad \text{for all } i \geqslant i',$$

where $B_i = [i+1, i+3] \times [0, h(t) - k]$. As these rectangles B_i overlap each other and C is independent of i, it follows from the above estimates that

$$|u|_{1+\alpha,\,[i_0,\infty)\times[0,\,h(t)-k]} \leqslant C, \tag{8.21}$$

where $i_0 = i' + 1$.

(c) The estimate in $[i_0, \infty) \times [h(t) - k, h(t)]$. We shall show

$$|u|_{1+\alpha,\,[i_0,\infty)\times[h(t)-k,\,h(t)]} \leqslant C. \tag{8.22}$$

Once this is done, joining this with (8.19) and (8.21) we can derive (8.13).

In order to attain the estimate (8.22), we introduce a variable transformation: $y = h(t) - x$, $w(t, y) = u(t, h(t) - y)$. Then w satisfies

$$\begin{cases} w_t - dw_{yy} + h'(t)w_y = F(t, y), & t > 0, \ 0 < y < h(t), \\ w(t, 0) = \mathscr{B}^*w(t, h(t)) = 0, & t > 0, \\ w(0, y) = u(0, h_0 - y), & 0 < y < h_0, \end{cases}$$

where $F(t, y) = f(t, h(t) - y, w(t, y))$, $\mathscr{B}^*w = \beta w + (1 - \beta)w_y$. As above, for every integer $i \geqslant i'$, let $w^i(t, y) = w(t + i, y)$ and $h^i(t) = h(t + i)$. Then w^i satisfies

$$\begin{cases} w_t^i - dw_{yy}^i + (h^i(t))'w_y^i = F^i(t, y), & 0 < t \leqslant 3, \ 0 < y < h^i(t), \\ w^i(t, 0) = \mathscr{B}^*w^i(t, h^i(t)) = 0, & 0 < t \leqslant 3, \\ w^i(0, y) = w(i, y), & 0 < y < h(i), \end{cases} \tag{8.23}$$

where $F^i(t, y) = F(t+i, y)$. According to Theorem 8.5, we know that w^i, $(h^i(t))'$ and F^i are bounded uniformly with respect to i. Set $\Gamma_i = [i+1, i+3] \times [0, k]$. Remember

$$h^i(t) - k \geqslant h(i') - k > 1 \quad \text{for all } i \geqslant i', \ 0 \leqslant t \leqslant 3.$$

Applying the interior L^p estimate to (8.23) firstly and the embedding theorem secondly, we conclude $|w^i|_{1+\alpha, \Gamma_0} \leqslant C$. Accordingly, $|w|_{1+\alpha, \Gamma_i} \leqslant C$. Since these rectangles Γ_i overlap each other and C is independent of i, it follows that

$$|w|_{1+\alpha, [i_0, \infty) \times [0, k]} \leqslant C, \quad i_0 = i' + 1.$$

Notice that $0 \leqslant y \leqslant k$ is equivalent to that $h(t) - k \leqslant x \leqslant h(t)$. By means of $0 < h'(t) \leqslant 2\mu MK$ and $u_x(t, x) = -w_y(t, y)$, the estimate (8.22) is followed.

Step 4. The estimate (8.16) can be proved by the similar way. □

8.4 SOME TECHNICAL LEMMAS

In this section, we will give four basic conclusions. The first one will be used to study asymptotic behaviour of u when $h_\infty < \infty$, which indicates that if species cannot spread into $[0, \infty)$, then it will die out in the long run. The second one concerns with the comparison principle, which, as well known, is an important tool. The last two are very useful in determining the criteria for spreading and vanishing.

Lemma 8.7 *Let d, C, μ and g_0 be positive constants, and $m(t, x)$ be a bounded function. Let $v \in W_p^{1,2}((0, T) \times (0, g(t)))$ for each $0 < T < \infty$, v, $v_x \in C(\overline{\mathbb{R}}_+ \times [0, g(t)])$ and $g \in C^1(\overline{\mathbb{R}}_+)$. Assume that (v, g) satisfies*

$$\begin{cases} v_t - dv_{xx} + m(t, x)v_x \geqslant -Cv, & t > 0, \quad 0 < x < g(t), \\ v \geqslant 0, & t > 0, \quad x = 0, \\ v = 0, \quad g'(t) \geqslant -\mu v_x, & t > 0, \quad x = g(t), \\ v(0, x) = v_0(x) \geqslant 0, \not\equiv 0, & x \in (0, g_0), \\ g(0) = g_0 \end{cases}$$

and

$$\lim_{t \to \infty} g(t) = g_\infty < \infty, \quad \lim_{t \to \infty} g'(t) = 0, \tag{8.24}$$

$$\|v(t, \cdot)\|_{C^1([0, g(t)])} \leqslant M \quad \text{for all } t \geqslant 1 \tag{8.25}$$

for some constant $M > 0$. Then

$$\lim_{t \to \infty} \max_{0 \leqslant x \leqslant g(t)} v(t, x) = 0. \tag{8.26}$$

Proof. Firstly, the maximum principle gives $v(t, x) > 0$ for $t > 0$ and $0 < x < g(t)$, hence $v_x(t, g(t)) \leqslant 0$. Furthermore, applying the Hopf boundary lemma to problem of $w(t, y) = v(t, g(t)y)$ we can get $w_y(t, 1) < 0$. It demonstrates $v_x(t, g(t)) < 0$, and then $g'(t) > 0$ for $t > 0$.

The idea of the following arguments comes from [112, 121, 122]. On the contrary, we assume that there exist $\sigma > 0$ and $\{(t_i, x_i)\}_{i=1}^{\infty}$, with $0 \leqslant x_i < g(t_i)$ and $t_i \to \infty$ as $i \to \infty$, so that

$$v(t_i, x_i) \geqslant 4\sigma, \quad i = 1, 2, \ldots. \tag{8.27}$$

Then there exist a sub-sequence of $\{x_i\}$, noted by itself, and $x_0 \in [0, g_\infty]$ such that $x_i \to x_0$ as $i \to \infty$. We claim $x_0 \neq g_\infty$. Otherwise, $x_i - g(t_i) \to 0$ as $i \to \infty$. By use of the inequality (8.27) firstly and the inequality (8.25) secondly, it derives that

$$4\sigma \leqslant |v(t_i, x_i) - v(t_i, g(t_i))| = |v_x(t_i, \bar{x}_i)||x_i - g(t_i)| \leqslant M|x_i - g(t_i)|,$$

where $\bar{x}_i \in (x_i, g(t_i))$. This is a contradiction since $x_i - g(t_i) \to 0$.

The strategy in the following arguments is to construct a series of lower solutions and use the comparison principle. In view of (8.25) and (8.27), there exists $\delta > 0$ so that $x_0 + \delta < g_\infty$ and

$$v(t_i, x) \geqslant 2\sigma \quad \text{for all } x \in [x_0, x_0 + \delta]$$

for all large i. Since $g(t_i) \to g_\infty$ as $i \to \infty$, without loss of generality, we may think that $g(t_i) > x_0 + \delta$ for all i. Let

$$r_i(t) = x_0 + \delta + t - t_i.$$

Then $r_i(t_i) = x_0 + \delta < g(t_i)$. Since $g'(t) > 0$ and $g_\infty < \infty = r_i(\infty)$, there exists a unique $\tau_i > t_i$ such that $g(\tau_i) = r_i(\tau_i)$. See Fig. 8.2.

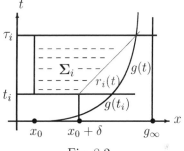

Fig. 8.2

Now, we define

$$\Sigma_i = \{(t, x) : t_i < t < \tau_i, \ x_0 < x < r_i(t)\},$$

$$v_i(t, x) = \sigma e^{-k(t-t_i)}(\cos y_i(t, x) + \cos \theta) \quad \text{for } (t, x) \in \overline{\Sigma}_i$$

with

$$y_i(t, x) = (\pi - \theta)\frac{2(x - x_0) - (\delta + t - t_i)}{\delta + t - t_i},$$

where θ $(0 < \theta < \pi/8)$ and k are positive constants to be chosen later. It is obvious that $|y_i(t, x)| \leqslant \pi - \theta$ for $(t, x) \in \overline{\Sigma}_i$, which implies $v_i(t, x) \geqslant 0$ in Σ_i. We want to compare $v(t, x)$ and $v_i(t, x)$ in $\overline{\Sigma}_i$. Firstly, it is clear that

$$v(t, x_0) \geqslant 0 = v_i(t, x_0), \quad v(t, r_i(t)) \geqslant 0 = v_i(t, r_i(t)) \quad \text{for all } t \in [t_i, \tau_i]$$

and

$$v(t_i, x) \geqslant 2\sigma > v_i(t_i, x) \quad \text{for all } x \in [x_0, x_0 + \delta].$$

Thus, if positive constants θ and k can be chosen independent of i such that

$$v_{it} - dv_{ixx} + m(t, x)v_{ix} \leqslant -Cv_i \quad \text{in } \Sigma_i, \tag{8.28}$$

then we can deduce $v \geqslant v_i$ in Σ_i by applying the comparison principle to v and v_i over Σ_i. Since $v(\tau_i, g(\tau_i)) = 0 = v_i(\tau_i, r_i(\tau_i))$ and $g(\tau_i) = r_i(\tau_i)$, it follows that

$$v_x(\tau_i, g(\tau_i)) \leqslant v_{ix}(\tau_i, r_i(\tau_i)).$$

Due to the fact that $\theta < \pi/8$ and $\delta + \tau_i - t_i < g_\infty$, it concludes

$$v_{ix}(\tau_i, r_i(\tau_i)) = -\frac{2\sigma(\pi - \theta)}{\delta + \tau_i - t_i} e^{-k(\tau_i - t_i)} \sin(\pi - \theta) \leqslant -\frac{7\sigma\pi}{4g_\infty} e^{-kg_\times} \sin\theta.$$

The free boundary condition $-\mu v_x(\tau_i, g(\tau_i)) \leqslant g'(\tau_i)$ implies

$$g'(\tau_i) \geqslant -\mu v_{ix}(\tau_i, r_i(\tau_i)) \geqslant \frac{7\mu\sigma\pi}{4g_\infty} e^{-kg_\times} \sin\theta.$$

This contradicts (8.24) since $\tau_i \to \infty$, and subsequently, (8.26) holds.

In the following, we prove that if θ and k satisfy

$$\theta < \frac{\pi}{8}, \quad 4\pi \left(\frac{g_\infty}{\delta^2} + \frac{\|m\|_\infty}{\delta} \right) \sin\theta < \frac{3d\pi^2}{g_\infty^2}, \tag{8.29}$$

$$k > C + d \left(\frac{2\pi}{\delta} \right)^2 + \frac{2\pi(g_\infty + \delta\|m\|_\infty)}{\delta^2(\cos\theta - \cos 2\theta)}, \tag{8.30}$$

then (8.28) holds for all large i. In fact, thanks to $0 \leqslant v_i \leqslant 2\sigma$ and $\delta + \tau_i - t_i < g_\infty$, a series of computations indicate that, for $(t, x) \in \Sigma_i$,

$$e^{k(t-t_i)}\left(v_{it} - dv_{ixx} + m(t, x)v_{ix} + Cv_i \right)$$

$$= e^{k(t-t_i)}\left[Cv_i - kv_i - \sigma e^{-k(t-t_i)}\left(y_{it} \sin y_i - dy_{ix}^2 \cos y_i + my_{ix} \sin y_i \right) \right]$$

$$= e^{k(t-t_i)}\left[(C + dy_{ix}^2 - k)v_i - \sigma e^{-k(t-t_i)}\left(dy_{ix}^2 \cos\theta + (y_{it} + my_{ix}) \sin y_i \right) \right]$$

$$\leqslant \sigma\left[C + d(2\pi/\delta)^2 - k \right](\cos y_i + \cos\theta)$$

$$+ 2\sigma\pi\left[(x - x_0)/\delta^2 + \|m\|_\infty/\delta \right]|\sin y_i| - \sigma d[2(\pi - \theta)/g_\infty]^2 \cos\theta$$

$$=: \sigma I(t, x).$$

By (8.30), $C + d(2\pi/\delta)^2 - k < 0$ holds. Since $-(\pi - \theta) \leqslant y_i \leqslant \pi - \theta$ when $(t, x) \in \overline{\Sigma}_i$, we can decompose $\Sigma_i = \Sigma_i^1 \bigcup \Sigma_i^2$, where

$$\Sigma_i^1 = \{(t, x) \in \Sigma_i : t_i < t < \tau_i, \ \pi - 2\theta < |y_i(t, x)| < \pi - \theta\},$$

$$\Sigma_i^2 = \{(t, x) \in \Sigma_i : t_i < t < \tau_i, \ |y_i(t, x)| \leqslant \pi - 2\theta\}.$$

It is obvious that $|\sin y_i| \leqslant \sin 2\theta$ when $(t, x) \in \Sigma_i^1$, and $\cos y_i \geqslant -\cos 2\theta$ when $(t, x) \in \Sigma_i^2$. Recall that $\theta < \pi/8$, $v_i(t, x) \geqslant 0$ and $x - x_0 \leqslant g_\infty$ in Σ_i. Taking advantage of (8.29) and (8.30), we conclude

$$I \leqslant 2\pi \left(\frac{g_\infty}{\delta^2} + \frac{\|m\|_\infty}{\delta} \right) \sin 2\theta - \frac{3d\pi^2}{g_\infty^2} \cos\theta \leqslant 0$$

when $(t, x) \in \Sigma_i^1$, and

$$I \leqslant (C + d(2\pi/\delta)^2 - k)(\cos\theta - \cos 2\theta) + 2\pi\delta^{-2}(g_\infty + \delta\|m\|_\infty) < 0$$

when $(t, x) \in \Sigma_i^2$. The proof is complete. □

Now, we give a comparison principle.

Lemma 8.8 (Comparison principle) *Assume that f satisfies conditions* **(J)** *and* (8.3). *Let $\bar{h} \in C^1([0, T])$, $\bar{h} > 0$ in $[0, T]$, and $\bar{u} \in C^{0,1}([0, T] \times [0, \bar{h}(t)]) \cap C^{1,2}([0, T] \times (0, \bar{h}(t)))$ satisfy*

$$\begin{cases} \bar{u}_t - d\bar{u}_{xx} \geqslant f(t, x, \bar{u}), & 0 < t \leqslant T, \ 0 < x < \bar{h}(t), \\ \mathscr{B}\bar{u}(t, 0) \geqslant 0, \ \bar{u}(t, \bar{h}(t)) = 0, & 0 < t \leqslant T, \\ \bar{h}'(t) \geqslant -\bar{\mu}\bar{u}_x(t, \bar{h}(t)), & 0 < t \leqslant T. \end{cases}$$

Let (u, h) be the unique solution of (8.1) *defined in $[0, T]$. If $\mu \leqslant \bar{\mu}$, $h_0 \leqslant \bar{h}(0)$, $\bar{u}(0, x) \geqslant 0$ in $[0, \bar{h}(0)]$, and $\bar{u}(0, x) \geqslant u_0(x) > 0$ in $(0, h_0)$. Then $h \leqslant \bar{h}$ in $[0, T]$ and $u \leqslant \bar{u}$ in $[0, T] \times [0, h(t)]$.*

Proof. The idea of this proof comes from [25], and the strategy is to use the classical maximum principle and comparison principle. For small $\varepsilon > 0$, let $(u_\varepsilon, h_\varepsilon)$ be the unique solution of (8.1) with $h_0, \mu, u_0(x)$ replaced by

$$h_0^\varepsilon = (1 - \varepsilon)h_0, \quad \mu_\varepsilon = (1 - \varepsilon)\mu, \quad u_0^\varepsilon(x) \in W_p^2([0, h_0^\varepsilon]),$$

respectively, where $u_0^\varepsilon(x)$ satisfies $\mathscr{B}u_0^\varepsilon(0) = u_0^\varepsilon(h_0^\varepsilon) = 0$ and

$$0 < u_0^\varepsilon(x) \leqslant u_0(x) \quad \text{in} \ (0, h_0^\varepsilon],$$

$$\lim_{\varepsilon \to 0} u_0^\varepsilon((1 - \varepsilon)x) = u_0(x) \quad \text{in} \ W_p^2([0, h_0]).$$

Then $h_\varepsilon \in C^{1+\alpha/2}([0, T])$.

We shall prove that $h_\varepsilon < \bar{h}$ in $(0, T]$. Obviously, this is reasonable for small $t > 0$. Arguing indirectly, it is assumed that we can find a first $0 < \tau \leqslant T$ such that $h_\varepsilon < \bar{h}$ in $(0, \tau)$ and $h_\varepsilon(\tau) = \bar{h}(\tau)$. Then

$$h_\varepsilon'(\tau) \geqslant \bar{h}'(\tau). \tag{8.31}$$

As $\bar{u}(0, x) \geqslant 0$ in $[0, \bar{h}(0)]$, and $\bar{u}(0, x) > 0$ in $(0, h_0)$, we have $\bar{u} > 0$ in $(0, \tau] \times (0, \bar{h}(t))$ by the maximum principle, and $\bar{u}_x(t, \bar{h}(t)) < 0$ by the Hopf boundary lemma. And then the comparison principle asserts $\bar{u} > u_\varepsilon$ in $(0, \tau] \times (0, h_\varepsilon(t))$ since $\bar{u}|_{x=h_\varepsilon(t)} > 0 = u_\varepsilon|_{x=h_\varepsilon(t)}$. Obviously, $\bar{u}(\tau, h_\varepsilon(\tau)) = u_\varepsilon(\tau, h_\varepsilon(\tau)) = 0$ since $h_\varepsilon(\tau) = \bar{h}(\tau)$, and then $\bar{u}_x(\tau, h_\varepsilon(\tau)) \leqslant u_{\varepsilon x}(\tau, h_\varepsilon(\tau))$. Notice $\bar{u}_x(\tau, h_\varepsilon(\tau)) < 0$ and $\mu_\varepsilon < \mu \leqslant \bar{\mu}$. It follows that $h_\varepsilon'(\tau) < \bar{h}'(\tau)$. This contradicts (8.31). As a result, $h_\varepsilon < \bar{h}$ in $(0, T]$. Apply the comparison principle to u_ε and $\bar{u}$ over $(0, T] \times (0, h_\varepsilon(t))$ to deduce $u_\varepsilon < \bar{u}$ in $(0, T] \times (0, h_\varepsilon(t))$.

As the unique solution of (8.1) depends continuously on parameters in (8.1), and $(u_\varepsilon, h_\varepsilon)$ converges to (u, h), the unique solution of (8.1), the desired result then follows by letting $\varepsilon \to 0$ in inequalities: $u_\varepsilon < \bar{u}$ and $h_\varepsilon < \bar{h}$. □

The pair $(\bar{u}, \bar{h})$ in Lemma 8.8 is usually called an upper solution of (8.1). We can define a lower solution by reversing all inequalities in the obvious places. Moreover, one can easily obtain an analogue of Lemma 8.8 for lower solutions.

Applying Lemma 8.8, we can prove that the solution (u, h) of (8.1) is strictly monotone increasing in $\mu > 0$. Details are left to readers as an exercise.

The next two lemmas are very useful in determining the criteria for spreading and vanishing.

Lemma 8.9 *Let $C > 0$ and $p > 3$ be constants. For any given constants $\bar{h}_0, H > 0$, and any function $\bar{u}_0 \in W_p^2([0, \bar{h}_0])$ satisfying $\mathscr{B}\bar{u}_0(0) = \bar{u}_0(\bar{h}_0) = 0$ and $\bar{u}_0 > 0$ in $(0, \bar{h}_0)$, there exists $\mu^0 > 0$ so that when $\mu \geqslant \mu^0$ and $(\bar{u}, \bar{h})$ satisfies*

$$
\begin{cases}
\bar{u}_t - d\bar{u}_{xx} \geqslant -C\bar{u}, & t > 0, \;\; 0 < x < \bar{h}(t), \\
\mathscr{B}\bar{u}(t, 0) = \bar{u}(t, \bar{h}(t)) = 0, & t > 0, \\
\bar{h}'(t) \geqslant -\mu\bar{u}_x(t, \bar{h}(t)), & t > 0, \\
\bar{u}(0, x) = \bar{u}_0(x), \;\; \bar{h}(0) = \bar{h}_0, & 0 < x < \bar{h}_0,
\end{cases}
$$

we must have $\lim_{t \to \infty} \bar{h}(t) > H$.

Proof. This proof is inspired by [90], and the strategy is to construct a lower solution and use the comparison principle. By results of §8.2 we see that the linear problem

$$
\begin{cases}
v_t - dv_{xx} = -Cv, & t > 0, \;\; 0 < x < r(t), \\
\mathscr{B}v(t, 0) = v(t, r(t)) = 0, & t > 0, \\
r'(t) = -\mu v_x(t, r(t)), & t > 0, \\
v(0, x) = \bar{u}_0(x), \;\; r(0) = \bar{h}_0, & 0 < x < \bar{h}_0
\end{cases}
\tag{8.32}
$$

admits a unique global solution (v, r), and $v \in C^{1+\alpha/2, 2+\alpha}(\mathbb{R}_+ \times [0, h(t)])$, $r'(t) > 0$ for $t > 0$. By Lemma 8.8,

$$
\bar{u}(t, x) \geqslant v(t, x), \quad \bar{h}(t) \geqslant r(t) \;\; \text{for all } t \geqslant 0, \; x \in [0, r(t)].
\tag{8.33}
$$

In the following, we are going to prove that for all large μ,

$$
r(2) \geqslant H.
\tag{8.34}
$$

To this end, we first choose a smooth function $\underline{r}(t)$ with $\underline{r}(0) = \bar{h}_0/2$, $\underline{r}'(t) > 0$ and $\underline{r}(2) = H$. We then discuss the following initial-boundary value problem

$$
\begin{cases}
\underline{v}_t - d\underline{v}_{xx} = -C\underline{v}, & t > 0, \;\; 0 < x < \underline{r}(t), \\
\mathscr{B}\underline{v}(t, 0) = \underline{v}(t, \underline{r}(t)) = 0, & t > 0, \\
\underline{v}(0, x) = \underline{v}_0(x), & 0 < x < \bar{h}_0/2.
\end{cases}
\tag{8.35}
$$

Here, for the smooth initial value $\underline{v}_0$, we require

$$
\begin{cases}
0 < \underline{v}_0(x) \leqslant \bar{u}_0(x), & x \in (0, \bar{h}_0/2), \\
\mathscr{B}\underline{v}_0(0) = \underline{v}_0(\bar{h}_0/2) = 0, \;\; \underline{v}_0'(\bar{h}_0/2) < 0.
\end{cases}
\tag{8.36}
$$

The standard theory ensures that (8.35) has a unique positive solution $\underline{v}$, and $\underline{v}_x(t, \underline{r}(t)) < 0$ for all $t \in [0, 2]$ due to the Hopf boundary lemma. According to our choices for $\underline{r}(t)$ and $\underline{v}_0(x)$, there is a constant $\mu^0 > 0$ such that, for all $\mu \geqslant \mu^0$,

$$\underline{r}'(t) \leqslant -\mu \underline{v}_x(t, \underline{r}(t)) \quad \text{for all } 0 \leqslant t \leqslant 2. \tag{8.37}$$

On the other hand, noting $\underline{r}(0) = \bar{h}_0/2 < r(0)$, and using (8.32) and (8.35)–(8.37), it is deduced by the comparison principle that

$$v(t, x) \geqslant \underline{v}(t, x), \quad r(t) \geqslant \underline{r}(t) \quad \text{for all } t \in [0, 2], \ x \in [0, \underline{r}(t)].$$

It particularly illustrates that $r(2) \geqslant \underline{r}(2) = H$, and then (8.34) holds. Making use of (8.33) and (8.34) we obtain

$$\lim_{t \to \infty} \bar{h}(t) \geqslant \lim_{t \to \infty} r(t) > r(2) \geqslant H.$$

The proof is complete. □

We are next to give a condition to judge vanishing. For any given $T_0 \geqslant 0$, consider problem

$$\begin{cases} u_t - du_{xx} \leqslant uf(t, x), & t > T_0, \ 0 < x < h(t), \\ u_x(t, 0) \geqslant 0, & t \geqslant T_0, \\ u = 0, \ h'(t) = -\mu u_x, & t \geqslant T_0, \ x = h(t), \\ u(T_0, x) > 0, & 0 \leqslant x < h(T_0), \end{cases} \tag{8.38}$$

where $f(t, x)$ and $h(t)$ are given continuous function and moving boundary, respectively.

Lemma 8.10 *Assume that (u, h) satisfies (8.38) and u is always positive for $t \geqslant T_0$. If f satisfies*

$$\limsup_{t \to \infty} \sup_{0 \leqslant x \leqslant h(t)} f(t, x) < 0,$$

then $h_\infty := \lim_{t \to \infty} h(t) < \infty$.

Proof. This proof comes from [51]. A direct calculation gives

$$\frac{\mathrm{d}}{\mathrm{d}t} \int_0^{h(t)} u(t, x) = \int_0^{h(t)} u_t(t, x) + h'(t)u(t, h(t))$$

$$\leqslant \int_0^{h(t)} du_{xx}(t, x) + \int_0^{h(t)} u(t, x)f(t, x)$$

$$\leqslant -\frac{d}{\mu} h'(t) + \int_0^{h(t)} u(t, x)f(t, x).$$

By the assumption, there exists $\tau > T_0$ such that $f(t, x) \leqslant 0$ for all $t \geqslant \tau$ and $0 \leqslant x \leqslant h(t)$. Integration of the above inequality from τ to t yields

$$0 \leqslant \int_0^{h(t)} u(t, x)$$

$$\leqslant \int_0^{h(\tau)} u(\tau, x) + \frac{d}{\mu}(h(\tau) - h(t)) + \int_\tau^t \int_0^{h(s)} u(s, x)f(s, x)\mathrm{d}s$$

$$\leqslant \int_0^{h(\tau)} u(\tau, x) + \frac{d}{\mu}(h(\tau) - h(t)),$$

which implies

$$h(t) \leqslant h(\tau) + \frac{\mu}{d} \int_0^{h(\tau)} u(\tau, x), \quad t \geqslant \tau.$$

Accordingly, $h_\infty < \infty$, and the proof is concluded. □

8.5 THE LOGISTIC EQUATION

Here, as an application of abstract conclusions obtained in above sections, we investigate a diffusive logistic equation with a free boundary and sign-changing intrinsic growth rate in heterogeneous time-periodic environment

$$\begin{cases} u_t - du_{xx} = a(t,x)u - b(t,x)u^2, & t > 0, \ 0 < x < h(t), \\ \mathscr{B}u(t,0) = 0, \quad u(t,h(t)) = 0, & t > 0, \\ h'(t) = -\mu u_x(t,h(t)), & t > 0, \\ h(0) = h_0, \quad u(0,x) = u_0(x), & 0 < x < h_0, \end{cases} \tag{8.39}$$

where $\mathscr{B}u = \beta u - (1-\beta)u_x$ with constant $0 \leqslant \beta \leqslant 1$, and $u_0(x)$ satisfies

$$u_0 \in C^2([0,h_0]), \quad u_0 > 0 \ \text{ in } (0,h_0), \quad \mathscr{B}u_0(0) = u_0(h_0) = 0.$$

Functions a and b satisfy

(K) $a, b \in (C^{\alpha/2,1} \cap L^\infty)([0,T] \times \overline{\mathbb{R}}_+)$ are time-periodic with period T, a is positive somewhere in $[0,T] \times \overline{\mathbb{R}}_+$, and $b_1 \leqslant b(t,x) \leqslant b_2$ in $[0,T] \times \overline{\mathbb{R}}_+$ for some positive constants b_1 and b_2.

By Theorem 8.5, problem (8.39) has a unique global solution (u,h). In this section, we will analyze longtime behaviours of (u,h), criteria for spreading and vanishing, and spreading-vanishing dichotomy. Contents of this part are due to [113].

8.5.1 Longtime behaviour of solution

In this subsection, we investigate *longtime behaviour* of solution. It will be shown that if $h_\infty < \infty$, i.e., the species u cannot spread over $[0,\infty)$, then it must die out in the long run; if $h_\infty = \infty$, i.e., the species u can spread over $[0,\infty)$, then it must tend to the unique positive solution U of (7.13) locally uniformly in $[0,T] \times [0,\infty)$ in the time-periodic sense.

The estimate (8.14) implies $\lim_{t\to\infty} h'(t) = 0$ when $h_\infty < \infty$. Applying the estimate (8.13) and Lemma 8.7, we obtain the following theorem immediately.

Theorem 8.11 (Vanishing) *Let (u,h) be the unique solution of (8.39). When $h_\infty < \infty$, we must have*

$$\lim_{t\to\infty} \max_{0 \leqslant x \leqslant h(t)} u(t,x) = 0.$$

Theorem 8.12 (Spreading) *Assume that the condition (7.16) holds. Let U be the unique positive solution of (7.13) and (u,h) be the unique solution of (8.39). If $h_\infty = \infty$, then $\lim_{i\to\infty} u(t+iT,x) = U(t,x)$ uniformly in $[0,T] \times [0,L]$ for any $L > 0$.*

Proof. In current circumstances, the condition **(I1)** is fulfilled. Let $\lambda(\ell; d, a)$ be the principal eigenvalue of (7.8). In view of Proposition 7.12, there is an $\ell_0 \gg 1$ such that $\lambda(\ell, d, a) < 0$ for all $\ell \geqslant \ell_0$. We divide this proof into two parts, i.e.,

$$\limsup_{i \to \infty} u(t + iT, x) \leqslant U(t, x) \quad \text{uniformly in} \ \ [0, T] \times [0, L] \tag{8.40}$$

and

$$\liminf_{i \to \infty} u(t + iT, x) \geqslant U(t, x) \quad \text{uniformly in} \ \ [0, T] \times [0, L]. \tag{8.41}$$

Step 1. Take $\theta = \|u_0\|_\infty + \frac{1}{b_1}\|a\|_\infty$, where b_1 is given by the condition **(K)**. Then $u < \theta$ by the maximum principle. For any fixed $\ell > \ell_0$, there exists an integer $m \gg 1$ such that

$$h(t) > \ell \quad \text{for all} \ \ t \geqslant mT.$$

Since $\lambda(\ell; d, \|a\|_\infty) \leqslant \lambda(\ell; d, a) < 0$, by the upper and lower solutions method for elliptic equations, we can prove that the boundary value problem

$$\begin{cases} -dV_{xx} = \|a\|_\infty V - b_1 V^2, & 0 < x < \ell, \\ \mathscr{B}V(0) = 0, \quad V(\ell) = \theta \end{cases}$$

has a unique positive solution $V(x)$, and $V(x) \leqslant \theta$ in $[0, \ell]$. In consideration of regularities of $u(mT, x)$ and $V(x)$ with respect to x, we can find a constant $k \geqslant 1$ such that $u(mT, x) \leqslant kV(x)$ for all $0 \leqslant x \leqslant \ell$. Noticing $u(t, \ell) < k\theta = kV(\ell)$, we can apply the comparison principle to u and kV, and then derive $u(t, x) \leqslant kV(x)$ for all $t \geqslant mT$ and $0 \leqslant x \leqslant \ell$. Since $k \geqslant 1$, it is easy to see that function $v := kV$ satisfies

$$-dv_{xx} \geqslant \|a\|_\infty v - b_1 v^2 \geqslant a(t, x)v - b(t, x)v^2.$$

Let w_ℓ be the unique solution of

$$\begin{cases} w_t - dw_{xx} = a(t, x)w - b(t, x)w^2, & t > mT, \ 0 < x < \ell, \\ \mathscr{B}w(t, 0) = 0, \quad w(t, \ell) = kV(\ell) = k\theta, & t > mT, \\ w(mT, x) = kV(x), & 0 < x < \ell. \end{cases}$$

The comparison principle gives $u \leqslant w_\ell \leqslant kV$ for $t \geqslant mT$ and $0 \leqslant x \leqslant \ell$. For any integer $i \geqslant m$, we define $w_\ell^i(t, x) = w_\ell(t + iT, x)$ for $(t, x) \in [0, T] \times [0, \ell]$. Similar to the proof of Theorem 7.10, it can be shown that $w_\ell^i \to W_\ell$ in $C^{1,2}([0, T] \times [0, \ell])$ as $i \to \infty$, where W_ℓ is a positive solution of

$$\begin{cases} W_t - dW_{xx} = a(t, x)W - b(t, x)W^2, & 0 < t \leqslant T, \ 0 < x < \ell, \\ \mathscr{B}W(t, 0) = 0, \quad W(t, \ell) = k\theta, & 0 < t \leqslant T, \\ W(0, x) = W(T, x), & 0 < x < \ell. \end{cases}$$

On account of $u(t + iT, x) \leqslant w_\ell(t + iT, x) = w_\ell^i(t, x)$ in $[0, T] \times [0, \ell]$, it yields

$$\limsup_{i \to \infty} u(t + iT, x) \leqslant W_\ell(t, x) \quad \text{uniformly in} \ \ [0, T] \times [0, \ell]. \tag{8.42}$$

Since $w_\ell \leqslant k\theta$ in $[0, \ell] \times [mT, \infty)$, we see that w_ℓ is monotonically decreasing in ℓ by the comparison principle, and hence W_ℓ is monotonically decreasing in ℓ. Remembering that U is the unique positive solution of (7.13), and arguing as Step 1 in proving Theorem 7.13, we can verify that $W_\ell \to U$ in $C^{1,2}([0, T] \times [0, L])$ as $\ell \to \infty$. This fact combined with (8.42) allows us to derive (8.40).

Step 2. Let $\ell > \ell_0$ and v_ℓ be the unique solution of

$$\begin{cases} v_t - dv_{xx} = a(t, x)v - b(t, x)v^2, & t > mT, \ 0 < x < \ell, \\ \mathscr{B}v(t, 0) = v(t, \ell) = 0, & t > mT, \\ v(mT, x) = u(mT, x), & 0 < x < \ell. \end{cases}$$

Then $u \geqslant v_\ell$ in $[mT, \infty) \times [0, \ell]$. Since $\lambda(\ell; d, a) < 0$, problem

$$\begin{cases} V_t - dV_{xx} = a(t, x)V - b(t, x)V^2, & 0 < t \leqslant T, \ 0 < x < \ell, \\ \mathscr{B}V(t, 0) = V(t, \ell) = 0, & 0 < t \leqslant T, \\ V(0, x) = V(T, x), & 0 < x < \ell \end{cases}$$

admits a unique positive solution V_ℓ. Noticing that $v_\ell(mT, x) = u(mT, x) > 0$ in $(0, \ell)$, it derives by Theorem 7.11 that $v_\ell(t + iT, x) \to V_\ell(t, x)$ in $C^{1,2}([0, T] \times [0, \ell])$ as $i \to \infty$.

Obviously, V_ℓ is monotonically increasing in ℓ. Similar to Step 1, we can deduce $V_\ell \to U$ in $C^{1,2}([0, T] \times [0, L])$. Consequently, the estimate (8.41) holds since $u \geqslant v_\ell$ in $[mT, \infty) \times [0, \ell]$ for all $\ell > \ell_0$. $\qquad\square$

8.5.2 Criteria and dichotomy for spreading and vanishing

We first give a necessary condition for vanishing. Let $\lambda(\ell; d, a)$ be the principal eigenvalue of (7.8).

Lemma 8.13 *If $h_\infty < \infty$, then $\lambda(h_\infty; d, a) \geqslant 0$. And hence, $\lambda(h_0; d, a) \leqslant 0$ implies $h_\infty = \infty$.*

Proof. We assume $\lambda(h_\infty; d, a) < 0$ to get a contradiction. By continuity of $\lambda(\ell; d, a)$ in ℓ and $h(t) \to h_\infty$, there exists $\tau \gg 1$ for which $\lambda(h(\tau); d, a) < 0$. Let z be the unique solution of

$$\begin{cases} z_t - dz_{xx} = a(t, x)z - b(t, x)z^2, & t > \tau, \ 0 < x < h(\tau), \\ \mathscr{B}z(t, 0) = z(t, h(\tau)) = 0, & t > \tau, \\ z(\tau, x) = u(\tau, x), & 0 < x < h(\tau). \end{cases}$$

Then $u \geqslant z$ in $[\tau, \infty) \times [0, h(\tau)]$ by the comparison principle. Since $\lambda(h(\tau); d, a) < 0$, it infers from Theorem 7.11 that $z(t + iT, x) \to Z(t, x)$ uniformly in $[0, T] \times [0, h(\tau)]$ as $i \to \infty$, where Z is the unique positive solution of time-periodic parabolic boundary value problem

$$\begin{cases} Z_t - dZ_{xx} = a(t, x)Z - b(t, x)Z^2, & 0 < t \leqslant T, \ 0 < x < h(\tau), \\ \mathscr{B}Z(t, 0) = Z(t, h(\tau)) = 0, & 0 < t \leqslant T, \\ Z(0, x) = Z(T, x), & 0 < x < h(\tau). \end{cases}$$

Since $u \geqslant z$ in $[\tau, \infty) \times [0, h(\tau)]$, it is deduced immediately that

$$\liminf_{i \to \infty} u(t + iT, x) \geqslant Z(t, x) \quad \text{for all } (t, x) \in [0, T] \times [0, h(\tau)].$$

This is a contradiction with Theorem 8.11, and the proof is complete. $\qquad\square$

Lemma 8.14 *If* $\lambda(h_0; d, a) > 0$, *then there exists* $\mu_0 > 0$ *such that* $h_\infty < \infty$ *for* $\mu \leqslant \mu_0$. *As a consequence,* $\lambda(h_\infty; d, a) \geqslant 0$ *for* $\mu \leqslant \mu_0$ *by Lemma 8.13.*

Proof. Let $\phi(t, x)$ be the positive eigenfunction of (7.8) with $\ell = h_0$ corresponding to $\lambda(h_0; d, a) =: \lambda_1$. Note that $\phi_x(t, h_0) < 0$, $\phi(t, 0) > 0$ in $[0, T]$ when $\beta < 1$, while $\phi(t, 0) = 0$ and $\phi_x(t, 0) > 0$ in $[0, T]$ when $\beta = 1$. In accordance with regularities of ϕ, there exists a constant $C > 0$ such that

$$x\phi_x(t, x) \leqslant C\phi(t, x) \quad \text{for all } (t, x) \in [0, T] \times [0, h_0], \tag{8.43}$$

please refer to the proof of Lemma 3.7. Let $0 < \delta, \sigma < 1$ and $k > 0$ be constants, which will be determined later. Set

$$s(t) = 1 + 2\delta - \delta e^{-\sigma t}, \quad \tau(t) = \int_0^t s^{-2}(\rho) \mathrm{d}\rho, \quad t \geqslant 0$$

and

$$v(t, x) = k e^{-\sigma t} \phi(\tau(t), y), \quad y = y(t, x) = \frac{x}{s(t)}, \quad 0 \leqslant x \leqslant h_0 s(t).$$

First of all, for any given $0 < \varepsilon \ll 1$, since a is uniformly continuous in $[0, T] \times [0, 3h_0]$ and time-periodic with period T, we can find $0 < \delta_0(\varepsilon) \ll 1$ such that, for all $0 < \delta \leqslant \delta_0(\varepsilon)$ and $0 < \sigma < 1$,

$$\left| s^{-2}(t) a\left(\tau(t), y(t, x)\right) - a(t, x) \right| \leqslant \varepsilon \quad \text{for all } t > 0, \ 0 \leqslant x \leqslant h_0 s(t). \tag{8.44}$$

Remembering (8.43), (8.44) and $\lambda_1 > 0$, a direct calculation alleges

$$v_t - d v_{xx} - a(t, x) v = v \left(\frac{a(\tau, y)}{s^2(t)} - \sigma - a(t, x) - \frac{y \phi_y(\tau, y)}{\phi(\tau, y)} \frac{\sigma \delta}{s(t)} e^{-\sigma t} + \frac{\lambda_1}{s^2(t)} \right)$$

$$\geqslant (\lambda_1/4 - \sigma - \varepsilon - C\sigma) v$$

$$> 0 \quad \text{for all } t > 0, \ 0 < x < h_0 s(t) \tag{8.45}$$

provided $0 < \sigma, \varepsilon \ll 1$. Evidently, $v(t, h_0 s(t)) = k e^{-\sigma t} \phi(\tau(t), h_0) = 0$. If either $\beta = 0$ or $\beta = 1$, then $\mathscr{B}v(t, 0) = 0$. If $0 < \beta < 1$, then $\beta\phi(\tau(t), 0) = (1 - \beta)\phi_y(\tau(t), 0)$, and $\phi_y(\tau(t), 0) > 0$. Therefore

$$\mathscr{B}v(t, 0) = (1 - \beta) k e^{-\sigma t} \phi_y(\tau(t), 0)[1 - 1/s(t)] > 0$$

due to $s(t) > 1$. In conclusion,

$$\mathscr{B}v(t, 0) \geqslant 0 \quad \text{and} \quad v(t, h_0 s(t)) = 0 \quad \text{for all } t > 0. \tag{8.46}$$

Fix $0 < \sigma, \varepsilon \ll 1$ and $0 < \delta \leqslant \delta_0(\varepsilon)$. Based on regularities of $u_0(x)$ and $\phi(0, x)$, we can choose a constant $k \gg 1$ so that

$$u_0(x) \leqslant k\phi\left(0, x/(1+\delta)\right) = v(0, x) \quad \text{for all } 0 \leqslant x \leqslant h_0. \tag{8.47}$$

As $h_0 s'(t) = h_0 \sigma \delta e^{-\sigma t}$ and $v_x(t, h_0 s(t)) = \frac{1}{s(t)} k e^{-\sigma t} \phi_y(\tau(t), h_0)$, there exists $\mu_0 > 0$ such that, for all $0 < \mu \leqslant \mu_0$,

$$-\mu v_x(t, h_0 s(t)) \leqslant h_0 s'(t) \quad \text{for all } t \geqslant 0. \tag{8.48}$$

Notice (8.45)–(8.48). Using Lemma 8.8 we conclude $h(t) \leqslant h_0 s(t)$ for all $t \geqslant 0$ and $0 < \mu \leqslant \mu_0$. Therefore, $h_\infty \leqslant h_0(1 + 2\delta)$, and the proof is complete. $\qquad\square$

In order to study the spreading phenomenon and establish sharp criteria, we shall introduce some sets and analyze their properties. For any given $d > 0$, we define

$$\Sigma_d = \{\ell > 0 : \lambda(\ell; d, a) = 0\}.$$

By the monotonicity of $\lambda(\ell; d, a)$ in ℓ, we see that Σ_d contains at most one element. For any given $\ell > 0$, we define

$$\Sigma_\ell^- = \{d > 0 : \lambda(\ell; d, a) \leqslant 0\}, \quad \Sigma_\ell^+ = \{d > 0 : \lambda(\ell; d, a) > 0\}.$$

Then $\Sigma_\ell^+ \neq \emptyset$ by Theorem 7.9 (3). Here we should remark that because $\lambda(\ell; d, a)$ is not monotone in d (cf. Theorem 2.2 of [52]), it is useless to define Σ_ℓ as the manner treating Σ_d.

Remark 8.15 *For any fixed $d > 0$, since*

$$\lim_{\ell \to 0^+} \lambda(\ell; d, a) = \infty, \quad \lim_{\ell \to \infty} \lambda(\ell; d, a) = \lambda(\infty; d, a) \quad \text{exists,}$$

we know that $\Sigma_d \neq \emptyset$ is equivalent to $\lambda(\infty; d, a) < 0$ by the monotonicity of $\lambda(\ell; d, a)$ in ℓ.

Proposition 8.16 *Assume that*

(I2) $\displaystyle\int_0^T a(t, \hat{x}) \mathrm{d}t > 0$ *for some $\hat{x} > 0$.*

Then the set Σ_ℓ^- is non-empty for all $\ell > \hat{x}$.

Proof. Since $\displaystyle\int_0^T a(t, \hat{x}) \mathrm{d}t > 0$, by the continuity of $a(t, x)$ we can find $0 < \varepsilon < \min\{\hat{x}, \ell - \hat{x}\}$ for which $\displaystyle\int_0^T a_\varepsilon(t) \mathrm{d}t > 0$, where $a_\varepsilon(t) = \min_{x \in \bar{I}_\varepsilon} a(t, x)$ and $I_\varepsilon = (\hat{x} - \varepsilon, \hat{x} + \varepsilon)$.

Let $\hat{\lambda}$ be the principal eigenvalue of the time-periodic parabolic eigenvalue problem

$$\begin{cases} \hat{\phi}_t - d\hat{\phi}_{xx} - a_\varepsilon(t)\hat{\phi} = \hat{\lambda}\hat{\phi}, & x \in I_\varepsilon, \ 0 < t \leqslant T, \\ \hat{\phi}(t, x) = 0, & x = \hat{x} \pm \varepsilon, \ 0 < t \leqslant T, \\ \hat{\phi}(0, x) = \hat{\phi}(T, x), & x \in \bar{I}_\varepsilon. \end{cases} \tag{8.49}$$

Then $\lambda(\ell; d, a) \leqslant \hat{\lambda}$ by the monotonicity of principal eigenvalue.

We calculate $\hat{\lambda}$ by decomposing $\hat{\phi}(t, x) = \xi(t)\varphi(x)$, where $\xi(t)$ is a function to be determined, and $\varphi(x) > 0$ is the principal eigenfunction of

$$
\begin{cases}
-\varphi'' = \gamma\varphi, & x \in I_\varepsilon, \\
\varphi(\hat{x} \pm \varepsilon) = 0
\end{cases}
$$

with principal eigenvalue $\gamma > 0$. It follows from (8.49) that

$$\xi'(t) = [a_\varepsilon(t) + \hat{\lambda} - d\gamma]\xi(t),$$

and consequently

$$\xi(t) = \xi_0 e^{(\hat{\lambda} - d\gamma)t} \exp\left(\int_0^t a_\varepsilon(s)\mathrm{d}s\right).$$

In order that $\hat{\phi}(t, x)$ is time-periodic with period T, we must have $\xi(0) = \xi(T)$, i.e.,

$$(\hat{\lambda} - d\gamma)T + \int_0^T a_\varepsilon(s)\mathrm{d}s = 0,$$

hence $\hat{\lambda} = d\gamma - \dfrac{1}{T}\displaystyle\int_0^T a_\varepsilon(s)\mathrm{d}s$. Accordingly

$$\lambda(\ell; d, a) \leqslant d\gamma - \frac{1}{T}\int_0^T a_\varepsilon(s)\mathrm{d}s$$

as $\lambda(\ell; d, a) \leqslant \hat{\lambda}$. Owing to $\displaystyle\int_0^T a_\varepsilon(s)\mathrm{d}s > 0$, we can find $d_0 > 0$ so that $\lambda(\ell; d, a) < 0$ for all $0 < d \leqslant d_0$. This implies $(0, d_0] \subset \Sigma_\ell^-$. □

We will study sharp criteria and dichotomy for spreading and vanishing in two cases.

Case 1: we fix d and consider h_0 and μ as varying parameters.

Firstly, recalling the estimate (8.12), as consequences of Lemmas 8.9, 8.13 and 8.14, we have

Corollary 8.17 *Assume $\Sigma_d \neq \emptyset$, and let $h_0^* = h_0^*(d) \in \Sigma_d$, i.e., $\lambda(h_0^*; d, a) = 0$. Then the following conclusions hold:*

(1) *if $h_\infty < \infty$, then $h_\infty \leqslant h_0^*$. Of course, $h_0 \geqslant h_0^*$ implies $h_\infty = \infty$ for all $\mu > 0$;*

(2) *when $h_0 < h_0^*$, there exist $\mu^0 > \mu_0 > 0$ such that $h_\infty \leqslant h_0^*$ for $\mu \leqslant \mu_0$, while $h_\infty = \infty$ for $\mu \geqslant \mu^0$.*

With regard to *sharp criteria* for spreading and vanishing we have the following theorem.

Theorem 8.18 *Assume $\Sigma_d \neq \emptyset$ and let $h_0^* = h_0^*(d) \in \Sigma_d$. Then we conclude*

(1) *$h_0 \geqslant h_0^*$ implies $h_\infty = \infty$ for all $\mu > 0$;*

(2) *if $h_0 < h_0^*$, then there exists $\mu^* > 0$ so that $h_\infty = \infty$ for $\mu > \mu^*$, while $h_\infty \leqslant h_0^*$ for $\mu \leqslant \mu^*$.*

Proof. The verdict (1) is exactly Corollary 8.17 (1). Now we prove the conclusion (2). Assume $h_0 < h_0^*$. Write (u^μ, h^μ) in place of (u, h) to clarify the dependence of the solution of (8.39) on μ. Define

$$\Sigma^* = \left\{ \mu > 0 : h_\infty^\mu \leqslant h_0^* \right\}.$$

According to Corollary 8.17 (2), $(0, \mu_0] \subset \Sigma^*$ and $\Sigma^* \cap [\mu^0, \infty) = \emptyset$. Therefore, $\mu^* := \sup \Sigma^* \in [\mu_0, \mu^0]$. In accordance with this definition and the monotonicity of (u^μ, h^μ) in μ, we find that $h_\infty^\mu = \infty$ when $\mu > \mu^*$, and so $\Sigma^* \subset (0, \mu^*]$.

Now, let us show that $\mu^* \in \Sigma^*$. Otherwise, $h_\infty^{\mu^*} = \infty$. There exists $T > 0$ so that $h^{\mu^*}(T) > h_0^*$. Utilizing the continuous dependence of (u^μ, h^μ) on μ, one can find an $\varepsilon > 0$ such that $h^\mu(T) > h_0^*$ for $\mu \in (\mu^* - \varepsilon, \mu^* + \varepsilon)$. It follows that, for all such μ,

$$\lim_{t \to \infty} h^\mu(t) \geqslant h^\mu(T) > h_0^*.$$

Therefore, $[\mu^* - \varepsilon, \mu^* + \varepsilon] \cap \Sigma^* = \emptyset$, and then $\sup \Sigma^* \leqslant \mu^* - \varepsilon$. This contradicts the definition of μ^*. □

Noting that (7.16) implies **(I1)**, and combining Theorems 8.11, 8.12 and 8.18, we immediately obtain the following *spreading-vanishing dichotomy* and *sharp criteria* for spreading and vanishing.

Theorem 8.19 *Assume (7.16) holds. Let h_0^* be the unique positive root of $\lambda(\ell; d, a) = 0$ and (u, h) the unique solution of (8.39). Then either*

(1) spreading: $h_\infty = \infty$ and $\lim_{i \to \infty} u(t + iT, x) = U(t, x)$ *uniformly in* $[0, T] \times [0, L]$ *for any $L > 0$, where $U(t, x)$ is the unique positive solution of (7.13);*

or

(2) vanishing: $h_\infty \leqslant h_0^*$ *and* $\lim_{t \to \infty} \max_{0 \leqslant x \leqslant h(t)} u(t, x) = 0$.

Moreover,

(3) $h_0 \geqslant h_0^*$ *implies* $h_\infty = \infty$ *for all $\mu > 0$;*

(4) *if $h_0 < h_0^*$, then there exists $\mu^* > 0$ so that $h_\infty = \infty$ for $\mu > \mu^*$, while $h_\infty \leqslant h_0^*$ for $\mu \leqslant \mu^*$.*

Case 2: we fix h_0, and regard d and μ as variable parameters.

Theorem 8.20 (1) *When $d \in \Sigma_{h_0}^-$, we have $h_\infty = \infty$ for all $\mu > 0$.*

(2) *For any fixed $d \in \Sigma_{h_0}^+$, there exists $\mu_0 = \mu_0(d) > 0$ such that $h_\infty < \infty$ as long as $0 < \mu \leqslant \mu_0$. If, in addition, $\Sigma_d \neq \emptyset$ for such d, then there exists $\mu^* > 0$, such that $h_\infty = \infty$ when $\mu > \mu^*$, and $h_\infty < \infty$ when $\mu \leqslant \mu^*$.*

Proof. (1) When $d \in \Sigma_{h_0}^-$, there holds $\lambda(h_0; d, a) \leqslant 0$, which implies $\Sigma_d \neq \emptyset$ and $h_0 \geqslant h_0^*(d)$. Theorem 8.18 (1) confirms $h_\infty = \infty$ for all $\mu > 0$.

(2) For the fixed $d \in \Sigma_{h_0}^+$, we have $\lambda(h_0; d, a) > 0$. By Lemma 8.14, there exists $\mu_0 > 0$ such that $h_\infty < \infty$ for $\mu \leqslant \mu_0$. If, in addition, $\Sigma_d \neq \emptyset$, then there exists $H \gg 1$ for which $\lambda(H; d, a) < 0$. Taking advantage of Lemma 8.9, there exists $\mu^0 > 0$ such that $h_\infty > H$ provided $\mu \geqslant \mu^0$, which implies $\lambda(h_\infty; d, a) < \lambda(H; d, a) < 0$. As a result, $h_\infty = \infty$ for $\mu \geqslant \mu^0$ by Lemma 8.13. The remaining proof is the same as Theorem 8.18. $\qquad\square$

Remark 8.21 (1) *When the condition* **(I1)** *holds, we have $\Sigma_d \neq \emptyset$ for all $d > 0$ by Theorem 7.9 (3) and Proposition 7.12.*

(2) *When the condition* **(I2)** *holds, from the proof of Proposition 8.16, we see that if $h_0 > \hat{x}$, then $\Sigma_{h_0}^- \neq \emptyset$ and there exists $d_0 > 0$ so that $d \in \Sigma_{h_0}^-$ for $0 < d \leqslant d_0$.*

Note that (7.16) implies **(I1)**. Invoking Theorems 8.11, 8.12 and 8.20, and Remark 8.21, we have the following *spreading-vanishing dichotomy* and *sharp criteria*.

Theorem 8.22 *Assume that conditions* (7.16) *and* **(I2)** *are satisfied and $h_0 > \hat{x}$. Let (u, h) be the unique solution of* (8.39). *Then either*

(1) *spreading: $h_\infty = \infty$ and $\lim_{i \to \infty} u(t+iT, x) = U(t, x)$ uniformly in $[0, T] \times [0, L]$ for any $L > 0$, where $U(t, x)$ is the unique positive solution of* (7.13);

or

(2) *vanishing: $h_\infty < \infty$ and $\lim_{t \to \infty} \max_{0 \leqslant x \leqslant h(t)} u(t, x) = 0$.*

Moreover,

(3) *$d \in \Sigma_{h_0}^-$ implies $h_\infty = \infty$ for all $\mu > 0$;*

(4) *if $d \in \Sigma_{h_0}^+$, then there exists $\mu^* > 0$ such that $h_\infty = \infty$ when $\mu > \mu^*$, while $h_\infty < \infty$ when $\mu \leqslant \mu^*$.*

8.6 SPREADING SPEED OF FREE BOUNDARY

This section concerns with spreading speed of free boundary. As an example, here we only treat the following problem.

$$\begin{cases} u_t - du_{xx} = au - bu^2, & t > 0, \ 0 < x < h(t), \\ \mathscr{B}u(t, 0) = 0, \ u(t, h(t)) = 0, & t > 0, \\ h'(t) = -\mu u_x(t, h(t)), & t > 0, \\ h(0) = h_0, \ u(0, x) = u_0(x), & 0 < x < h_0, \end{cases} \tag{8.50}$$

where a and b are positive constants. This problem has a unique global solution (u, h) by the conclusions of §8.2. It will be shown that when spreading occurs, the expanding front $x = h(t)$ moves at a constant speed for large time, namely

$$h(t) = (k_0 + o(1)) t \quad \text{as} \ t \to \infty.$$

This constant k_0 will be called the *asymptotic spreading speed*, and it is determined in Proposition 8.23 below. The fact $\lim_{t\to\infty} h(t)/t = k_0$ will be proved by using modifications of solution of the following elliptic problem (8.53). Contents of this part are taken from [12, 25].

We first remark that, since a and b are positive constants, the unique positive solution U of (7.13) does not depend on t, and

$$\lim_{x\to\infty} U(x) = a/b. \tag{8.51}$$

Moreover, the conclusion of Theorem 8.12 becomes

$$\lim_{t\to\infty} u(t,x) = U(x) \text{ uniformly in } [0, L] \text{ for any } L > 0. \tag{8.52}$$

Assume that (u, h) is the unique solution of (8.50) and $h(t) \to \infty$ as $t \to \infty$. Set $v(t, x) = u(t, h(t) - x)$. Then

$$\begin{cases} v_t - dv_{xx} + h'(t)v_x = v(a - bv), & t > 0, \ 0 < x < h(t), \\ v(t, 0) = 0, \quad h'(t) = \mu v_x(t, 0), & t > 0, \\ \beta v(t, h(t)) = (\beta - 1)v_x(t, h(t)), & t > 0, \\ h(0) = h_0, \quad v(0, x) = u_0(h_0 - x), \ 0 < x < h_0. \end{cases}$$

Note that $\lim_{t\to\infty} h(t) = \infty$. If $h'(t)$ approaches a constant k, and $v(t, x)$ approaches a positive function $U_k(x)$ as $t \to \infty$, then $U_k(x)$ must be a positive solution of

$$\begin{cases} -dU'' + kU' = aU - bU^2, & x > 0, \\ U(0) = 0 \end{cases} \tag{8.53}$$

and satisfies

$$U_k'(0) = k/\mu.$$

To study asymptotic behaviours of $h(t)$ we should investigate problem (8.53).

Proposition 8.23 *For any given $0 \leqslant k < 2\sqrt{ad}$, problem (8.53) has a unique positive solution, denoted by $U_k(x)$. Moreover, for each $\mu > 0$, there exists a unique $k_0 = k_0(\mu) \in (0, 2\sqrt{ad})$ such that $\mu U_{k_0}'(0) = k_0$.*

Proof. Since $0 \leqslant k < 2\sqrt{ad}$, it is well known that for any given constant $\theta \geqslant 0$ and all large $\ell > 0$, the problem

$$\begin{cases} -dU'' + kU' = aU - bU^2, & 0 < x < \ell, \\ U(0) = 0, \quad U(\ell) = \theta \end{cases} \tag{8.54}$$

has a unique positive solution, denoted by U_ℓ^θ. Similar to the proof of the existence in Theorem 7.13, we can show that problem (8.53) has at least one positive solution $U(x)$. The proof is left to the reader as an exercise. By the strong maximum principle for elliptic equations, we see that any non-trivial and non-negative solution of (8.53) satisfies

$$0 < U(x) < a/b \quad \text{for} \ \ x > 0.$$

Next, we claim that $U(x)$ is monotonically increasing and $\lim_{x \to \infty} U(x) = a/b$. Indeed, by use of (8.53) we conclude

$$(de^{-\frac{k}{d}x}U')' = -e^{-\frac{k}{d}x}(a - bU)U < 0 \quad \text{in} \ \ \mathbb{R}_+. \tag{8.55}$$

Since $U(x)$ is bounded, we can find a sequence $x_i \to \infty$ such that $U'(x_i) \to 0$ as $i \to \infty$. Hence, making use of (8.55),

$$e^{-\frac{k}{d}x}U'(x) > \lim_{i \to \infty} e^{-\frac{k}{d}x_i}U'(x_i) = 0 \quad \text{for} \ \ x > 0.$$

Of course $U'(x) > 0$, and then $\sigma = \lim_{x \to \infty} U(x)$ exists. In the light of (8.55) we easily find $\sigma = a/b$.

Let us now prove uniqueness. Suppose that both U_1 and U_2 are positive solutions of (8.53). Then for any $\varepsilon > 0$, function $w_i = (1 + \varepsilon)U_i$ satisfies

$$-dw_i'' + kw_i' > aw_i - bw_i^2 \quad \text{in} \ \ \mathbb{R}_+, \quad i = 1, 2.$$

On the basis of $\lim_{x \to \infty} w_i(x) = (1 + \varepsilon)a/b$ and $\lim_{x \to \infty} U_i(x) = a/b$, we can find $\ell_0 > 0$ large enough so that

$$w_1(\ell) > U_2(\ell), \quad w_2(\ell) > U_1(\ell) \quad \text{for all} \ \ell \geqslant \ell_0.$$

By means of the upper and lower solutions method and the uniqueness of positive solutions of (8.54), it can be proved that

$$(1 + \varepsilon)U_1(x) \geqslant U_2(x), \quad (1 + \varepsilon)U_2(x) \geqslant U_1(x) \quad \text{for all} \ 0 \leqslant x \leqslant \ell \ \text{and} \ \ell \geqslant \ell_0,$$

which implies $(1 + \varepsilon)U_1 \geqslant U_2$ and $(1 + \varepsilon)U_2 \geqslant U_1$ in $[0, \infty)$. Letting $\varepsilon \to 0$ we conclude $U_1 = U_2$, and the uniqueness is obtained.

Finally, if $0 \leqslant k_1 < k_2$, then, due to $U_{k_1}'(x) > 0$, one has

$$-dU_{k_1}'' + k_2 U_{k_1}' > -dU_{k_1}'' + k_1 U_{k_1}' = aU_{k_1} - bU_{k_1}^2, \quad x > 0.$$

Hence, for any $\varepsilon > 0$, function $w = (1 + \varepsilon)U_{k_1}$ satisfies

$$-dw'' + k_2 w' > aw - bw^2, \quad x > 0.$$

As above, $w \geqslant U_{k_2}$, i.e., $(1 + \varepsilon)U_{k_1} \geqslant U_{k_2}$, and then $U_{k_1} \geqslant U_{k_2}$ in $[0, \infty)$ by letting $\varepsilon \to 0$. Besides, applying the strong maximum principle for elliptic equations, we can still get $U_{k_1} > U_{k_2}$ in $\mathbb{R}_+$. Notice $U_{k_1}(0) = U_{k_2}(0) = 0$. Adopting the Hopf boundary lemma to differential equations satisfied by U_{k_1}, U_{k_2} and $U_{k_1} - U_{k_2}$, respectively, it can be derived that $U_{k_1}'(0) > 0$ and $U_{k_1}'(0) > U_{k_2}'(0)$. Hence, both sequences $\{U_k(x)\}$ and $\{U_k'(0)\}$ are monotonically decreasing in $k > 0$. Thus, for any fixed $\mu > 0$, function

$$\sigma(k) = k - \mu U_k'(0)$$

is strictly monotonically increasing in $k > 0$, and $\sigma(0) = -\mu U_0'(0) < 0$.

We claim that

$$\lim_{k \nearrow 2\sqrt{ad}} \sigma(k) = 2\sqrt{ad}. \tag{8.56}$$

Once this is done, there exists a unique $k_0 = k_0(\mu) \in (0, 2\sqrt{ad})$ such that $\sigma(k_0) = 0$.

To prove (8.56), we choose an arbitrary increasing sequence $\{k_i\}$ such that $k_i \to 2\sqrt{ad}$ as $i \to \infty$ and consider the corresponding sequence of solutions $\{U_{k_i}\}$ of (8.53). Since $0 < U_{k_i}(x) < a/b$ for $x > 0$, and $U_{k_i}(x)$ is decreasing in i, the limit

$$U_*(x) := \lim_{i \to \infty} U_{k_i}(x)$$

exists and satisfies $0 \leqslant U_*(x) < a/b$ for $x > 0$. Moreover, applying standard L^p estimates to the equation for U_{k_i} and then the embedding theorem (Theorem A.4 (3)), we easily see that $U_{k_i} \to U_*$ in $C^1_{\text{loc}}(\overline{\mathbb{R}_+})$. Hence U_* satisfies in the weak sense (and also classical sense)

$$\begin{cases} -dU_*'' + 2\sqrt{ad}U_*' = aU_* - bU_*^2, & 0 < x < \infty, \\ U_*(0) = 0. \end{cases}$$

Define

$$W_*(x) = \frac{b}{a}e^{-x}U_*\left(\sqrt{d/a}\,x\right).$$

Then one readily checks that

$$W_*'' = e^x W_*^2, \quad 0 \leqslant W_* < 1 \ \text{ in } (0, \infty), \quad \text{and} \quad W_*(0) = 0.$$

We claim that $U_* \equiv 0$. Otherwise, by the strong maximum principle, we have $U_*(x) > 0$ for $x > 0$ and $U_*'(0) > 0$. It follows that

$$W_*''(x) = e^x W_*^2(x) > 0 \ \text{ for } \ x > 0, \quad \text{and} \quad W_*'(0) > 0,$$

which implies $W_*(x) \to \infty$ as $x \to \infty$. This is a contradiction to the fact that $W_*(x) < 1$ for $x > 0$, and thus $U_* \equiv 0$.

Since $\{k_i\}$ is an arbitrary sequence increasing to $2\sqrt{ad}$, the above discussions show that $U_k \to 0$ as $k \to 2\sqrt{ad}$ in $C^1_{\text{loc}}(\overline{\mathbb{R}_+})$. In particular, $U_k'(0) \to 0$ as $k \to 2\sqrt{ad}$. Therefore

$$\lim_{k \nearrow 2\sqrt{ad}} \sigma(k) = 2\sqrt{ad},$$

as we wanted. $\qquad\qquad\qquad\qquad\qquad\qquad\qquad\qquad\qquad\qquad\qquad\qquad\qquad\qquad$ $\square$

Theorem 8.24 (Asymptotic spreading speed) *Let (u, h) be the unique solution of* (8.50). *If $h_\infty = \infty$, then $\lim_{t \to \infty} \dfrac{h(t)}{t} = k_0$, where k_0 is uniquely determined in Proposition* 8.23.

Proof. We first prove

$$\limsup_{t\to\infty} \frac{h(t)}{t} \leqslant k_0. \tag{8.57}$$

Clearly, $\limsup_{t\to\infty} u(t,x) \leqslant a/b$ uniformly for $x \geqslant 0$. Thus, for any given $\varepsilon > 0$ small, there exists $T = T_\varepsilon > 0$ such that

$$u(t,x) \leqslant a(1+\varepsilon)/b \quad \text{for all } t \geqslant T, \ x \geqslant 0.$$

Let $U_{k_0}(x)$ be the unique positive solution of (8.53) with $k = k_0$ and $k_0 = \mu U'_{k_0}(0)$. Since $U_{k_0}(x) \to a/b$ as $x \to \infty$, there exists $x_0 \gg 1$ so that

$$U_{k_0}(x) > \frac{a}{(1+\varepsilon)b} \quad \text{for all } x \geqslant x_0.$$

We now define

$$\xi(t) = (1+\varepsilon)^2 k_0 t + x_0 + h(T), \quad t \geqslant 0,$$
$$v(t,x) = (1+\varepsilon)^2 U_{k_0}(\xi(t) - x), \qquad t \geqslant 0, \ 0 \leqslant x \leqslant \xi(t).$$

Then

$$v(t, \xi(t)) = (1+\varepsilon)^2 U_{k_0}(0) = 0,$$
$$\xi'(t) = (1+\varepsilon)^2 k_0,$$
$$-\mu v_x(t, \xi(t)) = \mu(1+\varepsilon)^2 U'_{k_0}(0) = (1+\varepsilon)^2 k_0,$$

and so

$$\xi'(t) = -\mu v_x(t, \xi(t)).$$

Clearly,

$$\mathscr{B}v(t, 0) > 0$$

by noting that

$$v(t,0) = (1+\varepsilon)^2 U_{k_0}(\xi(t)) > 0, \quad v_x(t,0) = -(1+\varepsilon)^2 U'_{k_0}(\xi(t)) < 0.$$

Moreover, for $0 \leqslant x \leqslant h(T)$,

$$v(0,x) = (1+\varepsilon)^2 U_{k_0}(\xi(0) - x)$$
$$\geqslant (1+\varepsilon)^2 U_{k_0}(x_0)$$
$$\geqslant a(1+\varepsilon)/b \geqslant u(T,x),$$

and $v(0,x) > 0$ for $h(T) < x < \xi(0)$. A straightforward computation shows that

$$v_t - dv_{xx} = (1+\varepsilon)^2 (U'_{k_0}\xi' - dU''_{k_0})$$
$$= (1+\varepsilon)^2 ((1+\varepsilon)^2 k_0 U'_{k_0} - dU''_{k_0})$$
$$\geqslant (1+\varepsilon)^2 (k_0 U'_{k_0} - dU''_{k_0}) \quad \text{(due to } U'_{k_0} \geqslant 0)$$
$$= (1+\varepsilon)^2 (aU_{k_0} - bU^2_{k_0})$$
$$\geqslant av - bv^2 \quad \text{for } t > 0, \ 0 < x < \xi(t).$$

The comparison principle (Lemma 8.8) concludes $h(t + T) \leqslant \xi(t)$ for $t \geqslant 0$, and consequently

$$\limsup_{t \to \infty} \frac{h(t)}{t} \leqslant \lim_{t \to \infty} \frac{\xi(t - T)}{t} = k_0(1 + \varepsilon)^2.$$

Taking $\varepsilon \to 0$ we get (8.57).

In the following we will prove

$$\liminf_{t \to \infty} \frac{h(t)}{t} \geqslant k_0$$

by constructing a suitable lower solution. Consider

$$\begin{cases} -dV'' + k_0 V' = aV - bV^2, & 0 < x < \ell, \\ V(0) = V(\ell) = 0. \end{cases}$$

Similar to above, for all large ℓ, this problem has a unique positive solution, denoted by V_ℓ, and

$$V_\ell(x) < U_{k_0}(x) < a/b \quad \text{for } 0 < x \leqslant \ell.$$

Moreover, $V_\ell(x)$ is monotonically increasing in ℓ, and as $\ell \to \infty$, $V_\ell(x) \to V_\infty(x)$ which solves (8.53) with $k = k_0$. Then $V_\infty = U_{k_0}$ by the uniqueness of positive solutions of (8.53). Furthermore, a simple regularity and compactness consideration show that

$$\lim_{\ell \to \infty} V_\ell = U_{k_0} \quad \text{in } C^1_{\text{loc}}(\mathbb{R}_+).$$

Accordingly, for any given $0 < \varepsilon \ll 1$, we can find $\ell_0 = \ell_0(\varepsilon) \gg 1$ such that

$$V'_{\ell_0}(0) > (1 - \varepsilon)U'_{k_0}(0) = (1 - \varepsilon)k_0/\mu.$$

Define

$$V_0(x) = \begin{cases} V_{\ell_0}(x), & 0 \leqslant x \leqslant x_0, \\ V_{\ell_0}(x_0), & x > x_0, \end{cases}$$

where $x_0 \in (0, \ell_0)$ satisfies $V_{\ell_0}(x_0) = \max_{[0,\ell_0]} V_{\ell_0}(x)$. From the equation for V_{ℓ_0} we see that $e^{-\frac{k_0}{d}x}V'_{\ell_0}(x)$ is monotonically decreasing in $(0, \ell_0)$, see (8.55). Certainly, $V'_{\ell_0}(x)$ changes sign exactly once in this interval. It follows that such x_0 is unique, $V'_{\ell_0}(x) > 0$ for $x \in (0, x_0)$, and $V'_{\ell_0}(x) < 0$ for $x \in (x_0, \ell_0)$. As a consequence,

$$V_0(x) \leqslant V_{\ell_0}(x_0) < U_{k_0}(x_0) < a/b \quad \text{for } x \geqslant 0,$$
$$V'_0(x) = 0 \quad \text{for } x \geqslant x_0, \quad V'_0(x) > 0 \quad \text{for } 0 \leqslant x < x_0.$$

On the other hand, it is easy to check that V_0 satisfies (in the weak sense)

$$\begin{cases} -dV''_0 + k_0 V'_0 \leqslant aV_0 - bV_0^2, & x > 0, \\ V_0(0) = 0, \quad V_0(x) < a/b, & x \geqslant 0. \end{cases}$$

According to (8.51), there exists $L > 0$ such that $U(x) > (1 - \varepsilon/2)a/b$ for all $x \geqslant L$. Due to $h_\infty = \infty$, using (8.52) we can choose $T = T(\varepsilon, x_0, L)$ so large that

$$h(t) > L + x_0, \quad u(t, x) \geqslant (1 - \varepsilon)a/b \quad \text{for } t \geqslant T, \ x \in [L, L + x_0].$$

Define

$$\eta(t) = (1 - \varepsilon)^2 k_0 t + x_0, \quad t \geqslant 0,$$
$$w(t, x) = (1 - \varepsilon)V_0(\eta(t) - x), \quad t \geqslant 0, \ 0 \leqslant x \leqslant \eta(t),$$

and set

$$\bar{u}(t, x) = u(t + T, x + L), \quad \bar{h}(t) = h(t + T) - L.$$

Obviously,

$$\bar{u}_t - d\bar{u}_{xx} = a\bar{u} - b\bar{u}^2 \quad \text{for } t > 0, \ 0 < x < \bar{h}(t),$$

and

$$\eta(0) = x_0 < h(T) - L = \bar{h}(0),$$
$$w(0, x) = (1 - \varepsilon)V_0(x_0 - x) < (1 - \varepsilon)a/b \leqslant u(T, x + L) = \bar{u}(0, x), \quad 0 \leqslant x \leqslant x_0,$$
$$w(t, 0) = (1 - \varepsilon)V_0(\eta(t)) < (1 - \varepsilon)a/b \leqslant u(t + T, L) = \bar{u}(t, 0), \quad t \geqslant 0,$$
$$w(t, \eta(t)) = 0 = \bar{u}(t, \bar{h}(t)), \quad t \geqslant 0,$$
$$\eta'(t) = (1 - \varepsilon)^2 k_0 < \mu(1 - \varepsilon)V'_{\ell_0}(0) = -\mu w_x(t, \eta(t)), \quad t \geqslant 0,$$
$$\bar{h}'(t) = -\mu \bar{u}_x(t, \bar{h}(t)), \quad t \geqslant 0.$$

Moreover, due to $V'_0 \geqslant 0$, we have

$$
\begin{aligned}
w_t - dw_{xx} &= (1 - \varepsilon)[(1 - \varepsilon)^2 k_0 V'_0 - dV''_0] \\
&\leqslant (1 - \varepsilon)(k_0 V'_0 - dV''_0) \\
&\leqslant (1 - \varepsilon)(aV_0 - bV_0^2) \\
&\leqslant aw - bw^2, \quad t > 0, \ 0 < x < \eta(t).
\end{aligned}
$$

Hence, by the comparison principle,

$$u(t + T, x + L) \geqslant w(t, x) \quad \text{and} \quad h(t + T) - L \geqslant \eta(t) \quad \text{for } t \geqslant 0, \ 0 \leqslant x \leqslant \eta(t),$$

which leads to

$$\liminf_{t \to \infty} \frac{h(t)}{t} \geqslant \lim_{t \to \infty} \frac{\eta(t - T)}{t} = (1 - \varepsilon)^2 k_0.$$

The arbitrariness of ε implies

$$\liminf_{t \to \infty} \frac{h(t)}{t} \geqslant k_0,$$

and the proof is now complete. □

Next, we analyze the dependence of k_0 on parameters a, b, μ and d. From the proof of Proposition 8.23 we know that k_0 is the unique solution of

$$k - \mu U_k'(0) = 0.$$

Clearly, when all other parameters are fixed, k_0 increases with μ, $k_0 \to 0$ as $\mu \to 0$, and $k_0 \to 2\sqrt{ad}$ as $\mu \to \infty$. On the other hand, one easily sees by a comparison argument that for fixed k, $U_k(\cdot)$ increases with a and decreases with b, and it follows that $U_k'(0)$ increases with a and decreases with b. This implies that k_0 increases with a and decreases with b. Combining these analysis, we find that for fixed d, if k_0 is viewed as a function of (μ, a, b), namely $k_0 = k_0(\mu, a, b)$, then $\mu_1 \geqslant \mu_2$, $a_1 \geqslant a_2$ and $b_1 \leqslant b_2$ imply $k_0(\mu_1, a_1, b_1) \geqslant k_0(\mu_2, a_2, b_2)$, and the strict inequality holds when $(\mu_1, a_1, b_1) \neq (\mu_2, a_2, b_2)$.

Proposition 8.25 *Let k_0 be the spreading speed determined by Proposition 8.23. Then*

$$\lim_{\frac{a\mu}{bd} \to \infty} \frac{k_0}{\sqrt{ad}} = 2 \quad and \quad \lim_{\frac{a\mu}{bd} \to 0} \frac{k_0}{\sqrt{ad}} \frac{bd}{a\mu} = \frac{1}{\sqrt{3}}.$$

Proof. By Proposition 8.23 above, for any $\lambda \in [0, 2)$, the problem

$$-W'' + \lambda W' = W - W^2 \quad \text{in } (0, \infty), \quad W(0) = 0$$

has a unique positive solution denoted by W_λ, and for each $\alpha > 0$, there exists a unique $\lambda_0(\alpha) \in (0, 2)$ such that

$$\lambda_0(\alpha) = \alpha W_{\lambda_0(\alpha)}'(0)$$

and

$$W_0'(0) > 0, \quad \lim_{\lambda \to 2^-} W_\lambda'(0) = 0.$$

Clearly, the pair $(\lambda_0(\alpha), \lambda_0(\alpha)/\alpha)$ satisfies

$$\lim_{\alpha \to 0} (\lambda_0(\alpha), \lambda_0(\alpha)/\alpha) = (0, W_0'(0)) \quad \text{and} \quad \lim_{\alpha \to \infty} (\lambda_0(\alpha), \lambda_0(\alpha)/\alpha) = (2, 0). \quad (8.58)$$

A simple calculation confirms that for each $k > 0$,

$$W_{\frac{k}{\sqrt{ad}}}(x) = \frac{b}{a} U_k\left(\sqrt{d/a}\, x\right).$$

And consequently,

$$W_{\frac{k}{\sqrt{ad}}}'(0) = \frac{b}{a}\sqrt{\frac{d}{a}} U_k'(0)$$

and

$$\mu U_{k_0}'(0) = k_0 \iff \frac{a\mu}{bd} W_{\frac{k_0}{\sqrt{ad}}}'(0) = \frac{k_0}{\sqrt{ad}}.$$

It follows that

$$\frac{k_0}{\sqrt{ad}} = \lambda_0 \left(\frac{a\mu}{bd} \right).$$

This combined with (8.58) allows us to derive

$$\lim_{\frac{a\mu}{bd} \to \infty} \frac{k_0}{\sqrt{ad}} = 2 \quad \text{and} \quad \lim_{\frac{a\mu}{bd} \to 0} \frac{k_0}{\sqrt{ad}} \frac{bd}{a\mu} = W_0'(0).$$

We show next that $W_0'(0) = 1/\sqrt{3}$. Indeed, W_0 satisfies

$$-W_0'' = W_0 - W_0^2, \quad W_0 > 0 \quad \text{in} \quad (0, \infty), \quad W_0(0) = 0 \quad \text{and} \quad W_0(\infty) = 1.$$

It follows that

$$\frac{1}{2}(W_0'(0))^2 = -\int_0^\infty W_0'' W_0' \mathrm{d}x = \int_0^\infty (W_0 - W_0^2) W_0' \mathrm{d}x = \frac{1}{6},$$

which implies $W_0'(0) = 1/\sqrt{3}$, and the proof is complete. □

8.7 A SYSTEM COUPLED BY ONE PDE AND ONE ODE

In this section we introduce a free boundary problem of a system coupled by one PDE and one ODE

$$\begin{cases} u_t = f_1(t, x, u, v), & t > 0, \quad g(t) < x < h(t), \\ v_t = dv_{xx} + f_2(t, x, u, v), & t > 0, \quad g(t) < x < h(t), \\ u(t, x) = v(t, x) = 0, & t \geqslant 0, \quad x = g(t), \ h(t), \\ g'(t) = -\mu v_x(t, g(t)), & t \geqslant 0, \\ h'(t) = -\beta v_x(t, h(t)), & t \geqslant 0, \\ u(0, x) = u_0(x), \quad v(0, x) = v_0(x), & -h_0 \leqslant x \leqslant h_0, \\ h(0) = -g(0) = h_0. \end{cases} \quad (8.59)$$

This model describes the relationship between two species, such as viruses and healthy cells, in which the viruses have the ability of diffusion and the healthy cells do not. Contents of this section are taken from [79].

Let us define a couple of notations for the sake of arguments. Let $a < b$ and $T > 0$ be constants, and $g, h \in C([0, T])$ satisfy $g(t) < h(t)$ for all $0 \leqslant t \leqslant T$. We define

- $C^{1-}([a, b])$ to be the Lipschitz continuous function space in $[a, b]$;

- $D_{g,h}^T = \{0 < t \leqslant T, \ g(t) < x < h(t)\}$;

- function $u \in C^{1,1-}(\overline{D}_{g,h}^T)$ to mean that u is continuously differentiable in $t \in [0, T]$ and is Lipschitz continuous in $x \in [g(t), h(t)]$ for all $t \in [0, T]$;

- function $u \in C^{1-,1-}([0, T] \times [g(T), h(T)])$ to mean that u is Lipschitz continuous in $t \in [0, T]$ and in $x \in [g(T), h(T)]$.

We assume that initial functions u_0 and v_0 satisfy

- $(u_0, v_0) \in C^{1-}([-h_0, h_0]) \times W_p^2((-h_0, h_0))$ with $p > 3$, $u_0(\pm h_0) = v_0(\pm h_0) = 0$, $u_0, v_0 > 0$ in $(-h_0, h_0)$, and $v_0'(h_0) < 0, v_0'(-h_0) > 0$,

and denote by L_0 the Lipschitz constant of u_0 in x.

Let $\alpha = 1 - 3/p$. It is assumed that (f_1, f_2) satisfies

(L) $f_1(t, x, 0, v) \geqslant 0$ for all $v \geqslant 0$ and $f_2(t, x, u, 0) \geqslant 0$ for all $u \geqslant 0$. For any given $\tau, l, k_1, k_2 > 0$, $f_i(\cdot, 0, 0) \in L^\infty((0, \tau) \times (-l, l))$, and there exists a constant $L(\tau, l, k_1, k_2) > 0$ such that

$$|f_i(t, x, u_1, v_1) - f_i(t, y, u_2, v_2)| \leqslant L(\tau, l, k_1, k_2)(|x - y| + |u_1 - u_2| + |v_1 - v_2|)$$

for all $t \in [0, \tau]$, $x, y \in [-l, l]$, $u_1, u_2 \in [0, k_1]$, $v_1, v_2 \in [0, k_2]$ and $i = 1, 2$.

It is easy to see that the condition **(L)** implies $f_i \in L^\infty((0, \tau) \times (-l, l) \times (0, k_1) \times (0, k_2))$ for any given $\tau, l, k_1, k_2 > 0$.

Conditions that $u(t, g(t)) = u(t, h(t)) = 0$ in (8.59) look like boundary conditions of u, but they do actually play roles of initial conditions of u at points $x = g(t)$ and $x = h(t)$, respectively.

It can be seen from the following proof of Theorem 8.26 that the method and process of proving existence and uniqueness of solutions are quite different from problem (8.1).

Theorem 8.26 *Under above assumption* **(L)**, *there exists a $T > 0$ such that problem (8.59) has a unique solution (u, v, g, h) defined in $[0, T]$. Moreover,*

$$g, h \in C^{1+\alpha/2}([0, T]), \quad g'(t) < 0, \quad h'(t) > 0 \quad in \quad [0, T],$$

$$u \in C^{1,1-}(\overline{D}_{g,h}^T), \quad v \in W_p^{1,2}(D_{g,h}^T) \cap C^{(1+\alpha)/2, 1+\alpha}(\overline{D}_{g,h}^T), \quad u, v > 0 \quad in \quad D_{g,h}^T.$$

Proof. The proof process is very long, and we will divide it into several steps. As parameters d, μ and β are fixed, in the following arguments we do not emphasize the dependence of estimations on them.

Step 1. We denote

$$\sigma = \frac{1}{2} \min \{ -\beta v_0'(h_0), \, \mu v_0'(-h_0) \} > 0,$$

$$A = \max_{[-h_0, h_0]} u_0 + 1, \quad B = \max_{[-h_0, h_0]} v_0 + 1, \quad L_1 = L(1, 2h_0, A, B),$$

$$\mathscr{C} = \{h_0, A, B, \|v_0\|_{W_p^2((-h_0, h_0))}, v_0'(\pm h_0), \|f_2\|_{\infty, \Pi_1 \times (0, A) \times (0, B)}, L_1\},$$

and for any given $0 < T < \infty$, we define

$$\Pi_T = [0, T] \times [-2h_0, 2h_0], \quad \Delta_T^* = [0, T] \times [-1, 1],$$

$$\mathbb{X}_{u_0}^T = \{\phi \in C(\Pi_T) : \phi(0, x) = u_0(x), \, 0 \leqslant \phi \leqslant A\}.$$

We say $u \in C_x^{1-}(\Pi_T)$ if there is a constant $L(u, T)$ such that

$$|u(t, x_1) - u(t, x_2)| \leqslant L(u, T)|x_1 - x_2|$$

for all $x_1, x_2 \in [-2h_0, 2h_0]$ and $t \in [0, T]$. For the given $u \in \mathbb{X}_{u_0}^1 \cap C_x^{1-}(\Pi_1)$, we consider the following problem

$$\begin{cases} v_t = dv_{xx} + f_2(t, x, u(t, x), v), & 0 < t \leqslant 1, \ g(t) < x < h(t), \\ v(t, g(t)) = v(t, h(t)) = 0, & 0 \leqslant t \leqslant 1, \\ g'(t) = -\mu v_x(t, g(t)), & 0 \leqslant t \leqslant 1, \\ h'(t) = -\beta v_x(t, h(t)), & 0 \leqslant t \leqslant 1, \\ v(0, x) = v_0(x), & |x| \leqslant h_0, \\ h(0) = -g(0) = h_0. \end{cases} \qquad (8.60)$$

Due to properties of f_2 and u, using a similar argument in proving Theorem 8.1 we can show that there exists $0 < T_0 \leqslant 1$, independent of $L(u, 1)$, the Lipschitz constant of u in x, such that (8.60) has a unique solution $(v, g, h)^\ddagger$ which satisfies

$$g, h \in C^{1+\alpha/2}([0, T_0]), \quad v \in W_p^{1,2}(D_{g,h}^{T_0}) \cap C^{(1+\alpha)/2,\, 1+\alpha}(\overline{D}_{g,h}^{T_0}),$$

and $|g(t)|, h(t) \leqslant 2h_0$ in $[0, T_0]$ and

$$\|g, h\|_{C^{1+\alpha/2}([0, T_0])}, \ |v|_{1+\alpha, \overline{D}_{g,h}^{T_0}} \leqslant K(T_0^{-1}), \ \ 0 < v \leqslant B \ \ \text{in} \ D_{g,h}^{T_0}, \qquad (8.61)$$

where the constant $K(T_0^{-1})$ is independent of $L(u, 1)$.

Moreover, we can choose T_0 to be independent of $u \in \mathbb{X}_{u_0}^1 \cap C_x^{1-}(\Pi_1)$. In fact, let $f_2^*(t, x, v) = \max_{0 \leqslant u \leqslant A} f_2(t, x, u, v)$ and (v^*, g^*, h^*) be the unique solution of (8.60) with $f_2(t, x, u(t, x), v)$ replaced by $f_2^*(t, x, v)$. We denote T^* as the maximum existence time of (v^*, g^*, h^*). Then the comparison principle shows that $v \leqslant v^*$, $g \geqslant g^*$, $h \leqslant h^*$ and $T_0 \geqslant T^*$.

Take $T_0^* = \min\left\{T_0, \left(\sigma/K(T_0^{-1})\right)^{2/\alpha}\right\}$. Then T_0^* does not depend on $u \in \mathbb{X}_{u_0}^1 \cap C_x^{1-}(\Pi_1)$ and $L(u, 1)$, and (8.61) holds with T_0 replaced by T_0^*. For convenience, we also write T_0^* as T_0. Noting $-g'(0) \geqslant 2\sigma$ and $h'(0) \geqslant 2\sigma$. Using the $C^{\alpha/2}([0, T_0])$ estimates of $g'(t), h'(t)$ given in (8.61), we can find a $0 < T_* \leqslant T_0$, which depends only on $K(T_0^{-1})$, such that

$$-g'(t) \geqslant \sigma, \ \ h'(t) \geqslant \sigma \ \ \text{for all} \ 0 \leqslant t \leqslant T_*.$$

Clearly, T_* does not depend on $u \in \mathbb{X}_{u_0}^1 \cap C_x^{1-}(\Pi_1)$ and $L(u, 1)$, and (8.61) holds with T_0 replaced by T_*. We also write T_* as T_0.

Define $\tilde{u}_0(x) = u_0(x)$ when $|x| \leqslant h_0$, and $\tilde{u}_0(x) = 0$ when $|x| > h_0$. Then $\tilde{u}_0 \in C^{1-}([g(T_0), h(T_0)])$ since $u_0 \in C^{1-}([-h_0, h_0])$. For functions $g(t)$ and $h(t)$ obtained

---

$\ddagger$Using the Schauder fixed point theorem to determine T_0 and get the existence of solutions first, we can see that T_0 is independent of $L(u, 1)$. Then prove the uniqueness of solutions.

above, it is easy to see that their inverse functions $g^{-1}(x)$ and $h^{-1}(x)$ exist for $x \in [g(T_0), h(T_0)]$. We set (see Fig. 8.3)

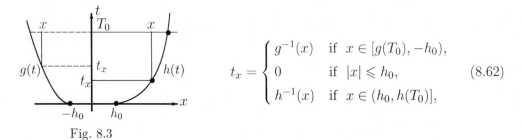

$$t_x = \begin{cases} g^{-1}(x) & \text{if } x \in [g(T_0), -h_0), \\ 0 & \text{if } |x| \leqslant h_0, \\ h^{-1}(x) & \text{if } x \in (h_0, h(T_0)], \end{cases} \tag{8.62}$$

Fig. 8.3

which is Lipschitz continuous in x. For the function $v(t, x)$ obtained above, and for every $g(T_0) < x < h(T_0)$, we consider the following initial value problem

$$\begin{cases} \tilde{u}_t = f_1(t, x, \tilde{u}, v(t, x)), & t_x < t \leqslant T_0, \\ \tilde{u}(t_x, x) = \tilde{u}_0(x). \end{cases} \tag{8.63}$$

Note properties of f_1 and v. By the standard theory for ODE, we can prove that there exists $0 < T < T_0$, which depends on L_1, A, B and K, such that for each $g(T) \leqslant x \leqslant h(T)$, the solution $\tilde{u}(t, x)$ of (8.63) exists uniquely in $[t_x, T]$ as a function of t. Consequently, we get a function $\tilde{u}$ of (t, x), which is defined in $\overline{D}_{g,h}^T$. Moreover, $\tilde{u} \in C^{1,1-}(\overline{D}_{g,h}^T)$, of which the Lipschitz constant will be calculated in the next step, and $\tilde{u}(t, g(t)) = \tilde{u}(t, h(t)) = 0$ for $0 \leqslant t \leqslant T$. Make zero extension of $\tilde{u}$ to $[0, t_x]$ for every $g(T) \leqslant x \leqslant h(T)$. Then $\tilde{u} \in C^{1-,1-}([0, T] \times [g(T), h(T)])$.

Step 2. Now, we intend to get an estimate of the Lipschitz constant of $\tilde{u}$ in x. Since $-g'(t), h'(t) \geqslant \sigma > 0$ for all $0 \leqslant t \leqslant T_0$, we have

$$\|(g^{-1})'\|_{C([g(T_0), -h_0])} \leqslant \sigma^{-1} \quad \text{and} \quad \|(h^{-1})'\|_{C([h_0, h(T_0)])} \leqslant \sigma^{-1}. \tag{8.64}$$

Set $F_1(s, x) = f_1(s, x, \tilde{u}(s, x), v(s, x))$. It follows from the first equation of (8.63) that, for $t_x < \tau \leqslant T$,

$$\tilde{u}(\tau, x) = \tilde{u}(t_x, x) + \int_{t_x}^{\tau} F_1(s, x) \mathrm{d}s.$$

For any given $(t, x_1), (t, x_2) \in \Pi_T$, we divide arguments into several cases.

Case 1: $(t, x_1), (t, x_2) \in \overline{D}_{g,h}^T$ with $-h_0 \leqslant x_2 < x_1$. Then $t \geqslant t_{x_1} \geqslant t_{x_2} \geqslant 0$, and consequently, for any $t_{x_1} \leqslant \tau \leqslant t$,

$$|\tilde{u}(\tau, x_1) - \tilde{u}(\tau, x_2)| \leqslant |\tilde{u}(t_{x_1}, x_1) - \tilde{u}(t_{x_2}, x_2)| + \int_{t_{x_2}}^{t_{x_1}} |F_1(s, x_2)| \mathrm{d}s$$

$$+ \int_{t_{x_1}}^{\tau} |F_1(s, x_1) - F_1(s, x_2)| \mathrm{d}s.$$

In view of the condition **(L)**, it is easy to derive that

$$|F_1(s, x_1) - F_1(s, x_2)| \leqslant L_1(|x_1 - x_2| + |\tilde{u}(s, x_1) - \tilde{u}(s, x_2)| + |v(s, x_1) - v(s, x_2)|),$$

where $L_1 = L(1, 2h_0, A, B)$. It yields

$$
\begin{aligned}
|\tilde{u}(\tau, x_1) - \tilde{u}(\tau, x_2)| &\leqslant T L_1 \|\tilde{u}(\cdot, x_1) - \tilde{u}(\cdot, x_2)\|_{C([t_{x_1}, t])} + |\tilde{u}(t_{x_1}, x_1) - \tilde{u}(t_{x_2}, x_2)| \\
&\quad + L_1|x_1 - x_2| + C_1|t_{x_1} - t_{x_2}| \\
&\quad + L_1 \int_{t_{x_1}}^{\tau} |v(s, x_1) - v(s, x_2)| ds
\end{aligned}
\tag{8.65}
$$

as $\tau \leqslant T \leqslant 1$, where $C_1 = \|f_1\|_{\infty, \Pi_1 \times (0,A) \times (0,B)}$.

By use of (8.61),

$$
\int_{t_{x_1}}^{\tau} |v(s, x_1) - v(s, x_2)| ds \leqslant T \|v_x\|_{\infty, D_{g,h}^T} |x_1 - x_2| \leqslant C_2 |x_1 - x_2|
$$

as $\tau \leqslant T \leqslant 1$. If $t_{x_1} > 0$ and $t_{x_2} > 0$, then $\tilde{u}(t_{x_1}, x_1) = \tilde{u}(t_{x_2}, x_2) = 0$, and

$$
|t_{x_1} - t_{x_2}| = |h^{-1}(x_1) - h^{-1}(x_2)| \leqslant \|(h^{-1})'\|_{C([h_0, h(T)])} |x_1 - x_2| \leqslant \sigma^{-1}|x_1 - x_2|
$$

by (8.64). If $t_{x_1} > 0$ and $t_{x_2} = 0$, then $x_2 \in [-h_0, h_0]$, $x_1 > h_0$ and $\tilde{u}(t_{x_1}, x_1) = 0$. Let L_0 be the Lipschitz constant of u_0 in x. It then follows that

$$
|t_{x_1} - t_{x_2}| = |h^{-1}(x_1) - h^{-1}(h_0)| \leqslant \sigma^{-1}|x_1 - h_0| \leqslant \sigma^{-1}|x_1 - x_2|,
$$
$$
|\tilde{u}(t_{x_1}, x_1) - \tilde{u}(t_{x_2}, x_2)| = |u_0(h_0) - u_0(x_2)| \leqslant L_0|h_0 - x_2| \leqslant L_0|x_1 - x_2|.
$$

If $t_{x_1} = t_{x_2} = 0$, i.e., $x_1, x_2 \in [-h_0, h_0]$, then

$$
|\tilde{u}(t_{x_1}, x_1) - \tilde{u}(t_{x_2}, x_2)| = |u_0(x_1) - u_0(x_2)| \leqslant L_0|x_2 - x_1|.
$$

Substituting these estimates into (8.65), we conclude

$$
|\tilde{u}(\tau, x_1) - \tilde{u}(\tau, x_2)| \leqslant T L_1 \|\tilde{u}(\cdot, x_1) - \tilde{u}(\cdot, x_2)\|_{C([t_{x_1}, t])} + M|x_1 - x_2|,
$$

where $M = L_0 + L_1 + C_1 \sigma^{-1} + C_2 L_1$. Take the maximum of $|\tilde{u}(\tau, x_1) - \tilde{u}(\tau, x_2)|$ in $[t_{x_1}, t]$ to yield

$$
\|\tilde{u}(\cdot, x_1) - \tilde{u}(\cdot, x_2)\|_{C([t_{x_1}, t])} \leqslant T L_1 \|\tilde{u}(\cdot, x_1) - \tilde{u}(\cdot, x_2)\|_{C([t_{x_1}, t])} + M|x_1 - x_2|,
$$

which implies

$$
|\tilde{u}(t, x_1) - \tilde{u}(t, x_2)| \leqslant \|\tilde{u}(\cdot, x_1) - \tilde{u}(\cdot, x_2)\|_{C([t_{x_1}, t])} \leqslant 2M|x_1 - x_2|
\tag{8.66}
$$

provided $0 < T \leqslant \min\{1, \frac{1}{2L_1}\}$.

Case 2: $(t, x_1), (t, x_2) \in \overline{D}_{g,h}^T$ with $x_2 < x_1 \leqslant h_0$. Similar to above, (8.66) holds.

Case 3: $(t, x_1), (t, x_2) \in \overline{D}_{g,h}^T$ with $x_2 < -h_0 < h_0 < x_1$. Then

$$
|\tilde{u}(t, x_1) - \tilde{u}(t, x_2)| \leqslant |\tilde{u}(t, x_1) - \tilde{u}(t, h_0)| + |\tilde{u}(t, h_0) - \tilde{u}(t, x_2)|
$$
$$
\leqslant 4M|x_1 - x_2|.
\tag{8.67}
$$

Case 4: $(t, x_1), (t, x_2) \notin \overline{D}_{g,h}^T$. Then $\tilde{u}(t, x_1) = \tilde{u}(t, x_2) = 0$.

Case 5: $(t, x_1) \notin \overline{D}_{g,h}^T$ and $(t, x_2) \in \overline{D}_{g,h}^T$. We may assume $x_1 > h(t)$. It is clear that $(t, h(t)), (t, x_2) \in \overline{D}_{g,h}^T$ and $\tilde{u}(t, x_1) = \tilde{u}(t, h(t)) = 0$. Hence, we have

$$|\tilde{u}(t, x_1) - \tilde{u}(t, x_2)| \leqslant |\tilde{u}(t, h(t)) - \tilde{u}(t, x_2)| \leqslant 4M|h(t) - x_2| \leqslant 4M|x_1 - x_2|.$$

In conclusion, the estimate (8.67) always holds provided $0 < T \leqslant \min\{1, \frac{1}{2L_1}\}$. Define

$$\mathbb{Y}_{u_0}^T = \{\phi \in \mathbb{X}_{u_0}^T : |\phi(t, x) - \phi(t, y)| \leqslant 4M|x - y| \text{ for } -2h_0 \leqslant x, y \leqslant 2h_0\}.$$

Obviously, $\mathbb{Y}_{u_0}^T$ is complete with metric $d(\phi_1, \phi_2) = \max_{\Pi_T} |\phi_1 - \phi_2|$. For any given $u \in \mathbb{Y}_{u_0}^T$, we extend u to $[T, 1] \times [-2h_0, 2h_0]$ by setting $u(t, x) = u(T, x)$. Then $u \in \mathbb{X}_{u_0}^1 \cap C_x^{1-}(\Pi_1)$. Define a mapping Γ by

$$\Gamma(u) = \tilde{u},$$

where $\tilde{u}$ is given in Step 1. Above discussions show that Γ maps $\mathbb{Y}_{u_0}^T$ into itself.

Step 3. We hope to show that Γ is a contraction mapping in $\mathbb{Y}_{u_0}^T$ for $0 < T \ll 1$. Let (v_i, g_i, h_i) be the unique local solution of (8.60) with $u = u_i \in \mathbb{Y}_{u_0}^T$, $i = 1, 2$, and define t_x^i by the manner (8.62) with $(g, h) = (g_i, h_i)$. Let $\tilde{u}_i$ be the unique solution of (8.63) with $t_x = t_x^i$, $v = v_i$ and $T_0 = T$. Then

$$\tilde{u}_i(t, x) = \tilde{u}_i(t_x^i, x) + \int_{t_x^i}^t f_1(s, x, \tilde{u}_i, v_i) ds \quad \text{for } x \in [g_i(T), h_i(T)].$$

Make zero extensions of $\tilde{u}_i$ and v_i in $([0, T] \times \mathbb{R}) \setminus D_{g_i, h_i}^T$, and set

$$U = u_1 - u_2, \ \widetilde{U} = \tilde{u}_1 - \tilde{u}_2, \ h = h_1 - h_2, \ g = g_1 - g_2, \ \Omega_T = D_{g_1, h_1}^T \cup D_{g_2, h_2}^T.$$

Fix $(t, x) \in \Omega_T$, we now estimate $|\widetilde{U}(t, x)|$ in all possible cases.

Case 1: $x \in (g_1(t), h_1(t)) \setminus (g_2(t), h_2(t))$. In such case, either $g_1(t) < x \leqslant g_2(t)$ or $h_2(t) \leqslant x < h_1(t)$, so we have $\tilde{u}_1(t_x^1, x) = 0$, $\tilde{u}_2(t, x) = 0$. Thus

$$|\widetilde{U}(t, x)| = |\tilde{u}_1(t, x)| = \left| \int_{t_x^1}^t f_1(s, x, \tilde{u}_1, v_1) ds \right| \leqslant C_1 |t - t_x^1|,$$

where $C_1 = \|f_1\|_{\infty, \Pi_1 \times (0, A) \times (0, B)}$.

When $h_2(t) \leqslant x < h_1(t)$, we have $0 < t_x^1 < t$ and $h_1(t) > h_1(t_x^1) = x \geqslant h_2(t)$. Therefore,

$$\begin{aligned}
|\widetilde{U}(t, x)| &\leqslant C_1 |t - t_x^1| = C_1 |h_1^{-1}(h_1(t)) - h_1^{-1}(h_1(t_x^1))| \\
&\leqslant C_1 \|(h_1^{-1})'\|_{C([0,T])} |h_1(t) - h_1(t_x^1)| \\
&\leqslant C_1 \sigma^{-1} |h_1(t) - h_1(t_x^1)| \\
&\leqslant C_1 \sigma^{-1} |h_1(t) - h_2(t)| \\
&\leqslant C_1 \sigma^{-1} \|h\|_{C([0,T])},
\end{aligned}$$

where $\sigma > 0$ is given in (8.64). When $g_1(t) < x \leqslant g_2(t)$, we can analogously obtain

$$|\widetilde{U}(t,x)| = |\tilde{u}_1(t,x)| \leqslant C_1\sigma^{-1}\|g\|_{C([0,T])}.$$

<u>Case 2</u>: $x \in (g_2(t), h_2(t)) \setminus (g_1(t), h_1(t))$. Similar to <u>Case 1</u>, we can get

$$|\widetilde{U}(t,x)| = |\tilde{u}_2(t,x)| \leqslant C_1\sigma^{-1}\|g,\,h\|_{C([0,T])}.$$

<u>Case 3</u>: $x \in (g_1(t), h_1(t)) \cap (g_2(t), h_2(t))$. When $x \in [-h_0, h_0]$, we have $t_x^1 = t_x^2 = 0$ and $\tilde{u}_1(t_x^1, x) = \tilde{u}_2(t_x^2, x) = \tilde{u}_0(x)$. Hence

$$\begin{aligned}
|\widetilde{U}(t,x)| &\leqslant \int_0^t |f_1(s,x,\tilde{u}_1,v_1) - f_1(s,x,\tilde{u}_2,v_2)|\mathrm{d}s \\
&\leqslant TL_1\left(\|\widetilde{U}\|_{C(\overline{\Omega}_T)} + \|v_1 - v_2\|_{C(\overline{\Omega}_T)}\right).
\end{aligned}$$

When $x \in (g_1(t), h_1(t)) \cap (g_2(t), h_2(t)) \setminus [-h_0, h_0]$, we have $t_x^1 > 0, t_x^2 > 0$ and $\tilde{u}_1(t_x^1, x) = \tilde{u}_2(t_x^2, x) = 0$. Without loss of generality, we assume $x > h_0$ and $t_x^2 > t_x^1 > 0$. Then $h_1(t_x^2) > h_1(t_x^1) = x = h_2(t_x^2)$, $x \in (g_1(s), h_1(s)) \cap (g_2(s), h_2(s))$ for all $t_x^2 < s \leqslant t$ and $x \in (g_1(t_x^2), h_1(t_x^2)) \setminus (g_2(t_x^2), h_2(t_x^2))$. Consequently

$$|\widetilde{U}(t_x^2, x)| = |\tilde{u}_1(t_x^2, x)| \leqslant C_1\sigma^{-1}\|g,\,h\|_{C([0,T])}$$

in line with the conclusion of <u>Case 1</u>. Integrating the differential equation of $\tilde{u}_i$ from t_x^2 to t, we obtain

$$\tilde{u}_1(t,x) = \tilde{u}_1(t_x^2, x) + \int_{t_x^2}^t f_1(s,x,\tilde{u}_1,v_1)\mathrm{d}s, \quad \tilde{u}_2(t,x) = \int_{t_x^2}^t f_1(s,x,\tilde{u}_2,v_2)\mathrm{d}s.$$

It follows that

$$\begin{aligned}
|\widetilde{U}(t,x)| &\leqslant |\tilde{u}_1(t_x^2, x)| + \int_{t_x^2}^t |f_1(s,x,\tilde{u}_1,v_1) - f_1(s,x,\tilde{u}_2,v_2)|\mathrm{d}s \\
&\leqslant |\widetilde{U}(t_x^2,x)| + \int_0^t |f_1(s,x,\tilde{u}_1,v_1) - f_1(s,x,\tilde{u}_2,v_2)|\mathrm{d}s \\
&\leqslant C_1\sigma^{-1}\|g,\,h\|_{C([0,T])} + TL_1\left(\|\widetilde{U}\|_{C(\overline{\Omega}_T)} + \|v_1 - v_2\|_{C(\overline{\Omega}_T)}\right).
\end{aligned}$$

In conclusion,

$$|\widetilde{U}(t,x)| \leqslant C_1\sigma^{-1}\|g,\,h\|_{C([0,T])} + TL_1\left(\|\widetilde{U}\|_{C(\overline{\Omega}_T)} + \|v_1 - v_2\|_{C(\overline{\Omega}_T)}\right). \tag{8.68}$$

We will show in the following Step 4 that if $0 < T \ll 1$ then there exists a positive constant C such that

$$\|g,\,h\|_{C^1([0,T])} \leqslant C\|U\|_{C(\Pi_T)} \quad \text{and} \quad \|v_1 - v_2\|_{C(\overline{\Omega}_T)} \leqslant C\|U\|_{C(\Pi_T)}. \tag{8.69}$$

Once this is done, noticing $g(0) = h(0) = 0$, the first inequality of (8.69) implies

$$\|g,\,h\|_{C([0,T])} \leqslant T\|g,\,h\|_{C^1([0,T])} \leqslant TC\|U\|_{C(\Pi_T)}.$$

Then combining with (8.68), we have

$$\|\tilde{U}\|_{C(\Pi_T)} \leqslant \frac{1}{3}\|U\|_{C(\Pi_T)} \quad \text{if } 0 < T \ll 1.$$

This demonstrates that Γ is a contraction mapping in $\mathbb{Y}_{u_0}^T$. Hence, Γ has a unique fixed point u in $\mathbb{Y}_{u_0}^T$ by the contraction mapping principle. Let (v, g, h) be the unique solution of (8.60) with such u. Then (u, v, g, h) is a solution of (8.59), and it is the unique one provided $u \in \mathbb{Y}_{u_0}^T$. Moreover, we can see that $u \in C^{1,1-}(\overline{D}_{g,h}^T)$ and $v \in C^{1,2}(D_{g,h}^T) \cap C^{(1+\alpha)/2, 1+\alpha}(\overline{D}_{g,h}^T)$.

Step 4. Now we show the estimate (8.69). Before our statement, some preparations are needed. Let

$$x_i(t, y) = \frac{1}{2}[(h_i(t) - g_i(t))y + h_i(t) + g_i(t)],$$

$$\xi_i(t) = \frac{2}{h_i(t) - g_i(t)}, \quad \zeta_i(t, y) = \frac{h_i'(t) + g_i'(t)}{h_i(t) - g_i(t)} + \frac{h_i'(t) - g_i'(t)}{h_i(t) - g_i(t)}y,$$

$$w_i(t, y) = u_i(t, x_i(t, y)), \quad z_i(t, y) = v_i(t, x_i(t, y))$$

and

$$f_2^i(t, y, w, z) = f_2(t, x_i(t, y), w, z)$$

for $i = 1, 2$. Then

$$\begin{cases} z_{i,t} = d\xi_i^2 z_{i,yy} + \zeta_i z_{i,y} + f_2^i(t, y, w_i(t, y), z_i(t, y)), & 0 < t \leqslant T, \ |y| < 1, \\ z_i(t, \pm 1) = 0, & 0 \leqslant t \leqslant T, \\ z_i(0, y) = v_0(h_0 y) =: z_{i0}(y), & |y| \leqslant 1. \end{cases}$$

Recalling (8.61), we have

$$\frac{1}{2h_0} \leqslant \xi_i(t) \leqslant \frac{1}{h_0}, \quad t \in [0, T], \tag{8.70}$$

$$\sup_{|t-s|<R} |\xi_i(t) - \xi_i(s)| = \sup_{|t-s|<R} \left| \frac{2[h_i(t) - h_i(s) - (g_i(t) - g_i(s))]}{[h_i(t) - g_i(t)][h_i(s) - g_i(s)]} \right|$$

$$\leqslant \frac{\|g_i, h_i\|_{C^1([0,T])}}{2h_0^2} R \to 0 \quad \text{as } R \to 0. \tag{8.71}$$

Moreover,

$$\|\zeta_i\|_{\infty, \Delta_T^*} \leqslant \frac{2K}{h_0} \quad \text{and} \quad \|f_2^i\|_{\infty, \Delta_T^*} \leqslant C_0. \tag{8.72}$$

Now, we can apply the L^p theory to obtain $\|z_i\|_{W_p^{1,2}(\Delta_T^*)} \leqslant C_1'$. Noting that the dimension of space is 1, and following discussions of [111, §5.4 and §5.5] for z_i and $z_{i,y}$ without extension of z_i to a large domain, we can obtain

$$[z_i]_{C^{\alpha/2, \alpha}(\Delta_T^*)}, \ [z_{i,y}]_{C^{\alpha/2, \alpha}(\Delta_T^*)} \leqslant \overline{C}, \tag{8.73}$$

where $\overline{C}$ is independent of T^{-1}. It follows that

$$\|z_{i,y}\|_{C(\Delta_T^*)} \leqslant \|z_{i0}'(y)\|_{C([-1,1])} + \overline{C}T^{\alpha/2} \leqslant \|z_{i0}'(y)\|_{C([-1,1])} + \overline{C} =: C_2'. \quad (8.74)$$

On the other hand, $z = z_1 - z_2$ satisfies

$$\begin{cases} z_t - d\xi_1^2 z_{yy} - \zeta_1 z_y - a(t,y)z = d(\xi_1 - \xi_2)z_{2,yy} \\ \qquad + (\zeta_1^2 - \zeta_2^2)z_{2,y} + b(t,y)(w_1 - w_2) + c(t,y), \quad 0 < t \leqslant T, \ |y| < 1, \\ z(t, \pm 1) = 0, \qquad\qquad\qquad\qquad\qquad\qquad\qquad\qquad 0 \leqslant t \leqslant T, \\ z(0,y) = 0, \qquad\qquad\qquad\qquad\qquad\qquad\qquad\qquad\qquad |y| \leqslant 1, \end{cases} \quad (8.75)$$

and $g(t) = g_1(t) - g_2(t)$, $h(t) = h_1(t) - h_2(t)$ satisfy

$$\begin{cases} g'(t) = -\mu\xi_1(t)z_y(t,-1) - \mu(\xi_1(t) - \xi_2(t))z_{2,y}(t,-1), \quad 0 < t \leqslant T, \\ h'(t) = -\beta\xi_1(t)z_y(t,1) - \beta(\xi_1(t) - \xi_2(t))z_{2,y}(t,1), \qquad 0 < t \leqslant T, \\ g(0) = h(0) = 0, \end{cases}$$

where

$$a(t,y) = \int_0^1 f_{2,v}^1(t,y,w_1,z_2 + (z_1 - z_2)\tau)d\tau,$$

$$b(t,y) = \int_0^1 f_{2,u}^2(t,y,w_2 + (w_1 - w_2)\tau, z_2)d\tau,$$

$$c(t,y) = f_2^1(t,y,w_1,z_2) - f_2^2(t,y,w_1,z_2).$$

Clearly, $\|a, b, c\|_{\infty, \Delta_T^*} \leqslant L_1$. Due to (8.70)–(8.72), (8.74) and

$$\begin{aligned} |c(t,y)| &= |f_2(t,x_1(t,y),w_1,z_2) - f_2(t,x_2(t,y),w_1,z_2)| \\ &\leqslant L|x_1(t,y) - x_2(t,y)| \\ &\leqslant C(|g(t)| + |h(t)|), \end{aligned}$$

an application of the L^p theory to (8.75) deduces

$$\|z\|_{W_p^{1,2}(\Delta_T^*)} \leqslant C_3 \left(\|g,h\|_{C^1([0,T])} + \|w_1 - w_2\|_{C(\Delta_T^*)} \right),$$

where C_3 depends on h_0, μ, β, A, B, and K. Same as the derivation of (8.73), we have

$$[z]_{C^{\alpha/2,\alpha}(\Delta_T^*)} + [z_y]_{C^{\alpha/2,\alpha}(\Delta_T^*)} \leqslant C_4 \left(\|g,h\|_{C^1([0,T])} + \|w_1 - w_2\|_{C(\Delta_T^*)} \right), \quad (8.76)$$

where $C_4 > 0$ is independent of T^{-1}. When $(t,y) \in \Delta_T^*$, we have

$$\begin{aligned} |w_1(t,y) - w_2(t,y)| &= |u_1(t,x_1(t,y)) - u_2(t,x_2(t,y))| \\ &\leqslant |u_1(t,x_1(t,y)) - u_2(t,x_1(t,y))| \\ &\quad + |u_2(t,x_1(t,y)) - u_2(t,x_2(t,y))| \\ &\leqslant \|U\|_{C(\Pi_T)} + L(u,1)|x_1(t,y) - x_2(t,y)| \\ &\leqslant C_5 \left(\|U\|_{C(\Pi_T)} + \|g,h\|_{C([0,T])} \right), \end{aligned}$$

where C_5 depends only on h_0 and the Lipschitz constant $L(u, 1)$ of u. Therefore,

$$\|w_1 - w_2\|_{C(\Delta_T^*)} \leqslant C_5 \left(\|U\|_{C(\Pi_T)} + \|g, h\|_{C([0,T])} \right).$$

This combined with (8.76) asserts

$$[z]_{C^{\alpha/2, \alpha}(\Delta_T^*)} + [z_y]_{C^{\alpha/2, \alpha}(\Delta_T^*)} \leqslant C_6 \left(\|g, h\|_{C^1([0,T])} + \|U\|_{C(\Pi_T)} \right). \tag{8.77}$$

Noticing $z_y(0, 1) = 0$, the above estimate implies

$$\|z_y(t, 1)\|_{C([0,T])} \leqslant C_6 T^{\alpha/2} \left(\|g, h\|_{C^1([0,T])} + \|U\|_{C(\Pi_T)} \right). \tag{8.78}$$

As $h(0) = g(0) = 0$, it is easy to see that

$$|h(t)| \leqslant T \|h'\|_{C([0,T])}, \quad |g(t)| \leqslant T \|g'\|_{C([0,T])}. \tag{8.79}$$

Making use of (8.74) and (8.78) we conclude

$$
\begin{aligned}
|h_1'(t) - h_2'(t)| &= \beta |v_{1,x}(t, h_1(t)) - v_{2,x}(t, h_2(t))| \\
&= 2\beta \left| \frac{[z_{1,y}(t, 1) - z_{2,y}(t, 1)]}{h_1(t) - g_1(t)} + \frac{z_{2,y}(t, 1)[g(t) - h(t)]}{[h_1(t) - g_1(t)][h_2(t) - g_2(t)]} \right| \\
&\leqslant \frac{\beta}{h_0} |z_y(t, 1)| + 2\beta |z_{2,y}(t, 1)| \frac{|h(t)| + |g(t)|}{4h_0^2} \\
&\leqslant C_7 T^{\alpha/2} \left(\|g, h\|_{C^1([0,T])} + \|U\|_{C(\Pi_T)} \right).
\end{aligned}
$$

Therefore, in the light of (8.79),

$$\|h'\|_{C([0,T])} \leqslant C_8 T^{\alpha/2} \left(\|g', h'\|_{C([0,T])} + \|U\|_{C(\Pi_T)} \right).$$

Analogously,

$$\|g'\|_{C([0,T])} \leqslant C_8' T^{\alpha/2} \left(\|g', h'\|_{C([0,T])} + \|U\|_{C(\Pi_T)} \right).$$

Consequently, $\|g', h'\|_{C([0,T])} \leqslant C_9 \|U\|_{C(\Pi_T)}$ provided that T is small enough. Recalling (8.79), we get the first inequality of (8.69), i.e.,

$$\|g, h\|_{C^1([0,T])} \leqslant C_9' \|U\|_{C(\Pi_T)}. \tag{8.80}$$

Moreover, due to $z(0, y) = 0$, we have

$$|z(t, y)| = |z(t, y) - z(0, y)| \leqslant t^{\alpha/2} [z]_{C^{\alpha/2, \alpha}(\Delta_T^*)} \quad \text{for all } (t, y) \in \Delta_T^*.$$

This combined with (8.77) allows us to derive

$$\|z\|_{C(\Delta_T^*)} \leqslant T^{\alpha/2} [z]_{C^{\alpha/2, \alpha}(\Delta_T^*)} \leqslant C_6 T^{\alpha/2} \left(\|g, h\|_{C^1([0,T])} + \|U\|_{C(\Pi_T)} \right). \tag{8.81}$$

Now, we are ready to estimate $\|v_1 - v_2\|_{C(\bar{\Omega}_T)}$. Fix $(t, x) \in \bar{\Omega}_T$, and let

$$y_i(t, x) = \frac{2x - g_i(t) - h_i(t)}{h_i(t) - g_i(t)}.$$

When $x \in [g_1(t), h_1(t)] \cap [g_2(t), h_2(t)]$, using (8.74), (8.80) and (8.81) in succession, we have

$$
\begin{aligned}
|v_1(t, x) - v_2(t, x)| &= |z_1(t, y_1) - z_2(t, y_2)| \\
&\leqslant |z_1(t, y_1) - z_2(t, y_1)| + |z_2(t, y_1) - z_2(t, y_2)| \\
&\leqslant \|z\|_{C(\Delta_T^*)} + \|z_{2,y}\|_{C(\Delta_T^*)}|y_1 - y_2| \\
&\leqslant \|z\|_{C(\Delta_T^*)} + \frac{2}{h_0}\|z_{2,y}\|_{C(\Delta_T^*)}\|g, h\|_{C([0,T])} \\
&\leqslant C_6 T^{\alpha/2}\left(\|g, h\|_{C^1([0,T])} + \|U\|_{C(\Pi_T)}\right) + \frac{2C_2'}{h_0}\|g, h\|_{C([0,T])} \\
&\leqslant C_{10}\|U\|_{C(\Pi_T)}.
\end{aligned}
\tag{8.82}
$$

When $x \in [g_1(t), h_1(t)] \setminus [g_2(t), h_2(t)]$, we have $v_2(t, x) = 0$. Without loss of generality, we may think that $x \in [g_1(t), g_2(t))$ and $g_2(t) \leqslant h_1(t)$. Take advantage of (8.82), it yields

$$
\begin{aligned}
|v_1(t, x) - v_2(t, x)| &= |v_1(t, x) - v_2(t, g_2(t))| \\
&\leqslant |v_1(t, x) - v_1(t, g_2(t))| + |v_1(t, g_2(t)) - v_2(t, g_2(t))| \\
&\leqslant \|v_{1,x}\|_{C(\bar{D}_{g_1,h_1}^T)}|g_1(t) - g_2(t)| + C_{10}\|U\|_{C(\Pi_T)} \\
&\leqslant C_{11}\|U\|_{C(\Pi_T)}.
\end{aligned}
$$

When $x \in [g_2(t), h_2(t)] \setminus [g_1(t), h_1(t)]$, similar to above we still have

$$|v_1(t, x) - v_2(t, x)| \leqslant C_{12}\|U\|_{C(\Pi_T)}.$$

In conclusion,

$$\|v_1 - v_2\|_{C(\bar{\Omega}_T)} \leqslant C\|U\|_{C(\Pi_T)}$$

if $0 < T \ll 1$. The estimate (8.69) is proved.

Step 5. Proof of the uniqueness. Let $(\tilde{u}, \tilde{v}, \tilde{g}, \tilde{h})$ be any solution of (8.59). It is easy to see from Step 2 that $\tilde{u} \in Y_{u_0}^T$ if $0 < T \ll 1$. Thus $(\tilde{u}, \tilde{v}, \tilde{g}, \tilde{h}) = (u, v, g, h)$, and the proof is fulfilled. $\qquad\square$

Theorem 8.27 *Suppose that the condition* **(L)** *holds, and*

(L1) (f_1, f_2) *does not depend on* (t, x), *i.e.,* $f_1(t, x, u, v) = f_1(u, v)$, $f_2(t, x, u, v) = f_2(u, v)$.

Assume also that (f_1, f_2) *is quasi-monotonically increasing for* $u, v \geqslant 0$. *If the initial value problem*

$$
\begin{cases}
\phi'(t) = f_1(\phi, \psi), \quad \psi'(t) = f_2(\phi, \psi), \quad t > 0, \\
\phi(0) = \max_{[-h_0, h_0]} u_0 > 0, \quad \psi(0) = \max_{[-h_0, h_0]} v_0 > 0
\end{cases}
$$

has a global solution (ϕ, ψ), then the unique solution (u, v, g, h) of (8.59) also exists globally.

Proof. Clearly, $\phi(t) > 0$, $\psi(t) > 0$ for all $t \geqslant 0$. Let $T_{\max}$ be the maximal existence time of (u, v, g, h). For any fixed $0 < T < T_{\max}$, applying the comparison principle in region $D_{g,h}^T$, we conclude:

$$
\begin{cases}
u(t, x) \leqslant \phi(t) \leqslant \max_{[0, T+1]} \phi(t) =: M(T) & \text{in } \overline{D}_{g,h}^T, \\
v(t, x) \leqslant \psi(t) \leqslant \max_{[0, T+1]} \psi(t) =: N(T) & \text{in } \overline{D}_{g,h}^T.
\end{cases}
\tag{8.83}
$$

We have known in proving Theorem 8.26 that $h'(t) > 0$. Set

$$
A = \sup_{[0, M(T)] \times [0, N(T)]} f_2(u, v).
$$

As in the proof of Lemma 8.2,

$$
h'(t) \leqslant 2\beta \max\left\{ \frac{2}{h_0}, \sqrt{\frac{AN(T)}{2d}}, -\min_{[0, h_0]} v_0'(x) \right\} \quad \text{for all } 0 \leqslant t \leqslant T,
$$

$$
g'(t) \geqslant -2\mu \max\left\{ \frac{2}{h_0}, \sqrt{\frac{AN(T)}{2d}}, \max_{[-h_0, 0]} v_0'(x) \right\} \quad \text{for all } 0 \leqslant t \leqslant T.
$$

Recalling the estimate (8.83) and above estimates, and using a similar method to the proof of Theorem 8.5 we conclude $T_{\max} = \infty$. □

Theorem 8.28 *Let conditions* **(L)** *and* **(L1)** *hold. Assume also that there exists $k_0 > 0$ such that $f_1(u, v) < 0$ for all $u > k_0$ and $v \geqslant 0$, and suppose that for any given $\eta > 0$, there exists $\Theta(\eta) > 0$ such that $f_2(u, v) < 0$ for $0 \leqslant u \leqslant \eta$, $v \geqslant \Theta(\eta)$, then the unique solution (u, v, g, h) of (8.59) exists globally.*

Proof. It is easy to see that

$$
u(t, x) \leqslant \max_{[-h_0, h_0]} u_0 + k_0 =: \eta, \quad v(t, x) \leqslant \max_{[-h_0, h_0]} v_0 + \Theta(\eta).
$$

The remaining proof is the same as that of Theorem 8.27. □

In the past decade, research on free boundary models originating from species invasion and epidemic spread has developed rapidly, and a large number of research results have appeared. This book only focuses on the reaction-diffusion equation and one system coupled by one PDE and one ODE. In these two respects, in addition to the work mentioned above, see also [23, 28, 27, 41, 3, 56, 66, 83, 20, 30, 54, 78, 69] and references therein.

With regard to the work of systems, we just refer the reader to [26, 121, 43, 44, 123, 29, 122, 119, 77] for the competition models, [112, 120] for the predator models, and references therein. In particular, papers [44, 119, 77, 120] focus on the case where two components have different free boundaries, and the paper [29] deals with the propagation speed of free boundary of a competitive model. There have also been many studies on epidemic free boundary models, readers can refer to [51, 39, 75] and references therein.

At present, there are also many advances in the study of free boundary problems of non-local diffusion equations, and systems with non-local vs. local diffusions. The processing method is different from free boundary problems of general reaction-diffusion equations. Interested readers can refer to [13, 70, 103, 104, 125] for non-local diffusion problems and references therein.

EXERCISES

8.1 Prove the estimate (8.16).

8.2 Prove that solution (u, h) of (8.1) is strictly monotonically increasing in $\mu > 0$.

8.3 Study the following free boundary problem with double free boundaries and sign-changing intrinsic growth rate in heterogeneous time-periodic environment

$$\begin{cases} u_t - du_{xx} = a(t,x)u - b(t,x)u^2, & t > 0, \ g(t) < x < h(t), \\ u(t, g(t)) = u(t, h(t)) = 0, & t \geqslant 0, \\ g'(t) = -\rho u_x(t, g(t)), & t \geqslant 0, \\ h'(t) = -\mu u_x(t, h(t)), & t \geqslant 0, \\ u(0, x) = u_0(x), & -h_0 \leqslant x \leqslant h_0, \\ g(0) = -h_0, \quad h(0) = h_0, \end{cases}$$

where $a, b \in (C^{\alpha/2,1} \cap L^\infty)([0,T] \times \mathbb{R})$ with $\alpha \in (0,1)$, and are time-periodic with period T, a is positive somewhere in $[0,T] \times \mathbb{R}$, and $b_1 \leqslant b \leqslant b_2$ in $[0,T] \times \mathbb{R}$ for some positive constants b_1 and b_2. Investigate the following problems:

(1) existence and uniqueness of global solution;

(2) regularity and uniform estimates;

(3) longtime behaviour of solution;

(4) criteria and dichotomy for spreading and vanishing.

8.4 Prove that problem (8.53) has at least one positive solution.

8.5 Prove that there exists $0 < T_0 \ll 1$ such that (8.60) has a unique solution (v, g, h) which satisfies

$$g, h \in C^{1+\alpha/2}([0, T_0]), \quad v \in W_p^{1,2}(D_{g,h}^{T_0}) \cap C^{\frac{1+\alpha}{2}, 1+\alpha}(\overline{D}_{g,h}^{T_0}).$$

8.6 Consider the following free boundary problem of predator-prey model with double free boundaries

$$
\begin{cases}
u_t - u_{xx} = u(1 - u + av), & t > 0, \quad g(t) < x < h(t), \\
u(t, x) \equiv 0, & t \geqslant 0, \quad x \notin (g(t), h(t)), \\
v_t - dv_{xx} = v(b - v - cu), & t > 0, \quad x \in \mathbb{R}, \\
u = 0, \quad g'(t) = -\mu u_x, & t \geqslant 0, \quad x = g(t), \\
u = 0, \quad h'(t) = -\mu u_x, & t \geqslant 0, \quad x = h(t), \\
g(0) = -h_0, \quad h(0) = h_0, & \\
u(0, x) = u_0(x), & x \in [-h_0, h_0], \\
v(0, x) = v_0(x), & x \in \mathbb{R},
\end{cases}
$$

where a, b, c, d, h_0 and μ are given positive constants. The initial functions $u_0(x)$, $v_0(x)$ satisfy: $u_0 \in C^2([-h_0, h_0])$, $u_0(\pm h_0) = 0$, $u_0 > 0$ in $(-h_0, h_0)$, $v_0 \in C_b(\mathbb{R})$ and $v_0 > 0$ in $\mathbb{R}$, here $C_b(\mathbb{R})$ is the space of continuous and bounded functions in $\mathbb{R}$. Same as Exercise 8.3, investigate the following problems:

(1) existence and uniqueness of global solution;

(2) regularity and uniform estimates;

(3) longtime behaviour of solution;

(4) criteria and dichotomy for spreading and vanishing.

CHAPTER 9

Semigroup Theory and Applications

Semigroup theory provides a unified and powerful tool for the study of differential equations on Banach spaces covering systems described by ordinary differential equations, partial differential equations, functional differential equations, and their combinations.

The basic idea of the semigroup method is to rewrite an evolutionary partial differential equation as an abstract ordinary differential equation, and to deal with the evolutionary partial differential equation by imitating the initial value problem of the ordinary differential equation.

Considering an initial value problem of an ordinary differential system

$$\begin{cases} y'(t) + Ay(t) = f(t, y), & t > 0, \\ y(0) = y_0 \in \mathbb{R}^m, \end{cases}$$

where A is a matrix of order m. Based on the variation-of-constants formula, its solution can be represented as

$$y(t) = \mathrm{e}^{-tA} y_0 + \int_0^t \mathrm{e}^{-(t-s)A} f(s, y(s)) \mathrm{d}s.$$

Given an initial-boundary value problem of parabolic equation (system), for example,

$$\begin{cases} u_t - \Delta u = f(x, t, u), & x \in \Omega, \quad t > 0, \\ u = 0, & x \in \partial\Omega, \ t > 0, \\ u(x, 0) = u_0(x), & x \in \Omega, \end{cases}$$

we can rewrite it as an initial value problem of abstract ordinary differential equation

$$\begin{cases} u'(t) + Au(t) = f(t, u), & t > 0, \\ u(0) = u_0, \end{cases}$$

where A is an operator from a subspace X_0 (of a Banach space X) to X. Similarly to the ordinary differential equation, its solution can be written formally as

$$u(t) = e^{-tA}u_0 + \int_0^t e^{-(t-s)A} f(s, u(s)) \mathrm{d}s.$$

The family of operators $\{e^{-tA}\}_{t \geqslant 0}$ is called a *semigroup*, and such a function u is called a *mild solution*. Then, by studying the properties of the semigroup, we can determine whether this mild solution satisfies the abstract ordinary differential equation.

This chapter is devoted to a brief review of basic results of semigroup theory and to establishing the existence, uniqueness, extension and regularity of mild solutions. An initial-boundary value problem of a system coupling parabolic equations and ordinary differential equations is systematically investigated.

For abstract integrals (Bochner integrals) and abstract derivatives (Fréchet derivatives and Gâteaux derivatives) of abstract functions (functions from a measurable space to a Banach space), readers can refer to textbooks on nonlinear functional analysis and evolution equations such as [15, 42]. Throughout this chapter, X is a Banach space equipped with the norm $\| \cdot \|$.

9.1 C_0 SEMIGROUP THEORY

9.1.1 Basic results

In this subsection, we review some basic results of C_0 semigroup theory. Their proofs will be omitted, and we refer to [89, 15, 82, 31, 109, 124] for details.

Let Ω be a smooth bounded domain and $\varphi \in C^2(\overline{\Omega})$, and let $u(x,t)$ be the unique solution of

$$\begin{cases} u_t - \Delta u = 0, & x \in \Omega,\ t > 0, \\ u = 0, & x \in \partial\Omega,\ t > 0, \\ u(x,0) = \varphi(x), & x \in \Omega. \end{cases}$$

Set $u(x,t) = Q(t)\varphi(x)$. Then $Q(t)$ has the following properties:

(a) $Q(0) = I$, the identity operator;

(b) $Q(t)Q(s) = Q(t+s)$ for all $t, s \geqslant 0$;

(c) $\lim_{t \to 0^+} \|Q(t)\varphi - \varphi\|_{C(\overline{\Omega})} = 0$ for all $\varphi \in C^2(\overline{\Omega})$;

(d) $\displaystyle \lim_{t \searrow 0} \frac{Q(t)\varphi - \varphi}{t} = \lim_{t \searrow 0} \frac{u(x,t) - u(x,0)}{t} = u_t(x,0) = \Delta u(x,0) = \Delta \varphi(x).$

Definition 9.1 *For $t > 0$, let $Q(t) : X \to X$ be continuously linear operators having above properties* (a)–(c) *with $\varphi \in C^2(\overline{\Omega})$ replaced by $\varphi \in X$ in* (c). *The family of operators $\{Q(t)\}_{t \geqslant 0}$ is called a* strongly continuous semigroup *or a C_0 semigroup* (*of operators*) *of X.*

For simplicity, we denote $Q(t) := \{Q(t)\}_{t \geqslant 0}$ henceforth.

Definition 9.2 *Let $Q(t)$ be a C_0 semigroup of X. We put*

$$D(A) = \left\{ v \in X : \lim_{t \to 0^+} \frac{Q(t)v - v}{t} \text{ exists} \right\}$$

and call the linear operator $-A : D(A) \to X$ defined by

$$-Av = \lim_{t \to 0^+} \frac{Q(t)v - v}{t}$$

the *infinitesimal generator of the semigroup $Q(t)$.*

For a given operator A, the resolvent set and spectrum of A are denoted by $\rho(A)$ and $\sigma(A)$, respectively.

Theorem 9.3 *Let $Q(t)$ be a C_0 semigroup of X and $-A$ the infinitesimal generator of $Q(t)$. Then the following conclusions are valid.*

(1) *There exist constants $M \geqslant 0$ and $a \in \mathbb{R}$ so that $\|Q(t)\| \leqslant Me^{-at}$ for all $t \geqslant 0$.*

(2) *For each $v \in X$, we have $Q(t)v \in C([0, \infty), X)$ and*

$$\int_0^t Q(s)v ds \in D(A), \quad Q(t)v - v = -A \int_0^t Q(s)v ds.$$

(3) *(Existence of solution) For $v \in X$, we define $u(t) = Q(t)v$, $t \geqslant 0$. When $v \in D(A)$, the derivative $u'(t)$ exists and*

$$u'(t) = -Au(t) = -Q(t)Av,$$

i.e., A commutes with $Q(t)$. In particular, if $v \in D(A)$ then so is $Q(t)v$. In that sense, there is no loss of regularity of $Q(t)v$ when compared with v. We can also write $Q'(t) = -Q(t)A$ in $D(A)$.

(4) *$(\lambda - A)^{-1} Q(t) = Q(t)(\lambda - A)^{-1}$ for all $\lambda \in \rho(A)$ and $t \geqslant 0$, where $\rho(A)$ is the resolvent set of A.*

(5) *A is closed and $D(A)$ is dense in X.*

(6) *(Uniqueness of solution) Let $\tau \in (0, \infty]$ and $u : [0, \tau) \to X$ be a continuous function. If for $0 < t < \tau$, $u(t) \in D(A)$, $u'(t)$ exists and $u'(t) = -Au(t)$, then $u(t) = Q(t)u(0)$ in $[0, \tau)$.*

(7) *(Uniqueness of semigroup) If $T(t)$ is a C_0 semigroup of X, and its infinitesimal generator is still $-A$, then $T(t) = Q(t)$ for all $t \geqslant 0$.*

The following two theorems give the necessary and sufficient conditions for a linear operator to be an infinitesimal generator of a C_0 semigroup.

Theorem 9.4 (Hille-Yosida Theorem) *Let $A : D(A) \to X$ be a closed and densely-defined linear operator. Then $-A$ is the infinitesimal generator of a C_0 semigroup $Q(t)$ satisfying $\|Q(t)\| \leqslant 1$ if and only if*

$$(-\infty, 0) \subset \rho(A) \quad \text{and} \quad \|(A - \lambda)^{-1}\| \leqslant \frac{1}{|\lambda|} \quad \text{for all } \lambda < 0. \tag{9.1}$$

A C_0 semigroup $Q(t)$ satisfying $\|Q(t)\| \leqslant 1$ is called a *contraction semigroup*.

Theorem 9.5 (Hille-Yosida Theorem) *Let $A : D(A) \to X$ be a linear operator. Then $-A$ is the infinitesimal generator of a C_0 semigroup $Q(t)$ satisfying $\|Q(t)\| \leqslant Me^{-at}$ for some constants M and a, if and only if*

(1) *$D(A)$ is dense in X;*

(2) *$(-\infty, a) \subset \rho(A)$ and*

$$\|(A - \lambda)^{-k}\| \leqslant M(a - \lambda)^{-k} \quad \text{for all } \lambda < a, \ k = 1, 2, \ldots. \tag{9.2}$$

Remark 9.6 (1) *Inequality (9.2) indicates that A is closed.*

(2) *Under conditions of Theorem 9.5 we have*

$$\lim_{k \to \infty} \sup_{t \in [0,T]} \left\| Q(t)v - \left(1 + \frac{t}{k} A \right)^{-k} v \right\| = 0 \quad \text{for } 0 < T < \infty, \ v \in X.$$

Based on this limit, it is reasonable to write $Q(t) = e^{-tA}$.

Here we consider an example.

Example 9.1 *Let Ω be of class C^2 and $X = \{u \in C(\overline{\Omega}) : u|_{\partial\Omega} = 0\}$. Define an operator A by*

$$D(A) = \{u \in H^2(\Omega) \cap X : \Delta u \in X\},$$

$$Au = -\Delta u \quad \text{for } u \in D(A).$$

Theorem 9.7 *The operator $-A$ is the infinitesimal generator of a contraction semigroup in X.*

Proof. Clearly, A is densely-defined because $C_0^\infty(\Omega)$ is dense in X. Now we prove that A is closed. Let $u_k \in D(A)$ and $u_k \to u$, $Au_k \to v$ in X. Then $u_k \in H^2(\Omega) \cap X$, $Au_k, u, v \in L^2(\Omega)$, and $u_k \to u$, $Au_k \to v$ in $L^2(\Omega)$. Thus, we have

$$\|u_k - u_l\|_{H^2(\Omega)} \leqslant C\|A(u_l - u_k)\|_2 \to 0 \quad \text{as } k, l \to \infty,$$

which implies $u_k \to u$ in $H^2(\Omega) \cap X$, and consequently

$$Au_k + \Delta u = -\Delta(u_k - u) \to 0 \quad \text{in } L^2(\Omega).$$

Since $Au_k \to v$ in X, we have $-\Delta u = v \in X$. This implies $u \in D(A)$ and $Au = v$. So, the operator A is closed.

In the following we shall prove that (9.1) holds. Given $f \in L^\infty(\Omega) \subset L^2(\Omega)$ and $\lambda < 0$. The boundary value problem

$$\begin{cases} -\Delta u - \lambda u = -\lambda f & \text{in } \Omega, \\ u = 0 & \text{on } \partial\Omega \end{cases}$$

has a unique solution $u \in H^2(\Omega) \cap H_0^1(\Omega)$. Let $M = \|f\|_\infty$. Then,

$$-\Delta(u - M) - \lambda(u - M) = -\lambda(f - M).$$

Note that $v = (u - M)^+ \in H_0^1(\Omega)$ and $\nabla v = \chi_{\{u>M\}}\nabla u$. Multiplying the above equation by v and integrating the result over Ω, we conclude

$$-\lambda \int_\Omega v^2 dx + \int_{\{u>M\}} |\nabla v|^2 dx = -\lambda \int_\Omega (f - M)v dx \leqslant 0.$$

This combined with the fact $\lambda < 0$ yields $v = 0$, and so $u \leqslant M$. Similarly, $u \geqslant -M$. Hence, $u \in L^\infty(\Omega)$, and $\|u\|_\infty \leqslant \|f\|_\infty$. Using $-\Delta u = \lambda(u - f)$ we also have $-\Delta u \in L^\infty(\Omega)$, which implies $u \in C(\overline{\Omega})$ by the imbedding theorem. Since $u \in H_0^1(\Omega)$, it yields $-\Delta u = \lambda(u - f) \in X$. Therefore, $u \in D(A)$. This indicates that $\lambda \in \rho(A)$.

On the other hand, the equation $-\Delta u - \lambda u = -\lambda f$ is equivalent to

$$u = (-\Delta - \lambda I)^{-1}(-\lambda f) = -\lambda(-\Delta - \lambda I)^{-1} f.$$

Hence we have

$$|\lambda| \|(-\Delta - \lambda I)^{-1} f\|_\infty = \|u\|_\infty \leqslant \|f\|_\infty,$$

which implies

$$\|(-\Delta - \lambda I)^{-1}\| \leqslant \frac{1}{|\lambda|}.$$

Consequently, (9.1) holds, and the conclusion is followed from Theorem 9.4. □

9.1.2 Solutions of linear problems

This subsection is concerned with the linear problem

$$\begin{cases} u'(t) + Au(t) = f(t), & 0 < t < T, \\ u(0) = u_0 \in X. \end{cases} \tag{9.3}$$

It is assumed throughout this subsection that

(i) $-A$ is the infinitesimal generator of a C_0 semigroup $Q(t)$ in X;

(ii) $0 < T < \infty$ and $f \in L^1((0, T), X)$.

Definition 9.8 *A function $u \in C([0, T], X)$ is called a* real solution *of (9.3) if the followings hold:*

(1) *for any fixed $0 < t < T$, $u'(t)$ exists, $u(t) \in D(A)$ and satisfies $u'(t) + Au(t) = f(t)$;*

(2) *$u(0) = u_0$, i.e., $\lim_{t\to 0+} u(t) = u_0$ in X.*

Theorem 9.9 *If u is a real solution of (9.3), then*

$$u(t) = Q(t)u_0 + \int_0^t Q(t - s)f(s)ds \quad \text{for all } t \in [0, T]. \tag{9.4}$$

Proof. Take $t \in (0, T)$ and denote

$$v(s) = Q(t-s)u(s), \quad 0 < s \leqslant t.$$

For $s, s+h \in (0, T]$ and $h > 0$, applying Theorem 9.3 (3) we conclude

$$\frac{v(s) - v(s-h)}{h} = \frac{I - Q(h)}{h} Q(t-s)u(s) + Q(t-s+h)\frac{u(s) - u(s-h)}{h}$$

$$\xrightarrow{h \to 0^+} AQ(t-s)u(s) + Q(t-s)u'(s)$$

$$= Q(t-s)f(s),$$

i.e., $v'(s) = Q(t-s)f(s)$. Integrating this differential equation from 0 to t and using $\lim_{s \to 0^+} v(s) = Q(t)u_0$ we get (9.4). □

Lemma 9.10 *Let* $F(t) = \int_0^t Q(t-s)f(s)\mathrm{d}s$. *Then* $F \in C([0, T), X)$.

Proof. Since $f \in L^1((0, T), X)$, function $F(t) = \int_0^t Q(t-s)f(s)\mathrm{d}s$ is well defined. For $t \in [0, T)$ and $0 < l \ll 1$, we have

$$F(t+l) - F(t) = \int_0^{t+l} Q(t+l-s)f(s)\mathrm{d}s - \int_0^t Q(t-s)f(s)\mathrm{d}s$$

$$= (Q(l) - I) \int_0^t Q(t-s)f(s)\mathrm{d}s + \int_t^{t+l} Q(t+l-s)f(s)\mathrm{d}s,$$

which implies

$$\lim_{l \to 0^+} [F(t+l) - F(t)] = 0.$$

For $t \in (0, T)$, $l < 0$ and $|l| \ll 1$, as

$$F(t+l) - F(t) = (I - Q(-l)) \int_0^{t+l} Q(t+l-s)f(s)\mathrm{d}s - \int_{t+l}^t Q(t-s)f(s)\mathrm{d}s,$$

it follows that

$$\lim_{l \to 0^-} [F(t+l) - F(t)] = 0.$$

Accordingly, $F \in C([0, T), X)$. □

Based on Lemma 9.10, the function u defined by (9.4) is in $C([0, T), X)$.

Definition 9.11 *The function* u *defined by* (9.4) *is called a* mild solution *of* (9.3).
A mild solution u *of* (9.3) *is called a* strong solution *if* $u \in L^1((0, T), D(A)) \cap W_1^1((0, T), X)$ *satisfies*

$$\begin{cases} u'(t) + Au(t) = f(t) & \text{a.e. in } (0, T), \\ u(0) = u_0. \end{cases} \tag{9.5}$$

A mild solution u *of* (9.3) *is called a* classical solution *of* (9.3) *if* $u \in C([0, T], D(A)) \cap C^1([0, T], X)$ *and satisfies* (9.3).

By this definition, a classical solution must be a real solution.

The following theorem gives an equivalent form of the mild solution.

Theorem 9.12 *Function u is a mild solution of (9.3) if and only if $u \in C([0,T), X)$ satisfies $\int_0^t u(s)ds \in D(A)$ and*

$$u(t) - u_0 + A \int_0^t u(s)ds = \int_0^t f(s)ds \quad \text{for all } t \in [0,T). \tag{9.6}$$

Proof. Let u be given by (9.4). For $t \in [0,T)$, one has

$$\begin{aligned}
\int_0^t u(r)dr &= \int_0^t Q(r)u_0 dr + \int_0^t dr \int_0^r Q(r-s)f(s)ds \\
&= \int_0^t Q(r)u_0 dr + \int_0^t ds \int_s^t Q(r-s)f(s)dr \\
&= \int_0^t Q(r)u_0 dr + \int_0^t ds \int_0^{t-s} Q(r)f(s)dr.
\end{aligned}$$

Making use of Theorem 9.3 (2) we conclude

$$\int_0^t Q(r)u_0 dr \in D(A) \quad \text{and} \quad A \int_0^t Q(r)u_0 dr = u_0 - Q(t)u_0,$$

$$\int_0^{t-s} Q(r)f(s)dr \in D(A) \quad \text{and} \quad A \int_0^{t-s} Q(r)f(s)dr = f(s) - Q(t-s)f(s).$$

Thus $\int_0^t u(r)dr \in D(A)$, and

$$\begin{aligned}
A \int_0^t u(r)dr &= u_0 - Q(t)u_0 + \int_0^t [f(s) - Q(t-s)f(s)]ds \\
&= u_0 - u(t) + \int_0^t f(s)ds.
\end{aligned}$$

This shows that every mild solution of (9.3) satisfies (9.6).

Conversely, let $v \in C([0,T), X)$ be a solution of (9.6) and set $w(t) = \int_0^t (v - u)$. Then $w' + Aw = 0$ and $w(0) = 0$. Theorem 9.3 (6) asserts $w \equiv 0$, i.e., $v \equiv u$. □

Corollary 9.13 *Suppose $u_0 \in X$ and $f \in L^1((0,T), X)$. If $u \in L^1((0,T), D(A))$ or $u \in W_1^1((0,T), X)$, then u is a mild solution of problem (9.3) if and only if u is a strong solution of (9.3).*

Proof. We first assume that u is a mild solution of (9.3), i.e., u is given by (9.4). Then u satisfies (9.6) by Theorem 9.12.

If $u \in L^1((0,T), D(A))$, then $A \int_0^t u(s)ds = \int_0^t Au(s)ds$. Thus we have in the light of (9.6),

$$u(t) - u_0 + \int_0^t Au(s)ds = \int_0^t f(s)ds \quad \text{for all } t \in [0,T).$$

Owing to the fact that $f, Au \in L^1((0,T), X)$, we see that $u(t)$ is differentiable almost everywhere, and the first equation of (9.5) holds. Accordingly, $u \in W_1^1((0,T), X)$.

If $u \in W_1^1((0,T), X)$, then, by use of (9.6),

$$F(t) = A \int_0^t u(s)\mathrm{d}s \in W_1^1((0,T), X).$$

Since A is a closed operator, we have that $\dfrac{1}{h} \displaystyle\int_t^{t+h} u(s)\mathrm{d}s \to u(t)$, and

$$Au(t) \leftarrow A \left(\frac{1}{h} \int_t^{t+h} u(s)\mathrm{d}s \right) = \frac{F(t+h) - F(t)}{h} \to F'(t) \quad \text{a.e. in } [0,T]$$

as $h \to 0$. It follows that $u(t) \in D(A)$ and $F'(t) = Au(t)$ almost everywhere in $[0,T]$. This fact and (9.6) affirm that u satisfies the first equation of (9.5). Moreover, $u \in L^1((0,T), D(A))$ since $u', f \in L^1((0,T), X)$.

Conversely, when u is a strong solution of (9.3), i.e., $u \in L^1((0,T), D(A)) \cap W_1^1((0,T), X)$ satisfies (9.5), similar to the proof of Theorem 9.9, we can show that u satisfies (9.4). □

Corollary 9.13 indicates that when the mild solution $u \in L^1((0,T), D(A))$ or $u \in W_1^1((0,T), X)$, it is exactly a strong solution.

Theorem 9.14 *Let $u_0 \in D(A)$, $f \in C([0,T], X)$ and u be given by (9.4). If either $f \in L^1((0,T), D(A))$ or $f \in W_1^1((0,T), X)$, then $u \in C([0,T], D(A)) \cap C^1([0,T], X)$ and satisfies (9.3). This indicates that u is a* classical solution.

Proof. Step 1. Define

$$v(t) = \int_0^t Q(t-s)f(s)\mathrm{d}s = \int_0^t Q(s)f(t-s)\mathrm{d}s, \quad t \in [0,T].$$

We shall show $v \in C^1([0,T], X)$. In fact, when $f \in L^1((0,T), D(A))$, for $t \in [0,T)$ and $h \in (0, T-t]$, we have

$$\frac{v(t+h) - v(t)}{h} = \int_0^t Q(t-s)\frac{Q(h) - I}{h}f(s)\mathrm{d}s + \frac{1}{h}\int_t^{t+h} Q(t+h-s)f(s)\mathrm{d}s$$

$$= \int_0^t Q(t-s)\frac{Q(h) - I}{h}f(s)\mathrm{d}s + \frac{1}{h}\int_0^h Q(s)f(t+h-s)\mathrm{d}s$$

$$= \frac{Q(h) - I}{h}\int_0^t Q(t-s)f(s)\mathrm{d}s + \frac{1}{h}\int_0^h Q(s)f(t+h-s)\mathrm{d}s.$$

Keeping $f(s) \in D(A)$ in mind, it is easy to see that

$$\lim_{h \to 0^+} \frac{Q(h) - I}{h}\int_0^t Q(t-s)f(s)\mathrm{d}s = -A\int_0^t Q(t-s)f(s)\mathrm{d}s$$

$$= -\int_0^t Q(t-s)Af(s)\mathrm{d}s.$$

Consequently, in accordance with $f \in C([0,T], X)$,

$$\frac{\mathrm{d}^+ v(t)}{\mathrm{d}t} = -\int_0^t Q(t-s)Af(s)\mathrm{d}s + f(t) \quad \text{for all } t \in [0,T).$$

Now we consider the case $f \in W_1^1((0,T), X)$. For $t \in [0,T)$ and $h \in (0, T-t]$, we have

$$\frac{v(t+h) - v(t)}{h} = \int_0^t Q(s)\frac{f(t+h-s) - f(t-s)}{h}\mathrm{d}s + \frac{Q(h)}{h}\int_0^h Q(t-s)f(s)\mathrm{d}s.$$

Notice that

$$\lim_{h \to 0^+} \frac{f(t+h-\cdot) - f(t-\cdot)}{h} = f'(t-\cdot).$$

Similar to above,

$$\frac{\mathrm{d}^+ v(t)}{\mathrm{d}t} = \int_0^t Q(s)f'(t-s)\mathrm{d}s + Q(t)f(0) \quad \text{for all } t \in [0,T).$$

In conclusion, the right derivative $\frac{\mathrm{d}^+ v(t)}{\mathrm{d}t} \in C([0,T), X)$ and hence $v \in C^1([0,T), X)$.

By the same way as above, the left derivative $\frac{\mathrm{d}^- v(T)}{\mathrm{d}t}$ exists and $\frac{\mathrm{d}^- v(T)}{\mathrm{d}t} = \lim_{t \to T^-} v'(t)$. Hence $v \in C^1([0,T], X)$.

Step 2. Let $t \in [0,T)$ and $h \in [0, T-t]$. Then

$$\frac{Q(h) - I}{h}v(t) = \frac{1}{h}\int_0^t Q(t+h-s)f(s)\mathrm{d}s - \frac{1}{h}\int_0^t Q(t-s)f(s)\mathrm{d}s$$

$$= \frac{v(t+h) - v(t)}{h} - \frac{1}{h}\int_t^{t+h} Q(t+h-s)f(s)\mathrm{d}s.$$

Taking $h \to 0^+$, it then yields $v(t) \in D(A)$ and

$$-Av(t) = v'(t) - f(t) \quad \text{for } t \in [0,T).$$

Since A is a closed operator, equation $-Av(T) = v'(T) - f(T)$ is valid. As a consequence, $v \in C([0,T], D(A))$.

Applying Theorem 9.3 (3) one has

$$u(t) = Q(t)u_0 + v(t) \in C([0,T], D(A)) \cap C^1([0,T], X),$$

$u(0) = u_0$ and

$$u'(t) = -AQ(t)u_0 - Av(t) + f(t) = -Au(t) + f(t) \quad \text{for all } t \in [0,T].$$

The proof is finished. □

9.1.3 Mild solutions of semilinear problems

Let $-A$ be the infinitesimal generator of a C_0 semigroup $Q(t)$ in X. Now, we consider the following *semilinear problem*

$$\begin{cases} u'(t) + Au(t) = F(t, u(t)), & t \in [0, T), \\ u(0) = u_0 \in X. \end{cases} \tag{9.7}$$

Assume that

(i) there exist $M \geqslant 1$ and $a \in \mathbb{R}$ such that $\|Q(t)\| \leqslant Me^{-at}$ for all $t \geqslant 0$;

(ii) $T \in (0, \infty]$, $\mathcal{O} \subset X$ is an open set, $u_0 \in \mathcal{O}$ and $F : [0, T) \times \mathcal{O} \to X$ is continuous. For any given $\tau \in (0, T)$ and $z \in \mathcal{O}$, there exist $\delta = \delta(\tau, z) > 0$ and $L = L(\tau, z) < \infty$ such that

$$\|F(t, u) - F(t, v)\| \leqslant L\|u - v\| \quad \text{for all } u, v \in B_\delta(z), \ t \in [0, \tau]. \tag{9.8}$$

Let $r \in (0, T]$. A function $u \in C([0, r), \mathcal{O})$ is referred to as a *mild solution* of (9.7) in $[0, r)$ if u solves the following integral equation

$$u(t) = Q(t)u_0 + \int_0^t Q(t - s)F(s, u(s))ds \quad \text{for all } t \in [0, r). \tag{9.9}$$

Clearly, the function $f(t) = F(t, u(t))$ satisfies $f \in L^1((0, T), X)$. Thanks to Theorem 9.12, u is a mild solution of (9.7) in $[0, r)$ if and only if $u \in C([0, r), \mathcal{O})$, $\int_0^t u(s)ds \in D(A)$ and

$$u(t) - u_0 + A \int_0^t u(s)ds = \int_0^t F(s, u(s))ds \quad \text{in } [0, r).$$

Let $\mathcal{S}_m(r)$ be a set of mild solutions of (9.7) in $[0, r)$. Sometimes, we write $\mathcal{S}_m(r, u_0)$ instead of $\mathcal{S}_m(r)$ to clarify dependence of $\mathcal{S}_m(r)$ on u_0.

Firstly, we give the local existence of mild solutions.

Theorem 9.15 (Local existence) *For any given $u_0 \in \mathcal{O}$, there exists $r \in (0, T)$ such that problem (9.7) has a mild solution $u \in \mathcal{S}_m(r)$. Moreover,*

$$\lim_{k \to \infty} \sup_{0 \leqslant t \leqslant r} \|u_k(t) - u(t)\| = 0,$$

where $u_k \in C([0, r], \mathcal{O})$ are defined by $u_0(t) \equiv u_0$ and

$$u_k(t) = Q(t)u_0 + \int_0^t Q(t - s)F(s, u_{k-1}(s))ds, \quad k = 1, 2, \ldots. \tag{9.10}$$

Proof. This proof is exactly the Picard iteration. We pick τ fixed such that $0 < \tau < T$. Since $\mathcal{O}$ is open, we can choose $0 < \delta \ll 1$ so that $\overline{B_\delta(u_0)} \subset \mathcal{O}$ and (9.8) is satisfied. Then there exists $M > 0$ for which

$$\|Q(t)\| \leqslant M, \quad \|F(t, u)\| \leqslant M \quad \text{for all } 0 \leqslant t \leqslant \tau \text{ and } u \in \overline{B_\delta(u_0)}. \tag{9.11}$$

According to the continuity of $Q(t)$, there exists $0 < t_0 \leqslant \tau$ such that

$$\|Q(t)u_0 - u_0\| \leqslant \delta/2 \quad \text{for all } 0 \leqslant t \leqslant t_0.$$

Then using (9.11), we derive that, when $0 \leqslant t \leqslant r = \min\{t_0, \delta M^{-2}/3\}$,

$$\|u_1(t) - u_0\| \leqslant \|Q(t)u_0 - u_0\| + \int_0^t \|Q(t-s)F(s, u_0)\|ds$$

$$\leqslant \delta/2 + M^2 t < \delta,$$

which implies $u_1 \in B_\delta(u_0)$. Analogously,

$$\|u_2(t) - u_0\| \leqslant \|Q(t)u_0 - u_0\| + \int_0^t \|Q(t-s)F(s, u_1(s))\|ds$$

$$\leqslant \delta/2 + M^2 t < \delta \quad \text{for } 0 \leqslant t \leqslant r.$$

Inductively, it can be deduced that $u_k(t) \in B_\delta(u_0)$ for all $k \geqslant 1$, $0 \leqslant t \leqslant r$.
Using (9.8) and (9.11) we conclude that, for $0 \leqslant t \leqslant r$,

$$\|u_2(t) - u_1(t)\| = \left\| \int_0^t Q(t-s)[F(s, u_1(s)) - F(s, u_0)]ds \right\|$$

$$\leqslant ML \int_0^t \|u_1(s) - u_0\|ds \leqslant ML\delta t,$$

$$\|u_3(t) - u_2(t)\| = \left\| \int_0^t Q(t-s)[F(s, u_2(s)) - F(s, u_1(s))]ds \right\|$$

$$\leqslant ML \int_0^t \|u_2(s) - u_1(s)\|ds$$

$$\leqslant (ML)^2 \delta \int_0^t sds = \frac{1}{2}(ML)^2 \delta t^2.$$

Taking advantage of the inductive method, we can prove that, for all $k \geqslant 1$,

$$\|u_{k+1}(t) - u_k(t)\| \leqslant \frac{1}{k!}(ML)^k \delta t^k \leqslant \frac{\delta}{k!}(MLr)^k, \quad 0 \leqslant t \leqslant r.$$

Hence for any given integer $N \geqslant 1$ and all $0 \leqslant t \leqslant r$, one has

$$\|u_{k+N}(t) - u_k(t)\| \leqslant \sum_{i=0}^{N-1} \|u_{k+i+1}(t) - u_{k+i}(t)\| \leqslant \delta \sum_{i=0}^{N-1} \frac{1}{(k+i)!}(MLr)^{k+i}.$$

According to Lemma 9.10, $u_k \in C([0, r], X)$. This and the above inequality assert that $\{u_k(t)\}$ is a Cauchy sequence in $C([0, r], X)$. Since $u_k \in C([0, r], \overline{B_\delta(u_0)})$, there is a $u \in C([0, r], \overline{B_\delta(u_0)})$ such that $u_k(t) \to u(t)$ in $C([0, r], \overline{B_\delta(u_0)})$. It then follows that u is a mild solution of (9.7) in $[0, r]$ by letting $k \to \infty$ in (9.10). $\square$

Secondly, we give the uniqueness of mild solution.

Theorem 9.16 (Uniqueness) *Let* $0 < \mu \leqslant r \leqslant T$, $w \in \mathcal{S}_m(\mu)$, $u \in \mathcal{S}_m(r)$. *Then* $w(t) = u(t)$ *for all* $0 \leqslant t < \mu$.

Proof. Noting that $w(t)$ and $u(t)$ are continuous and $w(0) = u(0) = u_0$, there exists $0 < t_1 \leqslant \mu$ so that $w(t), u(t) \in B_\delta(u_0)$ for all $0 \leqslant t \leqslant t_1$, where $\delta = \delta(t_1, u_0)$ is given by the condition (ii). Let $M = \max_{0 \leqslant t \leqslant t_1} \|Q(t)\|$. In view of (9.8) we have

$$\|u(t) - w(t)\| \leqslant \int_0^t \|Q(t-s)[F(s, u(s)) - F(s, w(s))]\| \mathrm{d}s$$

$$\leqslant ML \int_0^t \|u(s) - w(s)\| \mathrm{d}s, \quad 0 \leqslant t \leqslant t_1.$$

This leads to $w(t) = u(t)$ in $[0, t_1]$ because of $u(0) = w(0)$. Define

$$\tau = \sup \{s : 0 < s < \mu, \ u(t) = w(t) \ \text{in} \ [0, s]\}.$$

Clearly, $\tau \geqslant t_1$. If $\tau < \mu$, then $u(t) = w(t)$ in $[0, \tau]$ by continuities of $u(t)$ and $w(t)$. Regard τ and $u(\tau)$ as an initial time and an initial datum, respectively. Similarly to above it can be proved that $w(t) = u(t)$ in $[0, \tau + \varepsilon]$ for some $\varepsilon > 0$. This contradicts the definition of τ. Hence, $\tau = \mu$ and the proof is finished. □

Thirdly, we show the extension of mild solution.

Theorem 9.17 (Extension) *Let $u_0 \in \mathcal{O}$. Then there exist $T_{\max} := T_{\max}(u_0) \in (0, T]$ and a mild solution $u \in \mathcal{S}_m(T_{\max})$ with property: for any given $\mu \in (0, T]$, the set $\mathcal{S}_m(\mu) \neq \emptyset$ implies $T_{\max} \geqslant \mu$. This means that $T_{\max}$ is the maximal existence time.*

Furthermore, if $T_{\max} < T$, then either $\int_0^{T_{\max}} \|F(t, u(t))\| \mathrm{d}t = \infty$, or there exists $w \in \partial\mathcal{O}$ so that $\lim_{t \to T_{\max}} u(t) = w$. Such a solution u is called the maximal defined solution.

Proof. The first conclusion is followed from Theorems 9.15 and 9.16. Details are omitted here.

Now, we prove the second conclusion. Suppose $T_{\max} < T$ and

$$\int_0^{T_{\max}} \|F(t, u(t))\| \mathrm{d}t < \infty.$$

Denote $f(s) = F(s, u(s))$. The above inequality indicates $f \in L^1((0, T_{\max}), X)$, and then in line with Lemma 9.10,

$$\int_0^t Q(t-s)f(s)\mathrm{d}s \in C([0, T_{\max}], X).$$

Equation (9.9) implies $u \in C([0, T_{\max}], X)$. Accordingly, limit $\lim_{t \to T_{\max}} u(t) = v$ exists. Obviously, $v \in \bar{\mathcal{O}}$. We use a contradiction argument to show that $v \notin \mathcal{O}$. Otherwise, as $\mathcal{O}$ is an open set, there exist $t_0 > 0$ and $\delta = \delta(T_{\max} + t_0) > 0$ such that $T_{\max} + t_0 < T$ and $\overline{B_\delta(v)} \subset \mathcal{O}$. Let

$$K = \max_{0 \leqslant t \leqslant T_{\max} + t_0} \|Q(t)\| + \max_{\substack{0 \leqslant t \leqslant T_{\max} + t_0, \\ \|u - v\| \leqslant \delta}} \|F(t, u)\|, \quad \delta_0 = \frac{\delta}{2(K+2)},$$

and choose $t_i \nearrow T_{\max}$ such that $\|u(t_i) - v\| < \delta_0$. In line with the continuity of $Q(t)v$, there exists $\varepsilon > 0$ so small that

$$\|Q(t)v - v\| < \delta_0 \quad \text{for all } 0 \leqslant t \leqslant \varepsilon.$$

As a result,

$$\|Q(t)u(t_i) - u(t_i)\| \leqslant \|Q(t)(u(t_i) - v)\| + \|Q(t)v - v\| + \|u(t_i) - v\|$$
$$\leqslant \|Q(t)\|\|u(t_i) - v\| + \|Q(t)v - v\| + \delta_0,$$
$$\leqslant \delta/2 \quad \text{for all } 0 \leqslant t \leqslant \varepsilon, \ i \geqslant 1.$$

From the proof of Theorem 9.15 we know that there exist a positive constant r, depending only on ε, δ and K, and a function $u_i \in C([0, r], \mathcal{O})$, such that

$$u_i(t) = Q(t)u(t_i) + \int_0^t Q(t - s)F(t_i + s, u_i(s))ds, \quad t \in [0, r].$$

Define

$$w_i(t) = \begin{cases} u(t) & \text{for } 0 \leqslant t \leqslant t_i, \\ u_i(t - t_i) & \text{for } t_i \leqslant t \leqslant t_i + r. \end{cases}$$

It is obvious that $w_i(t)$ is a mild solution of (9.7) in $[0, t_i + r]$ and $w_i(t) \in \mathcal{S}_m(T_{\max} + r/2)$ when $i \gg 1$. According to Theorem 9.16, we have $u(t) = w_i(t)$ in $[0, T_{\max})$ for large i. This contradicts the first conclusion, and the proof is complete. □

Finally, we study the continuous dependence of mild solution on the initial data.

Theorem 9.18 (Continuous dependence) *Let $0 < \tau < \mu \leqslant T$ and $u \in \mathcal{S}_m(\mu)$. Then there exist $\varepsilon, C > 0$ such that for each $v_0 \in B_\varepsilon(u_0)$ we can find a mild solution $v \in \mathcal{S}_m(\tau, v_0)$ satisfying*

$$\|v(t) - u(t)\| \leqslant C\|v_0 - u_0\| \quad \text{for all } t \in [0, \tau]. \tag{9.12}$$

Proof. Firstly, for any fixed $t \in [0, \tau]$, according to the condition (ii) (think of $\tau = t, u(t) = z$), there exist $\delta(t) = \delta(\tau, u(t)) > 0$ and $L(t) = L(\tau, u(t)) < \infty$ so that $B_{\delta(t)}(u(t)) \subset \mathcal{O}$ and

$$\|F(s, x) - F(s, y)\| \leqslant L(t)\|x - y\| \quad \text{for all } x, y \in B_{\delta(t)}(u(t)), \ s \in [0, \tau].$$

Take $\mu(t) > 0$ such that $\|u(s) - u(t)\| < \delta(t)/2$ for $-\mu(t) + t < s < t + \mu(t)$.

Secondly, noting that $[0, \tau]$ is compact, there exists $t_i \in [0, \tau]$ such that

$$[0, \tau] \subset \bigcup_{i=1}^N [t_i - \mu(t_i), t_i + \mu(t_i)].$$

Let us define

$$L = \max_{1 \leqslant i \leqslant N} L(t_i) \quad \text{and} \quad \delta = \min_{1 \leqslant i \leqslant N} \frac{\delta(t_i)}{4}.$$

When $t \in [0, \tau]$ and $x, y \in B_{2\delta}(u(t))$, we have $x, y \in \mathcal{O}$, and subsequently

$$\|F(t, x) - F(t, y)\| \leqslant L\|x - y\| \quad \text{for all } t \in [0, \tau], \ x, y \in B_{2\delta}(u(t)). \tag{9.13}$$

Take

$$K = \max_{0 \leqslant t \leqslant \tau} \|Q(t)\| \quad \text{and} \quad 0 < \varepsilon < \frac{\delta}{K}e^{-KL\tau}.$$

For each $v_0 \in B_\varepsilon(u_0)$, in the light of Theorem 9.17, there exist $T_{\max} := T_{\max}(v_0) \in (0, T]$ and the maximal defined solution $v \in S_m(T_{\max}, v_0)$.

Thirdly, we define

$$\sigma = \sup \{s : 0 < s \leqslant \min\{\tau, T_{\max}\}, \ \|v(t) - u(t)\| < 2\delta \ \text{in} \ [0, s]\}.$$

Then $\sigma > 0$, and for any $0 < t < \sigma$, we have

$$\|v(t) - u(t)\| \leqslant \|Q(t)(v_0 - u_0)\| + \int_0^t \|Q(t - s)\| \|F(s, v(s)) - F(s, u(s))\| ds$$

$$\leqslant K\varepsilon + KL \int_0^t \|v(s) - u(s)\| ds.$$

Thus, by the Gronwall inequality,

$$\|v(t) - u(t)\| \leqslant \varepsilon K e^{KLt} \leqslant \varepsilon K e^{KL\tau} < \delta \quad \text{for all} \ 0 < t < \sigma.$$

As a result, this inequality holds when $t = \sigma$, and $v(t) \in B_\delta(u(t))$ for all $0 \leqslant t \leqslant \sigma$. Applying (9.13) and the definition of $T_{\max}$ we have that $T_{\max} > \tau$ and $\sigma = \tau$. Hence, for any $0 < t \leqslant \tau$,

$$\|v(t) - u(t)\| \leqslant \|Q(t)(v_0 - u_0)\| + \int_0^t \|Q(t - s)\| \|F(s, v(s)) - F(s, u(s))\| ds$$

$$\leqslant K\|v_0 - u_0\| + KL \int_0^t \|v(s) - u(s)\| ds.$$

As above, we can get (9.12). □

Here is an example of applications of the above theorems.

Example 9.2 *Let Ω be of class C^2 and $0 < T < \infty$. Consider the following initial-boundary value problem*

$$\begin{cases} u_t - \Delta u = f(x, t) & in \ Q_T, \\ u = 0 & on \ S_T, \\ u(x, 0) = u_0(x) & in \ \Omega. \end{cases} \tag{9.14}$$

Let $X = \{v \in C(\overline{\Omega}) : v|_{\partial\Omega} = 0\}$ and define the operator A as in Example 9.1. Then $-A$ is the infinitesimal generator of a contraction semigroup $Q(t)$ by Theorem 9.7. Assume that $f \in L^1((0, T), X)$. We have the following conclusions.

(1) If $u_0 \in X$, then problem (9.14) has a unique mild solution $u \in C([0, T), X)$ by Lemma 9.10.

(2) If $u_0 \in D(A) := \{v \in H^2(\Omega) \cap X : \Delta v \in X\}$, and either $f \in L^1((0, T), D(A))$ or $f \in W_1^1((0, T), X)$. Then problem (9.14) has a unique classical solution $u \in C([0, T], D(A)) \cap C^1([0, T], X)$ by Theorem 9.14.

(3) Consider the nonlinear version of (9.14), i.e.,

$$\begin{cases} u_t - \Delta u = \dfrac{u}{1+u^2} & \text{in } Q_T, \\ u = 0 & \text{on } S_T, \\ u(x,0) = u_0(x) & \text{in } \Omega. \end{cases} \tag{9.15}$$

If $u_0 \in X$, then problem (9.15) has a unique mild solution $u \in C([0,T], X)$, i.e., the integral equation

$$u(t) = Q(t)u_0 + \int_0^t Q(t-s)\frac{u(s)}{1+u^2(s)}ds \quad \text{for all } t \in [0,T]$$

has a unique solution $u \in C([0,T], X)$ by Theorems 9.15–9.17.

9.2 ANALYTIC SEMIGROUP THEORY

Let $Q(t)$ be a C_0 semigroup in X and $-A$ its infinitesimal generator. For any given $v \in D(A)$, we see from Theorem 9.3 (3) and (6) that problem

$$\begin{cases} u'(t) + Au(t) = 0, & t > 0, \\ u(0) = v \end{cases} \tag{9.16}$$

has a unique real solution $u(t) = Q(t)v$ (it is actually a classical solution). However, when $v \in X \setminus D(A)$, we do not know whether (9.16) has a real solution. To solve this problem, we introduce sectorial operator and analytic semigroup.

9.2.1 Basic results

In this subsection, we state some basic results of analytic semigroup theory. Their proofs will be omitted, and interested readers can refer to [89, 82, 124] for details.

Definition 9.19 *Let $Q(t)$ be a C_0 semigroup of X. We call $Q(t)$ a differentiable semigroup if for any given $x \in X$, function $Q(t)x$ is differentiable in $t > 0$.*

A C_0 semigroup $Q(t)$ of X is referred to as a real analytic semigroup *if for any given $x \in X$ and $\ell \in X^*$, function $\ell(Q(t)x)$ is analytic for $t > 0$, where X^* is the dual space of X.*

We make conventions: $\arg 0 = 0$, $\arg z \in (-\pi, \pi]$ when $z \in \mathbb{C} \setminus \{0\}$, and denote $z = |z|e^{i\arg z}$, where $\mathbb{C}$ is complex plane.

Definition 9.20 *Let $\varphi_1 < 0 < \varphi_2$, $\triangle := \{z : \varphi_1 < \arg z < \varphi_2\}$ and $Q(z)$ be a bounded linear operator for each $z \in \triangle$. The family $\{Q(z)\}_{z \in \triangle}$ is called an* analytic semigroup *if the followings hold:*

(1) for any given $x \in X$ and $\ell \in X^*$, function $\ell(Q(t)z)$ is analytic in $\triangle$;

(2) $Q(0) = I$ and $\lim_{\triangle \ni z \to 0} Q(z)v = v$ for every $v \in X$;

(3) $Q(z_1 + z_2) = Q(z_1)Q(z_1)$ *for* $z_1, z_2 \in \Delta$.

Theorem 9.21 *Let $Q(t)$ be a differentiable semigroup in X and $-A$ its infinitesimal generator. Then, for any $v \in X$, (9.16) has a unique solution $u(t) = Q(t)v$.*

Proof. Because $Q(t)$ is differentiable in $t > 0$, we know that, for any given $v \in X$, the limit

$$\lim_{h \to 0^+} \frac{1}{h}[Q(h)Q(t)v - Q(t)v] = \lim_{h \to 0^+} \frac{1}{h}[Q(t+h)v - Q(t)v]$$

exists. Therefore

$$u(t) = Q(t)v \in D(A) \quad \text{and} \quad \frac{\mathrm{d}^+ u(t)}{\mathrm{d}t} = -Au(t).$$

Analogously, $\frac{\mathrm{d}^- u(t)}{\mathrm{d}t} = -Au(t)$. Hence $u'(t) = -Au(t)$. The uniqueness is followed by Theorem 9.3 (6). $\qquad\square$

Corollary 9.22 *Assume that $-A$ is the infinitesimal generator of an analytic semigroup. Then, for each $v \in X$, problem (9.16) has a unique solution.*

Definition 9.23 *Suppose that $A : X \supset D(A) \to X$ is a linear operator. We call that A is a sectorial operator if the followings hold:*

(1) *A is closed and $D(A)$ is dense in X;*

(2) *there exist $a \in \mathbb{R}$ and $\theta \in (0, \pi/2)$ such that sector $S_{a,\theta} := \{\lambda \in \mathbb{C} : |\arg(\lambda - a)| \geqslant \theta\} \subset \rho(A)$;*

(3) *there exists $M \geqslant 1$ so that $\|(A - \lambda)^{-1}\| \leqslant M/|\lambda - a|$ for $\lambda \in S_{a,\theta}$.*

Definition 9.24 *Assume that A is a sectorial operator. For any complex number $z \in \mathbb{C}$ with $|\arg z| < \frac{\pi}{2} - \theta$, we define linear operators e^{-zA} in X as follows: $\mathrm{e}^{-zA} = I$ when $z = 0$, while*

$$\mathrm{e}^{-zA} = \frac{1}{2\pi \mathrm{i}} \int_{\mathbb{R}} (\phi(t) - A)^{-1} \mathrm{e}^{-\phi(t)z} \phi'(t) \mathrm{d}t$$

when $z \neq 0$, where

$$\phi(t) = b + |t| \cos \varphi - \mathrm{i}t \sin \varphi, \quad t \in \mathbb{R}$$

with $b < a$, $\varphi \in (\theta, \pi/2 - |\arg z|)$.

Theorem 9.25 *If A is a sectorial operator, then e^{-zA} is an analytic semigroup. Conversely, if $-A$ is the infinitesimal generator of an analytic semigroup $Q(z)$, then A is a sectorial operator and $Q(z) = \mathrm{e}^{-zA}$.*

Theorem 9.26 *Let A be a sectorial operator.*

(1) *For all integer $k \geqslant 0$ and $t > 0$, we have*

$1°$ $\mathrm{e}^{-tA}u \in D(A^k)$ *for every $u \in X$;*

$2°$ $A^k e^{-tA} = \left(-\frac{d}{dt}\right)^k e^{-tA}$;

$3°$ $A^k e^{-tA} u = e^{-tA} A^k u$ for every $u \in D(A^k)$.

(2) For any given $\tau \in \mathbb{R}_+$, there exists a constant $C = C(\tau) > 0$ such that

$$\|A e^{-tA}\| \leqslant C t^{-1}, \quad 0 < t \leqslant \tau, \tag{9.17}$$

$$\|e^{-tA} - e^{-sA}\| \leqslant C s^{-1}(t - s), \quad 0 < s \leqslant t \leqslant \tau, \tag{9.18}$$

$$\|A e^{-tA} - A e^{-sA}\| \leqslant C t^{-1} s^{-1}(t - s), \quad 0 < s \leqslant t \leqslant \tau. \tag{9.19}$$

Definition 9.27 Let $-A$ be the infinitesimal generator of a C_0 semigroup $Q(t)$ in X. If there exist positive constants a and M such that

$$\|Q(t)\| \leqslant M e^{-at} \quad \text{for all } t \geqslant 0,$$

then we call that $A \in \mathscr{B}(X)$. For every chosen $A \in \mathscr{B}(X)$ and $\alpha < 0$, we define a linear operator A^α in X by

$$A^\alpha v = \frac{1}{\Gamma(-\alpha)} \int_0^\infty t^{-\alpha-1} Q(t) v \, dt, \quad v \in X,$$

where

$$\Gamma(-\alpha) = a^{-\alpha} \int_0^\infty t^{-\alpha-1} e^{-at} dt.$$

For any given $A \in \mathscr{B}(X)$ and $\alpha \geqslant 0$, we define $A^\alpha = I$ when $\alpha = 0$, and A^α the inverse operator of $A^{-\alpha}$ when $\alpha > 0$.

Theorem 9.28 Suppose $A \in \mathscr{B}(X)$. Then the followings hold:

(1) A^α is closed and $D(A^\alpha)$ is dense in X. Moreover, $0 \in \rho(A^\alpha)$ and $(A^\alpha)^{-1} = A^{-\alpha}$ for all $\alpha > 0$;

(2) $A^\alpha Q(t) v = Q(t) A^\alpha v$ for all $\alpha \in \mathbb{R}$, $t \geqslant 0$ and $v \in D(A^\alpha)$;

(3) $D(A^\alpha) \subset D(A^\beta)$ when $\alpha > \beta$;

(4) if α, $\beta \in \mathbb{R}$ and $v \in D(A^\beta) \cap D(A^{\alpha+\beta})$, then $A^{\alpha+\beta} v = A^\alpha A^\beta v$;

(5) $\|A^\alpha v\| \leqslant 2(M+1)\|Av\|^\alpha \|v\|^{1-\alpha}$ for all $\alpha \in [0,1]$ and $v \in D(A)$;

(6) there exists a constant $C > 0$ such that

$$\|v - e^{\lambda t} Q(t) v\| \leqslant C(1 + e^{\lambda t})^{1-\alpha} \left((1 + |\lambda|) \frac{e^{\lambda t} - 1}{\lambda} \right)^\alpha \|A^\alpha v\|$$

for all $\lambda \in \mathbb{R}$, $\alpha \in [0,1]$, $v \in D(A^\alpha)$ and $t > 0$.

Theorem 9.29 *Let A be a sectorial operator, a and b be positive constants with $a \leqslant b$. If $\operatorname{Re}\lambda > b$ for all $\lambda \in \sigma(A)$, then $A - a \in \mathscr{B}(X)$, and for any given $k \geqslant 1$ we can find a positive constant $C(k)$ for which*

$$\|(A-a)^\alpha \mathrm{e}^{-tA}\| \leqslant C(k)t^{-\alpha}\mathrm{e}^{-bt} \text{ for all } t > 0, \ \alpha \in [0,k].$$

Definition 9.30 *Assume that A is a sectorial operator and a is a constant so that $\operatorname{Re}\lambda > a$ for all $\lambda \in \sigma(A)$. For $\alpha \geqslant 0$, we define $X^\alpha = D((A-a)^\alpha)$ and*

$$\|v\|_\alpha = \|(A-a)^\alpha v\|, \quad v \in X^\alpha.$$

Space X^α is referred to as a fractional power space *of X.*

Given $a_1 > a_2$. If $\operatorname{Re}\lambda > a_1$ for all $\lambda \in \sigma(A)$, then $D((A-a_1)^\alpha) = D((A-a_2)^\alpha)$ and $\|(A-a_1)^\alpha v\| = \|(A-a_2)^\alpha v\|$. These demonstrate that the fractional power space X^α and the norm $\|\cdot\|_\alpha$ do not depend on the choice of a.

Theorem 9.31 *Let A be a sectorial operator and integer $k \geqslant 1$. Then there exists a positive constant C such that*

$$\|\mathrm{e}^{-(t+h)A}v - \mathrm{e}^{-tA}v\|_\alpha \leqslant C\left(\mathrm{e}^{-a(t+h)} + \mathrm{e}^{-at}\right)h^\delta t^{\beta-\alpha-\delta}\|v\|_\beta$$

for all $h, t > 0$, $\delta \in [0,1]$, $\alpha \in [0,k]$, $\beta \in [0,\alpha+\delta]$ and $v \in X^\beta$.

Theorem 9.32 *Suppose that A is a sectorial operator in X, $A \in \mathscr{B}(X)$ and $0 < \alpha \leqslant 1$. If $v \in D(A^\alpha)$ then*

$$\|(\mathrm{e}^{-tA} - I)v\| \leqslant \frac{C}{\alpha}t^\alpha\|A^\alpha v\|.$$

Theorem 9.33 *Let $A \in \mathscr{B}(X)$ and $B : X \to Y$ be a closed linear operator with $D(B) \supset D(A)$. If there exist $\beta \in [0,1)$ and $C > 0$ so that*

$$\|Bx\|_Y \leqslant C\|Ax\|^\beta\|x\|^{1-\beta} \text{ for all } x \in D(A),$$

then, when $\alpha > \beta$, we have $D(B) \supset D(A^\alpha)$ and $BA^{-\alpha} \in \mathcal{L}(X,Y)$.

9.2.2 Regularity of function $\int_0^t \mathrm{e}^{-(t-s)A}f(s)\mathrm{d}s$

Let X be a Banach space, A be a sectorial operator in X and $\operatorname{Re}\lambda > a$ for all $\lambda \in \sigma(A)$. Assume that $T \in (0,\infty)$ and $f : [0,T] \to X$ is bounded. Set

$$v(t) := \int_0^t \mathrm{e}^{-(t-s)A}f(s)\mathrm{d}s.$$

Main aim of this part is to study properties of $v(t)$.

Lemma 9.34 $v \in C^{1-\alpha}([0,T], X^\alpha) \cap C^\alpha([0,T], X)$ *for all $\alpha \in (0,1)$.*

Proof. Rewrite

$$v(t+h) - v(t) = \int_0^t g_1(s,t)\mathrm{d}s + \int_t^{t+h} g_2(s,t)\mathrm{d}s, \quad 0 \leqslant t \leqslant t+h \leqslant T,$$

where

$$g_1(s,t) = \left(e^{-(t+h-s)A} - e^{-(t-s)A}\right) f(s) \quad \text{and} \quad g_2(s,t) = e^{-(t+h-s)A} f(s).$$

Let $A_1 = A - a$. Then we have

$$g_1(s,t) = \int_{t-s}^{t+h-s} \frac{\mathrm{d}}{\mathrm{d}\tau} e^{-\tau A} f(s)\mathrm{d}\tau$$

$$= - \int_{t-s}^{t+h-s} A e^{-\tau A} f(s)\mathrm{d}\tau$$

$$= -A \int_{t-s}^{t+h-s} e^{-\tau A} f(s)\mathrm{d}\tau,$$

$$A_1^\alpha g_1(s,t) = -A A_1^{-1} \int_{t-s}^{t+h-s} A_1^{1+\alpha} e^{-\tau A} f(s)\mathrm{d}\tau.$$

As f is bounded, applying Theorem 9.29 we declare

$$\|A_1^\alpha g_1(s,t)\| \leqslant C_1 \int_{t-s}^{t+h-s} \tau^{-(1+\alpha)}\mathrm{d}\tau = \frac{C_1}{\alpha}[(t-s)^{-\alpha} - (t+h-s)^{-\alpha}],$$

$$\int_0^t \|A_1^\alpha g_1(s,t)\|\mathrm{d}s \leqslant \frac{C_1}{\alpha(1-\alpha)} h^{1-\alpha}.$$

Similarly,

$$\int_t^{t+h} \|A_1^\alpha g_2(s,t)\|\mathrm{d}s \leqslant C_2 \int_t^{t+h} (t+h-s)^{-\alpha}\mathrm{d}s = \frac{C_2}{1-\alpha} h^{1-\alpha}.$$

Thus we have

$$\|v(t+h) - v(t)\|_\alpha = \|A_1^\alpha[v(t+h) - v(t)]\|$$

$$\leqslant \int_0^t \|A_1^\alpha g_1(s,t)\|\mathrm{d}s + \int_t^{t+h} \|A_1^\alpha g_2(s,t)\|\mathrm{d}s$$

$$\leqslant C h^{1-\alpha},$$

which implies $v \in C^{1-\alpha}([0,T], X^\alpha)$ for all $0 < \alpha < 1$. Thanks to

$$\|v(t+h) - v(t)\| \leqslant \|A_1^{\alpha-1}\| \times \|v(t+h) - v(t)\|_{1-\alpha},$$

it yields $v \in C^\alpha([0,T], X)$. □

If we define

$$G(t) = \int_0^t e^{-(t-s)A}[f(s) - f(t)]\mathrm{d}s, \quad t \in [0,T],$$

then

$$v(t) = G(t) + \int_0^t e^{-(t-s)A} f(t)\mathrm{d}s = G(t) + \int_0^t e^{-sA} f(t)\mathrm{d}s. \tag{9.20}$$

Lemma 9.35 *Let $0 < \mu < 1, 0 \leqslant \alpha < \mu$ and $f \in C^\mu([0,T],X)$. Then $AG(t) \in C^{\mu-\alpha}([0,T],X^\alpha)$.*

Proof. Take $0 \leqslant t < t + h \leqslant T$. Then we have

$$G(t+h) - G(t) = \int_0^t g_1(s,t)\mathrm{d}s + \int_0^t g_2(s,t)\mathrm{d}s + \int_t^{t+h} g_3(s,t)\mathrm{d}s, \qquad (9.21)$$

where

$$g_1(s,t) = \left(\mathrm{e}^{-(t+h-s)A} - \mathrm{e}^{-(t-s)A}\right)(f(s) - f(t)),$$

$$g_2(s,t) = \mathrm{e}^{-(t+h-s)A}(f(t) - f(t+h)),$$

$$g_3(s,t) = \mathrm{e}^{-(t+h-s)A}(f(s) - f(t+h)).$$

Write

$$A\left(\mathrm{e}^{-(t+h-s)A} - \mathrm{e}^{-(t-s)A}\right) = \mathrm{e}^{-(t-s)A/2}A\left(\mathrm{e}^{-(h+(t-s)/2)A} - \mathrm{e}^{-(t-s)A/2}\right),$$

then it follows from Theorem 9.29 and (9.19) that

$$\|Ag_1(s,t)\|_\alpha \leqslant C_2 h(t - s + 2h)^{-1}(t - s)^{\mu-\alpha-1}.$$

This in turn gives

$$\int_0^t \|Ag_1(s,t)\|_\alpha \mathrm{d}s \leqslant C_2 h^{\mu-\alpha} \int_0^{t/h} (2+r)^{-1} r^{\mu-\alpha-1} \mathrm{d}r \leqslant C_3 h^{\mu-\alpha}. \qquad (9.22)$$

Noticing

$$\int_0^t g_2(s,t)\mathrm{d}s = \int_0^t \mathrm{e}^{-(\tau+h)A}(f(t) - f(t+h))\mathrm{d}\tau,$$

and taking advantage of Theorem 9.3 (2), we have

$$A\int_0^t g_2(s,t)\mathrm{d}s = \mathrm{e}^{-hA}(1 - \mathrm{e}^{-tA})(f(t) - f(t+h)),$$

and thereafter, by taking $\delta = \beta = 0$ in Theorem 9.31, we have

$$\left\|A\int_0^t g_2(s,t)\mathrm{d}s\right\|_\alpha \leqslant Ch^{-\alpha}\|f(t) - f(t+h)\| \leqslant C_4 h^{\mu-\alpha}. \qquad (9.23)$$

On the other hand, thanks to Theorem 9.29 and (9.17), it deduces that

$$\int_t^{t+h} \|Ag_3(s,t)\|_\alpha \mathrm{d}s \leqslant C_5 \int_t^{t+h} (t+h-s)^{\mu-\alpha-1}\mathrm{d}s = C_5(\mu-\alpha)^{-1}h^{\mu-\alpha}.$$

We combine this with (9.21)–(9.23) to derive

$$G(t+h) - G(t) \in D(A) \quad \text{and} \quad AG \in C^{\mu-\alpha}([0,T],X^\alpha).$$

This completes the proof. □

Theorem 9.36 *Let $\mu \in (0,1)$, $\alpha \in [0,\mu)$, $\varepsilon \in (0,T)$ and $f \in C^\mu([0,T],X)$. Then the function $v(t) = \int_0^t e^{-(t-s)A} f(s)\mathrm{d}s$ has the followings properties.*

(1) $v'(t) + Av(t) = f(t)$ in $[0,T]$ and $v' \in C([0,T],X)$;

(2) $v \in C([0,T],X^1)$ and $Av \in C^\mu([\varepsilon,T],X)$;

(3) $v : (0,T) \to X^\alpha$ is differentiable and $v' \in C^{\mu-\alpha}([\varepsilon,T],X^\alpha)$.

Proof. Invoking Lemma 9.35, Theorem 9.3 (2) and (9.20), we conclude that $v(t) \in D(A)$ and

$$Av(t) = AG(t) + f(t) - e^{-tA}f(t) \quad \text{for all } t \in [0,T]. \tag{9.24}$$

As a result, $Av \in C([0,T],X)$ and $A\int_0^t v(s)\mathrm{d}s = \int_0^t Av(s)\mathrm{d}s$. According to the definition of $v(t)$, it is not difficult to see that $v(t)$ solves

$$\begin{cases} v'(t) + Av(t) = f(t), & 0 < t \leqslant T, \\ v(0) = 0. \end{cases}$$

Therefore, in line with Theorem 9.12,

$$v(t) + \int_0^t Av(s)\mathrm{d}s = v(t) + A\int_0^t v(s)\mathrm{d}s = \int_0^t f(s)\mathrm{d}s,$$

hence $v'(t) + Av(t) = f(t)$ in $[0,T]$ and $v' \in C([0,T],X)$. Conclusion (1) is accomplished.

As $f \in C^\mu([0,T],X)$, inequality (9.18) implies $e^{-tA}f(t) \in C^\mu([\varepsilon,T],X)$. Taking $\alpha = 0$ in Lemma 9.35, we have $AG(t) \in C^\mu([0,T],X)$ and then, by (9.24), $Av \in C^\mu([\varepsilon,T],X)$. It has been shown that $Av \in C([0,T],X)$. Write $A_1 v = Av - av$, then $A_1 v \in C([0,T],X)$, i.e., $v \in C([0,T],X^1)$. Conclusion (2) is proved.

The conclusion (1) and (9.24) lead to

$$v'(t) + AG(t) = e^{-tA}f(t). \tag{9.25}$$

It follows from Lemma 9.35 that $AG(t) \in C^{\mu-\alpha}([0,T],X^\alpha)$. We claim $e^{-tA}f(t) \in C^{\mu-\alpha}([\varepsilon,T],X^\alpha)$. Once this is done, then

$$v' \in C^{\mu-\alpha}([\varepsilon,T],X^\alpha)$$

by (9.25). Let $A_1 = A - a$ as above. We have

$$A_1^\alpha v(t+h) - A_1^\alpha v(t) = A_1^\alpha[v(t+h) - v(t)] = A_1^\alpha \int_t^{t+h} v'(s)\mathrm{d}s = \int_t^{t+h} A_1^\alpha v'(s)\mathrm{d}s,$$

and thus

$$\lim_{h \to 0} \frac{A_1^\alpha v(t+h) - A_1^\alpha v(t)}{h} = \lim_{h \to 0} \frac{1}{h} \int_t^{t+h} A_1^\alpha v'(s)\mathrm{d}s = A_1^\alpha v'(t),$$

i.e., $v : (0, T) \to X^\alpha$ is differentiable. The conclusion (3) is obtained.

Now we prove $e^{-tA} f(t) \in C^{\mu-\alpha}([\varepsilon, T], X^\alpha)$. For $h > 0$, we have

$$\|e^{-(t+h)A} f(t+h) - e^{-tA} f(t)\|_\alpha$$
$$\leqslant \|e^{-(t+h)A}[f(t+h) - f(t)]\|_\alpha + \|(e^{-(t+h)A} - e^{-tA}) f(t)\|_\alpha.$$

In view of Theorem 9.29, it follows that

$$\|e^{-(t+h)A}[f(t+h) - f(t)]\|_\alpha \leqslant C(t+h)^{-\alpha}\|f(t+h) - f(t)\| \leqslant Ch^{\mu-\alpha}.$$

Choose $\delta = \mu - \alpha$ and $\beta = 0$ in Theorem 9.31, then, for $\varepsilon \leqslant t \leqslant T$,

$$\|(e^{-(t+h)A} - e^{-tA}) f(t)\|_\alpha \leqslant Ch^{\mu-\alpha} t^{-\mu}\|f(t)\| \leqslant C\varepsilon^{-\mu} \max_{0 \leqslant t \leqslant T} \|f(t)\| h^{\mu-\alpha} \leqslant Ch^{\mu-\alpha}.$$

Therefore, $\|e^{-(t+h)A} f(t+h) - e^{-tA} f(t)\|_\alpha \leqslant Ch^{\mu-\alpha}$. The proof is complete. □

9.2.3 Real solutions of semilinear problems

Let X be a Banach space and A be a sectorial operator in X. We continue to discuss the semilinear problem (9.7). Assume that

(i) $\alpha \in [0, 1)$ and $\mathcal{O} \subset X^\alpha$ is a nonempty open set;

(ii) $\kappa \in (0, 1]$, $T \in (0, \infty]$, $F : [0, T) \times \mathcal{O} \to X$ and satisfies: for any given $t \in [0, T)$ and $u \in \mathcal{O}$, there exist positive constants $L = L(t, u)$ and $\delta = \delta(t, u)$ such that

$$\|F(t_2, u_2) - F(t_1, u_1)\| \leqslant L(|t_2 - t_1|^\kappa + \|u_2 - u_1\|_\alpha) \tag{9.26}$$

for all $t_i \in [0, T)$ and $u_i \in \mathcal{O}$ satisfying $|t_i - t| < \delta$ and $\|u_i - u\|_\alpha < \delta$.

For $r \in (0, T]$, let $RS(r)$ be the set of all functions $u \in C([0, r), \mathcal{O})$ solving (9.7) in the sense of Definition 9.8. When $u \in RS(r)$, we call that u is a *real solution* of (9.7) defined in $[0, r)$.

Theorem 9.37 *Let $r \in (0, T]$. Then $u \in RS(r)$ if and only if $u \in C([0, r), \mathcal{O})$ and satisfies*

$$u(t) = e^{-tA} u(0) + \int_0^t e^{-(t-s)A} F(s, u(s)) ds \quad \text{for all } t \in [0, r). \tag{9.27}$$

Moreover, if $u \in RS(r)$, then the following assertions are valid.

(1) *Functions $u'(t), Au(t), F(t, u(t)) : (0, r) \to X$ are locally Hölder continuous.*

(2) *If the constant κ given in (9.26) satisfies $\kappa \geqslant 1 - \alpha$, then for any $0 \leqslant \eta < 1 - \alpha$, function $u : (0, r) \to X^\eta$ is differentiable and its derivative $u' : (0, r) \to X^\eta$ is locally Hölder continuous.*

(3) *If $\gamma \in [\alpha, 1]$ and $u(0) \in X^\gamma$, then $u \in C([0, r), X^\gamma)$.*

Proof. Step 1. Obviously, $u \in C([0,r), \mathcal{O})$ implies

$$F(t, u(t)) =: f(t) \in C([0,r), X).$$

Applying Theorem 9.9 we see that (9.27) holds when $u \in RS(r)$.

Conversely, we assume that $u \in C([0,r), \mathcal{O})$ and satisfies (9.27). Rewrite (9.27) as

$$u(t) = e^{-tA}u(0) + g(t), \tag{9.28}$$

where

$$g(t) = \int_0^t e^{-(t-s)A} f(s) ds, \quad f(t) = F(t, u(t)).$$

For any given $0 < a < b < r$, since $f \in C([0,r), X)$, taking advantage of Lemma 9.34, we produce

$$g \in C^{1-\beta}([0,b], X^\beta) \cap C^\beta([0,b], X) \text{ for all } \beta \in (0,1). \tag{9.29}$$

On the other hand, for all $a \leqslant s \leqslant t \leqslant b$ and $\beta \in (0,1)$, Theorem 9.31 alleges

$$\|e^{-tA}u(0) - e^{-sA}u(0)\|_\beta \leqslant Ct^{\alpha-1}(t-s)^{1-\beta}\|u(0)\|_\alpha \leqslant Ca^{\alpha-1}(t-s)^{1-\beta}.$$

Combining this with (9.28) and (9.29), we get $u \in C^{1-\beta}([a,b], X^\beta)$, which in turn asserts that f is Hölder continuous in $[a,b]$, i.e., $f \in C^\tau([a,b], X)$ for some $\tau \in (0,1)$. Then $g(t)$ satisfies conclusions (1)–(3) of Theorem 9.36 with $[0,T]$, μ and α replaced respectively by $[a,b]$, τ and $0 < \alpha' < \tau$. As $j(t) := e^{-tA}u(0)$ satisfies $j'(t) + Aj(t) = 0$ by Theorem 9.26, we have $u'(t) + Au(t) = F(t, u(t))$ for $t \in [a,b]$. The arbitrariness of a, b implies $u \in RS(r)$.

Step 2. Assume $u \in RS(r)$. Then, for $a \leqslant s \leqslant t \leqslant b$, we have

$$\|Ae^{-tA}u(0) - Ae^{-sA}u(0)\| \leqslant Ct^{-1}s^{-1}(t-s)\|u(0)\| \leqslant Ca^{-2}\|u(0)\|(t-s)$$

by (9.19). This combines with Theorem 9.36 (2) shows that $Au \in C^\tau([a,b], X)$. We have known $F(t, u(t)) =: f(t) \in C^\tau([a,b], X)$. The arbitrariness of a, b implies that the conclusion (1) holds.

If the number κ given in (9.26) satisfies $\kappa \geqslant 1 - \alpha$, then for any $\tau \in (0,1)$ with $\tau \leqslant 1 - \alpha$, there holds $f \in C^\tau([a,b], X)$. Let us define

$$w(t) = e^{-tA}u(a),$$

$$v(t) = w(t) + \int_0^t e^{-(t-s)A} f(a+s) ds, \quad t \in [0, b-a].$$

Then in accordance with Theorem 9.26, we have that $w'(t)$ exists, $w(t) \in D(A)$ and satisfies $w'(t) + Aw(t) = 0$. According to Theorem 9.36, $v'(t)$ exists, $v(t) \in D(A)$,

$$v'(t) + Av(t) = f(a+t) \quad \text{in } [0, b-a],$$

and

$$Aw \in C^\tau([\varepsilon, b - a], X) \cap C^{\tau-\eta}([\varepsilon, b - a], X^\eta),$$

$$Av \in C^\tau([\varepsilon, b - a], X), \quad v' \in C^{\tau-\eta}([\varepsilon, b - a], X^\eta)$$

for any $0 < \varepsilon < b - a$, $0 \leqslant \eta < \tau$. Remembering (9.27), it is easy to see that $v(t) = u(t + a)$. The arbitrariness of a and b implies the conclusion (2).

When $\gamma \in (\alpha, 1)$, in view of (9.28), (9.29) and $e^{-tA}u(0) \in C((0, r), X^\gamma)$, we see that the conclusion (3) holds.

When $\gamma = 1$, taking $\beta = 1$ and $\delta = 1 - \alpha$ in Theorem 9.31, we deduce $e^{-tA}u(0) \in C^{1-\alpha}([0, b], X^\alpha)$. Making use of (9.28) and (9.29), $u \in C^\beta([0, b], X^\alpha)$ for some $\beta \in (0, 1)$. Thus $f \in C^\tau([0, b], X)$ for some $\tau \in (0, 1)$. Theorem 9.36 avers $Au \in C([0, b], X)$. Based on $u \in RS(r)$ and the arbitrariness of b we conclude $u \in C([0, r), X)$. The conclusion (3) holds for $\gamma = 1$. □

Now we give the local existence and uniqueness of real solutions.

Theorem 9.38 (Local existence and uniqueness of real solutions) *If $u_0 \in \mathcal{O}$, then there exists $r \in (0, T)$ such that problem (9.7) has a unique real solution u defined in $[0, r)$, i.e., $u \in RS(r)$. Furthermore,*

$$\lim_{k \to \infty} \max_{0 \leqslant t \leqslant r} \|u_k(t) - u(t)\|_\alpha = 0,$$

where $u_k \in C([0, r), \mathcal{O})$ is defined as follows: $u_1(t) = u_0$,

$$u_{k+1}(t) = e^{-tA}u_0 + \int_0^t e^{-(t-s)A}F(s, u_k(s))ds, \quad t \in [0, r), \quad k = 1, 2, \ldots.$$

Proof. Noticing that $\mathcal{O}$ is open, there exists $\delta > 0$ so that $B_\delta(u_0) \subset \mathcal{O}$. We define

$$Y := \{v : v \in C([0, r], X^\alpha)\}, \quad \|v\|_Y := \max_{0 \leqslant t \leqslant r} \|v(t)\|_\alpha$$

and

$$V := \{v : v \in Y, \|v(t) - u_0\|_Y \leqslant \delta\}.$$

For any given $v \in V$, let us define

$$G(v)(t) = e^{-tA}u_0 + \int_0^t e^{-(t-s)A}F(s, v(s))ds.$$

Obviously, $v \in X^\alpha$ implies $G(v) \in X^\alpha$. Since $u_0 \in X^\alpha$, it yields $e^{-tA}u_0 \in C([0, r], X^\alpha)$. In view of Theorem 9.29 and the condition (9.26), we can find a $0 < r \ll 1$ such that for any $v \in V$ and $0 \leqslant t \leqslant r$,

$$\|G(v)(t) - u_0\|_\alpha \leqslant \|e^{-tA}u_0 - u_0\|_\alpha + \int_0^t \|(A - a)^\alpha e^{-(t-s)A}\|\|F(s, v(s))\|ds$$

$$\leqslant \frac{\delta}{2} + C \max_{[0,r] \times B_\delta(u_0)} \|F(s, u)\| \int_0^t (t - s)^{-\alpha}ds$$

$$= \frac{\delta}{2} + \frac{C}{1 - \alpha} \max_{[0,r] \times B_\delta(u_0)} \|F(s, u)\| t^{1-\alpha}$$

$$\leqslant \delta,$$

and for any $u, v \in V$,

$$\|G(u)(t) - G(v)(t)\|_\alpha \leqslant \int_0^t \|(A-a)^\alpha e^{-(t-s)A}\|\|F(s, u(s)) - F(s, v(s))\|\mathrm{d}s$$

$$\leqslant CL \int_0^t (t-s)^{-\alpha}\|u(s) - v(s)\|_\alpha \mathrm{d}s$$

$$\leqslant CL\|u(s) - v(s)\|_Y \int_0^t (t-s)^{-\alpha}\mathrm{d}s$$

$$\leqslant \frac{1}{2}\|u - v\|_Y.$$

This shows $G : V \to V$, and G is a contraction map. Hence, by the contraction mapping principle, there exists a unique $u \in V$ such that $G(u) = u$. And so, $u \in RS(r)$ by Theorem 9.37.

Above arguments also indicate that $u_1 := G(u_0) \in V$, and then $u_{k+1} := G(u_k) \in V$ for all $k \geqslant 0$. Moreover, as u satisfies (9.27), it is deduced that

$$\|G(u_k)(t) - u(t)\|_\alpha \leqslant \int_0^t \|(A-a)^\alpha e^{-(t-s)A}\| \times \|F(s, u_k(s)) - F(s, u(s))\|\mathrm{d}s$$

$$\leqslant CL \int_0^t (t-s)^{-\alpha}\|u_k(s) - u(s)\|_\alpha \mathrm{d}s$$

$$\leqslant CL \max_{0 \leqslant s \leqslant r} \|u_k(s) - u(s)\|_\alpha \int_0^t (t-s)^{-\alpha}\mathrm{d}s$$

$$\leqslant \frac{1}{2} \max_{0 \leqslant s \leqslant r} \|u_k(s) - u(s)\|_\alpha \quad \text{for all } 0 \leqslant t \leqslant r$$

by use of Theorem 9.29 and the condition (9.26). Hence

$$\max_{0 \leqslant s \leqslant r} \|u_{k+1}(s) - u(s)\|_\alpha \leqslant \frac{1}{2} \max_{0 \leqslant s \leqslant r} \|u_k(s) - u(s)\|_\alpha \quad \text{for all } k \geqslant 0.$$

This leads to

$$\max_{0 \leqslant s \leqslant r} \|u_k(s) - u(s)\|_\alpha \leqslant 2^{-k} \max_{0 \leqslant s \leqslant r} \|u_0(s) - u(s)\|_\alpha \to 0$$

as $k \to \infty$. The proof is complete. □

Same as the mild solution, we have the following extension result for the real solution.

Theorem 9.39 (Extension) *Let $u_0 \in \mathcal{O}$. Then there exist $T_{\max} := T_{\max}(u_0) \in (0, T]$ and a unique $u \in RS(T_{\max})$ with the property: for any given $\mu \in (0, T]$, if there exists a $v \in RS(\mu)$ satisfying $v(0) = u_0$, then we must have $T_{\max} \geqslant \mu$. Furthermore, $T_{\max} < T$ implies that either $\int_0^{T_{\max}} \|F(t, u(t))\|_\alpha \mathrm{d}t = \infty$, or there exists $w \in \partial\mathcal{O}$ such that $\lim_{t \to T_{\max}} \|u(t) - w\|_\alpha = 0$. Such a time $T_{\max}$ is called the* maximal existence time, *and such a solution u is called the* maximal defined solution.

The proof of this theorem is the same as Theorem 9.17, and details are left to readers as an exercise.

9.3   SEMIGROUPS DETERMINED BY $-\Delta$ IN $L^p(\Omega)$

As the application, here we consider an example. Let Ω be of class $C^{2+\gamma}$ with $\gamma \in (0,1)$, and $1 < p < \infty$. Define $X = L^p(\Omega)$ and

$$D(A_p) = \{u \in \overset{\circ}{W}^1_p(\Omega) \cap W^2_p(\Omega)\},$$
$$A_p u = -\Delta u \ \text{ for } u \in D(A_p)$$

for the homogenous Dirichlet boundary condition, and

$$D(A_p) = \left\{u \in W^2_p(\Omega) : \partial_{\boldsymbol{n}} u = 0 \ \text{ on } \partial\Omega\right\},$$
$$A_p u = (-\Delta + 1)u \ \text{ for } u \in D(A_p)$$

for the homogenous Neumann boundary condition.

It will be proved that A_p is a sectorial operator. To this aim, we first give a theorem to judge whether an operator is a sectorial operator.

Theorem 9.40   *Let B be a closed linear operator and $D(B)$ dense in X. We denote the adjoint operator of B by B^*. Assume that there exist $R_0, C > 0$ and $\theta \in (0, \pi/2)$ such that for each*

$$\lambda \in \mathcal{S}_\theta := \{\lambda \in \mathbb{C} : |\lambda| > R_0, |\arg \lambda| \geqslant \theta\},$$

there holds

$$\|u\| \leqslant \frac{C}{|\lambda|}\|(\lambda - B)u\| \ \text{ for all } u \in D(B), \tag{9.30}$$

and equation $(\bar{\lambda} - B^*)v = 0$ only has zero solution, where $\bar{\lambda}$ is the conjugate number of λ. Then B is a sectorial operator.

Before proving Theorem 9.40, we first present a lemma.

Lemma 9.41   *Let A be a closed linear operator and $D(A)$ dense in X. If the equation $A^*v = 0$ only has zero solution, then range $\mathcal{R}(A)$ of A is dense in X.*

Proof. Firstly, $\mathcal{R}(A)$ is a linear space of X. If $\mathcal{R}(A)$ is not dense in X, then there exists $u_0 \in X$ so that $d = d(u_0, \mathcal{R}(A)) > 0$. According to the Hahn-Banach theorem, there exists $f \in X^*$ satisfying

$$\langle f, u_0 \rangle = d, \quad \|f\| = 1 \ \text{ and } \ \langle f, u \rangle = 0 \ \text{ for all } u \in \mathcal{R}(A).$$

This implies $\langle A^* f, u \rangle = \langle f, Au \rangle = 0$ for all $u \in D(A)$. Because $\overline{D(A)} = X$ and f is continuous, we have $\langle A^* f, u \rangle = 0$ for any $u \in X$. This leads to $A^* f = 0$, and then $f = 0$ by our assumption. It is a contradiction, and the proof is finished.   □

Proof of Theorem 9.40. Given $\lambda \in \mathcal{S}_\theta$. We first show $\mathcal{R}(\lambda - B) = X$. Since $\lambda - B$ is closed and equation $(\lambda - B)^*v = (\bar{\lambda} - B^*)v = 0$ only has zero solution, Lemma 9.41

indicates that $\mathcal{R}(\lambda - B)$ is dense in X. So, $\mathcal{R}(\lambda - B) = X$ once we prove $\mathcal{R}(\lambda - B)$ is closed. Assume $v_i \in \mathcal{R}(\lambda - B)$ and $v_i \to v$. Take $u_i \in D(B)$ satisfying $(\lambda - B)u_i = v_i$. According to (9.30), one has

$$\|u_i - u_j\| \leqslant \frac{C}{|\lambda|}\|v_i - v_j\| \to 0$$

as $i, j \to \infty$. Thus, $\{u_i\}$ is a Cauchy sequence, and $u_i \to u \in X$. Remembering that $\lambda - B$ is closed, it yields $u \in D(B)$ and $v = (\lambda - B)u$, i.e., $v \in \mathcal{R}(\lambda - B)$. Therefore, $\mathcal{R}(\lambda - B)$ is closed, and then $\mathcal{R}(\lambda - B) = X$.

Using (9.30) again, we know that $\lambda \in \mathcal{S}_\theta$ implies inverse $(\lambda - B)^{-1}$ exists and

$$\|(\lambda - B)^{-1}u\| \leqslant \frac{C}{|\lambda|}\|u\| \quad \text{for all } u \in D((\lambda - B)^{-1}) = \mathcal{R}(\lambda - B) = X,$$

which implies $\|(\lambda - B)^{-1}\| \leqslant C/|\lambda|$. Obviously, there exists $a \in \mathbb{R}$ such that

$$\mathcal{S}_{a,\theta} := \{\lambda \in \mathbb{C} : |\arg(\lambda - a)| \geqslant \theta\} \subset \mathcal{S}_\theta.$$

Thus, for $\lambda \in \mathcal{S}_{a,\theta}$,

$$\|(\lambda - B)^{-1}\| \leqslant \frac{C}{|\lambda|} = \frac{C}{|\lambda - a|}\frac{|\lambda - a|}{|\lambda|} \leqslant \frac{M}{|\lambda - a|}.$$

Hence, Theorem 9.40 is proved. □

In order to apply Theorem 9.40, we state two estimates.

Lemma 9.42 (*L^p estimate for elliptic equations*) *Let $1 < p < \infty$. Then there exists a constant $C > 0$ so that*

$$\|u\|_{W_p^2(\Omega)} \leqslant C\|A_p u\|_p \quad \text{for all } u \in D(A_p). \tag{9.31}$$

This lemma is exactly Lemma 9.17 of [40] for the homogeneous Dirichlet boundary condition, and Theorem 6.29 of [73] for the homogeneous Neumann boundary condition.

Lemma 9.43 ([2, Theorem 4.4]) *Let $1 < p < \infty$. Then there exist positive constants C, R and $0 < \theta < \pi/2$ such that, when $|\lambda| \geqslant R$ and $|\arg \lambda| \geqslant \theta$, there holds*

$$\|u\|_p \leqslant \frac{C}{|\lambda|}\|(\lambda - A_p)u\|_p \quad \text{for all } u \in D(A_p).$$

Now we prove that A_p is closed and A_q is adjoint operator of A_p for $q = p/(p-1)$.

Lemma 9.44 *If $1 < p < \infty$, then A_p is closed and $D(A_p)$ is dense in $L^p(\Omega)$.*

Proof. Since $C_0^\infty(\Omega) \subset D(A_p)$ and $\overline{C_0^\infty(\Omega)} = L^p(\Omega)$, it follows that $\overline{D(A_p)} = L^p(\Omega)$, i.e., $D(A_p)$ is dense.

Next, we prove that A_p is closed. Let $u_l \in D(A_p)$ and $u_l \to u, A_p u_l \to v$ in $L^p(\Omega)$. Thanks to Lemma 9.42,

$$\|u_l - u_k\|_{W_p^2(\Omega)} \leqslant C \|A_p(u_l - u_k)\|_p \to 0,$$

which implies $u_l \to u$ in $W_p^2(\Omega)$, and hence $u \in D(A_p)$. As

$$\|A_p(u_l - u)\|_p \leqslant \|u_l - u\|_{W_p^2(\Omega)} \to 0,$$

it follows that $A_p u = v$, and then A_p is closed. $\qquad\square$

Lemma 9.45   *If $1 < p < \infty$, then A_q is adjoint operator of A_p for $q = p/(p-1)$.*

Proof. Integrating by parts,

$$\langle A_p u, v \rangle = \langle u, A_q v \rangle \quad \text{for all } u \in D(A_p), \ v \in D(A_q). \tag{9.32}$$

So, $D(A_q) \subset D((A_p)^*)$ and $A_q v = (A_p)^* v$ for $v \in D(A_q)$.

Let $v \in D((A_p)^*)$ and $w = (A_p)^* v$. Then $v \in L^q(\Omega)$ and

$$\langle A_p u, v \rangle = \langle u, w \rangle \quad \text{for all } u \in D(A_p). \tag{9.33}$$

Noticing $\overline{D(A_q)} = L^q(\Omega)$, we can find a sequence $\{v_k\} \subset D(A_q)$ so that $\|v_k - v\|_q \to 0$ as $k \to \infty$, which implies $\langle A_p u, v_k \rangle \to \langle A_p u, v \rangle$. This combined with (9.32) and (9.33) allows us to derive

$$\langle u, A_q v_k \rangle = \langle A_p u, v_k \rangle \to \langle A_p u, v \rangle = \langle u, w \rangle \quad \text{for all } u \in D(A_p). \tag{9.34}$$

By virtue of $\overline{D(A_p)} = L^p(\Omega)$, it follows from (9.34) that $A_q v_k \rightharpoonup w$. Therefore $v \in D(A_q)$, i.e., $D((A_p)^*) \subset D(A_q)$ since A_q is closed. Thus $(A_p)^* = A_q$, and the proof is finished. $\qquad\square$

We shall prove that A_p is a sectorial operator and $-A_p$ generates an analytic semigroup.

Theorem 9.46   *Assume $1 < p < \infty$. Then A_p is a sectorial operator in $L^p(\Omega)$ and $-A_p$ is the infinitesimal generator of an analytic semigroup $Q_p(t) = e^{-tA_p}$.*

Proof. Utilizing Lemmas 9.44 and 9.45, we know that A_p is closed, $D(A_p)$ is dense in $L^p(\Omega)$ and $(A_p)^* = A_q$. Applying Lemma 9.43 to A_p and A_q, respectively, there exist $\theta \in (0, \pi/2)$ and positive constants R, C such that when $|\lambda| \geqslant R$ and $|\arg \lambda| \geqslant \theta$, there hold

$$\|u\|_p \leqslant C|\lambda|^{-1} \|(\lambda - A_p)u\|_p \quad \text{for all } u \in D(A_p),$$

$$\|v\|_q \leqslant C|\lambda|^{-1} \|(\bar{\lambda} - A_q)v\|_q \quad \text{for all } v \in D(A_q).$$

The second inequality implies that equation $(\bar{\lambda} - (A_p)^*)v = 0$ only has zero solution. It follows from Theorem 9.40 that A_p is a sectorial operator. $\qquad\square$

In the following we shall investigate properties of the semigroup $Q_p(t)$. Making use of $L^r(\Omega) \hookrightarrow L^p(\Omega)$ for $r \geqslant p$, we see that A_p is an extension of A_r for $r \geqslant p$ and

$$A_r u = A_p u \qquad \text{for } u \in D(A_r),$$
$$Q_r(t)u = Q_p(t)u \quad \text{for } u \in L^r(\Omega).$$

Taking advantage of Theorem 9.26 (1), $Q_p(t)u \in D(A_p^m)$ for all $u \in L^p(\Omega)$, $m \geqslant 1$ and $t > 0$. Exploiting the semigroup property and iteration method, we can deduce $Q_p(t)u \in C^{2+\alpha}(\overline{\Omega})$ for all $u \in L^p(\Omega)$ and $t > 0$. This leads to that for each $1 < q \leqslant \infty$,

$$Q_p(t)u \in D(A_q), \quad Q_p(t)u = Q_q(t - \varepsilon)Q_p(\varepsilon)u \ \text{ for all } \ u \in L^p(\Omega), \ t \geqslant \varepsilon > 0.$$

Let λ^* be the smallest positive eigenvalue of A_2. Then $\text{Re}\lambda \geqslant \lambda^*$ for all $\lambda \in \sigma(A_p)$. Accordingly, we can define the fractional power space:

$$X^\alpha = D(A_p^\alpha), \quad 0 \leqslant \alpha \leqslant 1.$$

Now, we prove the following embedding results.

Theorem 9.47 *Let Ω be of class C^1 and $0 \leqslant \alpha \leqslant 1$. Then we have*

(1) $X^\alpha \hookrightarrow W_q^1(\Omega)$ *when* $\alpha > \dfrac{1}{2} + \dfrac{n}{2}\left(\dfrac{1}{p} - \dfrac{1}{q}\right)$ *and* $p \leqslant q$;

(2) $X^\alpha \hookrightarrow L^q(\Omega)$ *when* $\alpha > \dfrac{n}{2}\left(\dfrac{1}{p} - \dfrac{1}{q}\right)$ *and* $p \leqslant q$;

(3) $X^\alpha \hookrightarrow C^\mu(\overline{\Omega})$ *when* $0 \leqslant \mu < 2\alpha - n/p$.

Moreover, all above embedding are continuous.

Proof. We first prove the conclusion (1). Taking $m = 2$ and $k = 1$ in Theorem A.9 (1) and noticing (9.31), we have that if $1/2 \leqslant \theta \leqslant 1$ and

$$q \geqslant p, \quad 1 - n/q = \theta(2 - n/p) - n(1 - \theta)/p = 2\theta - n/p, \qquad (9.35)$$

then

$$\|Bu\|_q = \|u\|_{W_q^1(\Omega)} \leqslant C\|u\|_{W_p^2(\Omega)}^\theta \|u\|_p^{1-\theta} \leqslant C\|A_p u\|_p^\theta \|u\|_p^{1-\theta}$$

for all $u \in D(A_p)$, where

$$B = \sum_{|\beta| \leqslant 1} D^\beta, \quad D(B) = W_q^1(\Omega) \subset X = L^p(\Omega).$$

Clearly, B is a closed linear operator and $D(B) \supset D(A_p)$. According to the hypothesis we can choose $1/2 \leqslant \theta < \alpha$ such that (9.35) is fulfilled. It then follows by Theorem 9.33 that

$$D(B) \supset D(A_p^\alpha), \quad BA_p^{-\alpha} \in \mathcal{L}(L^p(\Omega), L^q(\Omega)).$$

Thanks to
$$X^\alpha = D(A_p^\alpha) = \mathcal{R}(A_p^{-\alpha}), \quad BA_p^{-\alpha} \in \mathcal{L}(L^p(\Omega), L^q(\Omega)),$$
we have $X^\alpha \hookrightarrow D(B) = W_q^1(\Omega)$. Notice
$$\|Bu\|_q = \|BA_p^{-\alpha}A_p^\alpha u\|_q \leqslant C\|A_p^\alpha u\|_p = C\|u\|_\alpha \quad \text{for all } u \in X^\alpha.$$

It is clear that the embedding $X^\alpha \hookrightarrow W_q^1(\Omega)$ is continuous.

Similarly, we can prove conclusions (2) and (3). □

Lemma 9.48 *For each $u_0 \in L^2(\Omega)$, there holds*
$$\|Q_2(t)u_0\|_2 \leqslant \|u_0\|_2 e^{-\lambda^* t} \quad \text{for all } t \geqslant 0.$$

Proof. Denote
$$u(t) = Q_2(t)u_0 \quad \text{and} \quad f(t) = e^{2\lambda^* t}\|u(t)\|_2^2, \quad t \geqslant 0.$$

We first handle the homogeneous Dirichlet boundary condition and assume $u_0 \in H_0^1(\Omega) \cap H^2(\Omega)$. Utilizing the Poinceré inequality and straightforward calculations we arrive at

$$\begin{aligned}
e^{-2\lambda^* t} f'(t) &= 2\lambda^* \int_\Omega u^2(t) + 2\int_\Omega u(t)u'(t)\\
&= 2\lambda^* \int_\Omega u^2(t) + 2\int_\Omega u(t)\Delta u(t)\\
&= 2\lambda^* \int_\Omega u^2(t) - 2\int_\Omega |\nabla u(t)|^2\\
&\leqslant 0,
\end{aligned}$$

which implies

$$\|Q_2(t)u_0\|_2 = \|u(t)\|_2 \leqslant \|u_0\|_2 e^{-\lambda^* t} \quad \text{for all } u_0 \in H_0^1(\Omega) \cap H^2(\Omega),\ t \geqslant 0.$$

This estimate still holds for all $u_0 \in L^2(\Omega)$ based on denseness.

For the homogeneous Neumann boundary condition, we first assume $u_0 \in H^2(\Omega)$ and $\partial_n u_0 = 0$ on $\partial\Omega$. As above,

$$\begin{aligned}
e^{-2t} f'(t) &= 2\int_\Omega u^2(t) + 2\int_\Omega u(t)u'(t)\\
&= 2\int_\Omega u^2(t) + 2\int_\Omega u(t)[\Delta u(t) - u(t)]\\
&= -2\int_\Omega |\nabla u(t)|^2 \leqslant 0,
\end{aligned}$$

and then
$$\|Q_2(t)u_0\|_2 \leqslant \|u_0\|_2 e^{-t} \quad \text{for all } u_0 \in L^2(\Omega),\ t \geqslant 0.$$

The proof is complete. □

Lemma 9.49 ([95]) *Semigroups $Q_p(t)$ have the following properties.*

(1) *$Q_p(t)$ is positivity preserving, i.e., $u \geqslant 0$ in Ω implies $Q_p(t)u \geqslant 0$ in Ω for all $t > 0$.*

(2) *Assume $1 < p \leqslant q \leqslant \infty$. Then for any given $\delta > 0$, there exists a constant $C = C(n, p, q, \delta, \Omega)$ such that*

$$\|Q_p(t)u\|_q \leqslant C\left(1 + t^{-\delta - n(1/p - 1/q)/2}\right)e^{-\lambda^* t}\|u\|_p \quad \text{for all } u \in L^p(\Omega). \quad (9.36)$$

For heat equations with homogeneous Dirichlet boundary conditions, one may choose $\delta = 0$.

(3) *For any given $\mu \in (0, 2)$ and $\delta > 0$, there exists a constant $C = C(n, \mu, \delta, \Omega)$ such that*

$$\|Q_p(t)u\|_{C^\mu(\overline{\Omega})} \leqslant C\left(1 + t^{-\delta - \mu/2}\right)e^{-\lambda^* t}\|u\|_\infty \quad \text{for all } u \in L^\infty(\Omega).$$

Proof. (1) This is the weak maximum principle.

(2) In the following, constants C and C_i depend on exponents involved and Ω. If $1 < p \leqslant q \leqslant \infty$ satisfy $n(1/p - 1/q)/2 < 1$, then by Theorem 9.47 (2),

$$\|w\|_q \leqslant C_1\|A_p^\alpha w\|_p \quad \text{for all } w \in D(A_p^\alpha) \quad (9.37)$$

provided $n(1/p - 1/q)/2 < \alpha \leqslant 1$. Taking $a = b = 0$ in Theorem 9.29 one has

$$\|A_p^\alpha Q_p(t)u\|_p \leqslant C_2 t^{-\alpha}\|u\|_p \quad \text{for all } t > 0. \quad (9.38)$$

Because of $\delta > 0$, we can choose α such that $n(1/p - 1/q)/2 < \alpha \leqslant 1$ and $\alpha < \delta + n(1/p - 1/q)/2$. Then combining (9.37) and (9.38), we discover

$$\|Q_p(t)u\|_q \leqslant C_3 t^{-\delta - n(1/p - 1/q)/2}\|u\|_p \quad \text{for all } 0 < t \leqslant 3. \quad (9.39)$$

The restriction $n(1/p - 1/q)/2 < 1$ can be removed by exploiting the semigroup property and iteration. We left its proof to readers as an exercise. Hence (9.39) holds for all $1 < p \leqslant q \leqslant \infty$ and $\delta > 0$. Certainly,

$$\|Q_p(t)u\|_q \leqslant C\left(1 + t^{-\delta - n(1/p - 1/q)/2}\right)e^{-\lambda^* t}\|u\|_p \quad \text{for all } 0 < t \leqslant 3. \quad (9.40)$$

Now we treat the case $t \geqslant 3$. If $p > 2$, then $Q_p(t)v = Q_2(t)v$ for $v \in L^p(\Omega)$ and $t > 0$. Notice $q \geqslant p > 2$. In view of (9.39) and Lemma 9.48 it avers

$$\begin{aligned}
\|Q_p(t)u\|_q &= \|Q_2(t)u\|_q = \|Q_2(1)Q_2(t - 1)u\|_q \\
&\leqslant C_4\|Q_2(t - 1)u\|_2 \leqslant C_5 e^{-\lambda^*(t-1)}\|u\|_2 \\
&\leqslant C e^{-\lambda^* t}\|u\|_p.
\end{aligned}$$

When $1 < p \leqslant 2$, we have $Q_p(1)u \in L^2(\Omega)$. As above,

$$
\begin{aligned}
\|Q_p(t)u\|_q &= \|Q_p(1)Q_p(t-2)Q_p(1)u\|_q \leqslant C_6\|Q_p(t-2)Q_p(1)u\|_p \\
&= C_6\|Q_2(t-2)Q_p(1)u\|_p \leqslant C_7\|Q_2(t-2)Q_p(1)u\|_2 \\
&\leqslant C_8 e^{-\lambda^*(t-2)}\|Q_p(1)u\|_2 \leqslant C e^{-\lambda^* t}\|u\|_p.
\end{aligned}
$$

Anyhow, for each $p > 1$, there holds

$$
\|Q_p(t)u\|_q \leqslant C e^{-\lambda^* t}\|u\|_p \quad \text{for all } t \geqslant 3. \tag{9.41}
$$

Combining (9.41) and (9.40), we obtain (9.36).

For the case of the homogeneous Dirichlet boundary condition, we define

$$
v_0(x) = \begin{cases} |u(x)|, & x \in \Omega, \\ 0, & x \in \mathbb{R}^n \setminus \overline{\Omega}, \end{cases}
$$

and let $v(x,t)$ be the unique solution of

$$
\begin{cases} v_t - \Delta v = 0, & x \in \mathbb{R}^n, \ t > 0, \\ v(x,0) = v_0(x), & x \in \mathbb{R}^n. \end{cases}
$$

Then

$$
v(x,t) = \int_{\mathbb{R}^n} \frac{1}{(4\pi t)^{n/2}} \exp\left(-\frac{|x-y|^2}{4t}\right) v_0(y) \mathrm{d}y,
$$

and an explicit calculation shows

$$
\|v(\cdot,t)\|_q \leqslant C t^{-n(1/p - 1/q)/2}\|v_0\|_p, \quad 1 \leqslant p \leqslant q \leqslant \infty.
$$

Taking advantage of a comparison argument, we can get

$$
|Q(t)u| \leqslant v e^{-\lambda^* t} \quad \text{for all } x \in \Omega, \ t \geqslant 0,
$$

and hence (9.36) is valid for $\delta = 0$.

(3) When $\mu < 2\alpha - n/p$ and $w \in D(A_p^\alpha)$, we have $\|w\|_{C^\mu(\overline{\Omega})} \leqslant C_9\|A_p^\alpha w\|_p$ by Theorem 9.47. This leads to $\|Q_p(t)u\|_{C^\mu(\overline{\Omega})} \leqslant C_9\|A_p^\alpha Q(t)u\|_p$. Because of $0 < \mu < 2$ and $\delta > 0$, we can take $\alpha < 1$ and $p \gg 1$ such that $\mu < 2\alpha - n/p$ and $\alpha < \delta + \mu/2$. Then, by use of (9.38),

$$
\|Q_p(t)u\|_{C^\mu(\overline{\Omega})} \leqslant C_9\|A_p^\alpha Q_p(t)u\|_p \leqslant C_{10} t^{-\alpha}\|u\|_p. \tag{9.42}
$$

And so, for $0 < t \leqslant 4$,

$$
\|Q_p(t)u\|_{C^\mu(\overline{\Omega})} \leqslant C_{10} t^{-(\mu/2+\delta)}\|u\|_p \leqslant C t^{-(\mu/2+\delta)} e^{-\lambda^* t}\|u\|_\infty.
$$

When $t \geqslant 4$, it follows from (9.42) and (9.41) that

$$
\begin{aligned}
\|Q_p(t)u\|_{C^\mu(\overline{\Omega})} &= \|Q_p(1)Q_p(t-1)u\|_{C^\mu(\overline{\Omega})} \leqslant C_{11}\|Q_p(t-1)u\|_p \\
&\leqslant C_{12} e^{-\lambda^*(t-1)}\|u\|_p \leqslant C e^{-\lambda^* t}\|u\|_\infty.
\end{aligned}
$$

The conclusion (3) is obtained. $\qquad\qquad\qquad\qquad\qquad\qquad\qquad\qquad\quad \square$

9.4　AN EXAMPLE

As an application of the semigroup theory, in this section, we study the following initial-boundary value problem ([95])

$$
\begin{cases}
u_{it} - d_i \Delta u_i = f_i(x,t,u), & x \in \Omega, \quad t > 0, \quad 1 \leqslant i \leqslant k, \\
u_{jt} = f_j(x,t,u), & x \in \Omega, \quad t > 0, \quad k+1 \leqslant j \leqslant m, \\
u_i = 0 \ \text{ or } \ \partial_n u_i = 0, & x \in \partial\Omega, \ t > 0, \ 1 \leqslant i \leqslant k, \\
u_l(x,0) = u_{l0}(x), & x \in \Omega, \quad 1 \leqslant l \leqslant m,
\end{cases}
\tag{9.43}
$$

where, for each $1 \leqslant i \leqslant k$, $d_i > 0$ is a constant, and for each $1 \leqslant l \leqslant m$, $u_{l0} \in C^\gamma(\overline{\Omega})$ for some $0 < \gamma < 1$, $f_l : \overline{\Omega} \times [0,\infty) \times \mathbb{R}^m \to \mathbb{R}$ is measurable and satisfies

(M) For any bounded set $B \subset \overline{\Omega} \times [0,\infty) \times \mathbb{R}^m$, f_l is bounded in B and

$$
|f_l(x,t,u) - f_l(x,t,v)| \leqslant L(B)|u - v| \ \text{ for all } \ (x,t,u), (x,t,v) \in B.
$$

Let $Q_p(t) = e^{-tA_p}$, where A_p is defined at the beginning of §9.3. Define $\Gamma_i(t) = Q_p(d_i t)|_{L^\infty(\Omega)}$ for $1 \leqslant i \leqslant k$, and $\Gamma_i(t) = I$ for $k+1 \leqslant i \leqslant m$. Then operators $\Gamma(t) = (\Gamma_1(t), \ldots, \Gamma_m(t)) : [L^\infty(\Omega)]^m \to [L^\infty(\Omega)]^m$ satisfy $\|\Gamma(t)\| \leqslant 1$ for all $t \geqslant 0$.

For $1 \leqslant i \leqslant k$, we set $\tilde{f}_i(x,t,u) = f_i(x,t,u) + d_i u_i$ when the boundary is $\partial_n u_i|_{\partial\Omega} = 0$, while $\tilde{f}_i(x,t,u) = f_i(x,t,u)$ when the boundary is $u_i|_{\partial\Omega} = 0$. For $k+1 \leqslant i \leqslant m$, we set $\tilde{f}_i(x,t,u) = f_i(x,t,u)$.

Let $0 < T \leqslant \infty$. A mild solution of the initial-boundary value problem (9.43) is a measurable function $u : \Omega \times [0,T) \to \mathbb{R}^m$ satisfying $u(\cdot,t) \in [L^\infty(\Omega)]^m$ and

$$
u(\cdot,t) = \Gamma(t)u_0 + \int_0^t \Gamma(t-s)\tilde{f}(\cdot,s,u(\cdot,s))ds, \quad t \in [0,T),
$$

or precisely, for $t \in [0,T)$,

$$
\begin{cases}
u_i(\cdot,t) = \Gamma_i(t)u_{i0} + \displaystyle\int_0^t \Gamma_i(t-s)\tilde{f}_i(\cdot,s,u(\cdot,s))ds, & 1 \leqslant i \leqslant k, \\
u_i(\cdot,t) = u_{i0} + \displaystyle\int_0^t f_i(\cdot,s,u(\cdot,s))ds, & k+1 \leqslant i \leqslant m.
\end{cases}
\tag{9.44}
$$

Theorem 9.50 ([95]) *Under the condition* **(M)**, *we have the following conclusions.*

(1) *For each initial function $u_0 \in [L^\infty(\Omega)]^m$, there exists $T_{\max} > 0$ such that (9.43) has a unique mild solution in $[0,T_{\max})$.*

(2) *The "existence time $T_{\max}$" can be chosen maximal: either $T_{\max} = \infty$, or $T_{\max} < \infty$ and*

$$
\lim_{t \to T_{\max}} \|u(\cdot,t)\|_\infty = \infty.
\tag{9.45}
$$

(3) *Consider the existence time $T_{\max}$ as a functional of the initial data u_0. Then $\inf\{T_{\max}(u_0) : \|u_0\| \leqslant C\} > 0$ for any $C > 0$.*

Proof. The existence part of (1) and the conclusion (3) can be proved by the standard Picard iteration, and we just refer the reader to the proof of Theorem 9.15. The uniqueness part of (1) can be proved by a similar way as that of Theorem 9.16. They are left to the reader as an exercise.

Now, we prove the conclusion (2). Assume that the maximal existence time $T_{\max}$ is finite. First we show that assumption

$$\sup_{0 \leqslant t < T_{\max}} \|u(\cdot, t)\|_\infty \leqslant C < \infty \tag{9.46}$$

will lead to a contradiction. For simplicity, the spatial variable x is omitted. Choose an arbitrary $\mu \in (0, 2)$. Applying Lemma 9.49 (3) to the first equation of (9.44), one has

$$\sup_{\sigma \leqslant t < T_{\max}} \|u_i(t)\|_{C^\mu(\overline{\Omega})} < \infty, \quad 1 \leqslant i \leqslant k$$

for any $\sigma > 0$.

Applying Theorem 9.29 (take $a = b = 0$ there) to the first equation of (9.44), we find that, for $p \in (1, \infty)$ and $\alpha, \beta \in (0, 1)$ satisfying $\alpha + \beta < 1$,

$$\sup_{\sigma \leqslant t < T_{\max}} \|A_p^{\alpha+\beta} u_i(t)\|_p < \infty, \quad 1 \leqslant i \leqslant k. \tag{9.47}$$

Fix parameters μ, α and β to satisfy

$$0 < \mu < 2, \quad \alpha, \beta > 0, \quad \alpha + \beta < 1, \quad 2\alpha > \mu + n/p.$$

Using Theorem 9.47 (3), we know that the new norm

$$\|u\|_* := \|u\|_\infty + \sum_{i=1}^m d_i \|u_i\|_{C^\mu(\overline{\Omega})} \leqslant \|u\|_\infty + C \sum_{i=1}^m d_i \|A_p^\alpha u_i\|_p.$$

We have known that

$$\sup_{\sigma \leqslant t < T_{\max}} \|u(\cdot, t)\|_* \leqslant M < \infty.$$

Next, we hope to show

$$\lim_{t \to T_{\max}} u(t) \text{ exists in the norm } \|\cdot\|_*.$$

To this aim, we denote

$$\Sigma_C := \{s : 0 \leqslant s < T_{\max}, \ \|u(s)\|_\infty \leqslant C\},$$

and let a sequence $\{t_j\}$ be monotonically increasing and $t_j \to T_{\max}$ as $j \to \infty$. Take $t_l > t_j$. It follows from the second equation of (9.44) that

$$\|u_i(t_l) - u_i(t_j)\|_\infty \leqslant |t_l - t_j| \sup_{\Sigma_C} \|f_i(s, u)\|_\infty \quad \text{for } k+1 \leqslant i \leqslant m,$$

which implies, as $l, j \to \infty$,

$$\|u_i(t_l) - u_i(t_j)\|_\infty \to 0 \quad \text{for } k+1 \leqslant i \leqslant m. \tag{9.48}$$

For $1 \leqslant i \leqslant k$. Using the first equation of (9.44), we achieve

$$u_i(t_l) - u_i(t_j) = (\Gamma_i(t_l - t_j) - 1)u_i(t_j)$$
$$+ \int_0^{t_l - t_j} \Gamma_i(t_l - t_j - s)\tilde{f}_i(t_j + s, u(t_j + s))ds,$$

and then

$$\|A_p^\alpha(u_i(t_l) - u_i(t_j))\|_p \leqslant \|A_p^{-\beta}(\Gamma_i(t_l - t_j) - 1)\|_{p,p}\|A_p^{\alpha+\beta}u_i(t_j)\|_p$$
$$+ \sup_{\Sigma_C} \|\tilde{f}_i(s, u)\|_\infty \int_0^{t_l - t_j} \|A_p^\alpha \Gamma_i(s)\|_{p,p}ds, \qquad (9.49)$$

where $\| \cdot \|_{p,p} = \| \cdot \|_{\mathcal{L}(L^p, L^p)}$. Based on Theorem 9.28 (2) and Theorem 9.32,

$$\|A_p^{-\beta}(\Gamma_i(t) - 1)v\|_p = \|(\Gamma_i(t) - 1)A_p^{-\beta}v\|_p \leqslant \frac{C}{\beta}t^\beta\|v\|_p \text{ for all } v \in L^p(\Omega),$$

which implies $\|A_p^{-\beta}(\Gamma_i(t) - 1)\|_{p,p} \leqslant Kt^\beta$, and consequently

$$\|A_p^{-\beta}(\Gamma_i(t_l - t_j) - 1)\|_{p,p} \leqslant K(t_l - t_j)^\beta \to 0 \quad \text{as } l, j \to \infty. \qquad (9.50)$$

Taking advantage of Theorem 9.29, we conclude $\|A_p^\alpha \Gamma_i(s)\|_{p,p} \leqslant Cs^{-\alpha}$, and then

$$\int_0^{t_l - t_j} \|A_p^\alpha \Gamma_i(s)\|_{p,p}ds \leqslant C(t_l - t_j)^{1-\alpha} \to 0 \quad \text{as } l, j \to \infty \qquad (9.51)$$

because of $0 < \alpha < 1$. Remembering (9.47), it follows from (9.49)–(9.51) that

$$\|A_p^\alpha(u_i(t_l) - u_i(t_j))\|_p \to 0 \quad \text{for } 1 \leqslant i \leqslant k$$

as $l, j \to \infty$. Hence, by use of Theorem 9.47 (3), it derives

$$\|u_i(t_l) - u_i(t_j)\|_{C^\mu(\overline{\Omega})} \leqslant C\|A_p^\alpha(u_i(t_l) - u_i(t_j))\|_p \to 0 \quad \text{for } 1 \leqslant i \leqslant k. \qquad (9.52)$$

Recalling the definition of $\|u\|_*$, it follows from (9.48) and (9.52) that $\|u(t_l) - u(t_j)\|_* \to 0$ as $l, j \to \infty$. Thus, $\lim_{t \to T_{\max}} u(t)$ exists in the norm $\| \cdot \|_*$. Therefore, the mild solution can be prolonged to an interval $[0, T_{\max} + \varepsilon)$, which contradicts the maximality of $T_{\max}$. This indicates that (9.46) cannot hold, and thereby

$$\limsup_{t \to T_{\max}} \|u(t)\|_\infty = \infty. \qquad (9.53)$$

Let us assume

$$\liminf_{t \to T_{\max}} \|u(t)\|_\infty < r < \infty, \qquad (9.54)$$

and we are going to derive a contradiction. According to (9.53) and (9.54), there exist two sequences $\{t_l\}$ and $\{s_l\}$ with properties: $t_l < s_l$, $\lim_{l \to \infty} t_l = \lim_{l \to \infty} s_l = T_{\max}$ and

$$\|u(t_l)\|_\infty = r, \quad \|u(s_l)\|_\infty = r + 1, \quad \|u(t)\|_\infty \leqslant r + 1 \quad \text{in } [t_l, s_l].$$

From the integral equation

$$u(s_l) = \Gamma(s_l - t_l)u(t_l) + \int_0^{s_l - t_l} \Gamma(s_l - t_l - \tau)\tilde{f}(t_l + s, u(t_l + s))ds,$$

one gets

$$\|u(s_l)\|_\infty \leqslant \|u(t_l)\|_\infty + |s_l - t_l| \sup_{\Sigma_{r+1}} \|\tilde{f}(\tau, u(\tau))\|_\infty,$$

where $\Sigma_{r+1} = \{\tau : 0 \leqslant \tau < T_{\max}, \|u(\tau)\|_\infty \leqslant r+1\}$, and hence $r+1 \leqslant r$ due to the fact that $s_l - t_l \to 0$, which is a contradiction. Thus (9.54) cannot hold, and (9.45) is obtained. □

Note: Before closing this chapter, we point out that conclusions and solution spaces in this chapter are different from those obtained by the L^p theory and Schauder theory. For example, consider the initial-boundary value problem (9.15).

- The mild solution u of (9.15) is only the continuous solution in time t of integral equation

$$u(t) = e^{t\Delta}u_0 + \int_0^t e^{(t-s)\Delta}\frac{u(s)}{1 + u^2(s)}ds \text{ for all } t \in [0, T),$$

and its weak derivatives u_t and $D^2 u$ may not exist. For the strong solution of (9.15) here, although weak derivatives u_t and $D^2 u$ exist, they may not belong to $L^p(Q_T)$. That is, the strong solution u of (9.15) may not belong to $W_p^{2,1}(Q_T)$. For the classical solution of (9.15) here, although weak derivatives u_t and $D^2 u$ exist and are continuous from $[0, T]$ to $X = L^p(\Omega)$, they may not belong to $C(\overline{Q_T})$, i.e., the classical solution u may not belong to $C^{2,1}(\overline{Q_T})$.

- When we use the L^p theory or Schauder theory to discuss the existence and uniqueness of solutions of (9.15), we should assume that $u_0 \in W_p^2(\Omega) \cap \overset{\circ}{W}{}_p^1(\Omega)$ for the L^p theory and $u_0 \in C^{2+\alpha}(\overline{\Omega})$ for the Schauder theory. The corresponding solution spaces are $W_p^{2,1}(Q_T)$ and $C^{2+\alpha, 1+\alpha/2}(\overline{Q_T})$, respectively.

  When we use the semigroup theory to discuss the existence and uniqueness of solutions of (9.15), we can take $X = L^p(\Omega)$ and $D(-\Delta) = W_p^2(\Omega) \cap \overset{\circ}{W}{}_p^1(\Omega)$. Then $-\Delta$ is a sectorial operator in $L^p(\Omega)$ and Δ is the infinitesimal generator of an analytic semigroup $Q_p(t) = e^{t\Delta}$ by Theorem 9.46. For any $0 \leqslant \alpha < 1$ and $u_0 \in X^\alpha$, problem (9.15) has a unique solution $u \in C([0, T], X^\alpha)$ by Theorem 9.38.

  Because there is no necessary embedding relationship between spaces $W_p^2(\Omega) \cap \overset{\circ}{W}{}_p^1(\Omega)$ and $(L^p(\Omega))^\alpha$, the L^p theory (Schauder theory) and semigroup method can not replace each other.

  In order to establish the existence and uniqueness of solutions, generally speaking, the semigroup method has a lower requirement for initial values, for example, $u_0 \in L^p(\Omega)$, while L^p theory and Schauder theory have a higher requirement for initial values.

- Generally speaking, solutions obtained by the semigroup method are not necessarily weak solutions in the sense of distribution. Therefore, we cannot use standard interior regularity methods to improve the interior smoothness of this solution.

- Sometimes some people call the solution obtained by semigroup method as a *semigroup solution.*

EXERCISES

9.1 Take $X = L^2(\Omega)$ and $D(A) = H^2(\Omega) \cap H_0^1(\Omega)$ in Example 9.1. Prove that the operator A is the infinitesimal generator of a contraction semigroup in $L^2(\Omega)$.

9.2 Let Ω be of class C^2 and $p > \max\{1, n/2\}$. Define an operator C by

$$D(C) = \left\{ u \in W_p^2(\Omega) \cap \overset{\circ}{W}{}_p^1(\Omega) \cap C(\overline{\Omega}) : \Delta u \in C(\overline{\Omega}) \right\},$$

$$Cu = -\Delta u \quad \text{for} \quad u \in D(A).$$

Prove that the operator $-C$ is the infinitesimal generator of a contraction semigroup in $C(\overline{\Omega})$.

9.3 Consider the initial-boundary value problem (9.15). Let $D(C)$ be as above and assume $u_0 \in D(C)$. Prove that problem (9.15) has a unique global classical solution $u \in C([0, \infty), D(C)) \cap C^1([0, \infty), C(\overline{\Omega}))$.

9.4 Prove Theorem 9.39.

9.5 Prove that (9.39) holds for all $1 < p \leq q \leq \infty$ and $\delta > 0$ without the restriction $n(1/p - 1/q)/2 < 1$.

9.6 Let $\Omega = (0, 1)$ and $X = L^2(\Omega)$. Define a linear operator A:

$$Au = -u'' \quad \text{for } u \in D(A) := H^2(\Omega) \cap H_0^1(\Omega).$$

Prove

(1) $D(A^{1/2}) = H_0^1(\Omega)$ and $\|A^{1/2}u\| = \|u'\|$ for all $u \in H_0^1(\Omega)$;
(2) if $\alpha > 1/4$, then $X^\alpha \hookrightarrow L^\infty(\Omega)$, i.e., there exists constant $C > 0$ such that $\|u\|_\infty \leq C\|A^\alpha u\|$ for all $u \in D(A^\alpha)$.

9.7 Let A be a sectorial operator and $\mathrm{Re}(\sigma(A)) > 0$. Prove that the followings are equivalent:

(1) for any given $\alpha > 0$, operator $A^{-\alpha}$ is compact;
(2) for any given $t > 0$, operator e^{-tA} is compact.

9.8 Prove conclusions (2) and (3) of Theorem 9.47.

9.9 Prove conclusions (1) and (3) of Theorem 9.50.

9.10 Let Ω be of class $C^{2+\alpha}$. Consider the problem

$$\begin{cases} u_t - \Delta u = u|\nabla u|^2, & x \in \Omega, \quad t > 0, \\ u(x,t) = 0, & x \in \partial\Omega, \ t > 0, \\ u(x,0) = u_0(x), & x \in \Omega. \end{cases}$$

Give an appropriate condition regarding initial datum $u_0(x)$ and study the local existence, uniqueness and regularity of solutions.

Appendix

In this appendix, we review the fixed point theorem, embedding theorems and interpolation inequalities.

Theorem A.1 (Contraction mapping principle) *Let X be a nonempty complete metric space, and let $\mathcal{F} : X \to X$ be a strict contraction mapping, i.e.,*

$$d(\mathcal{F}(u), \mathcal{F}(v)) \leqslant k d(u, v) \quad \text{for all } u, v \in X \text{ with } k < 1.$$

Then $\mathcal{F}$ has a unique fixed point, $\mathcal{F}(u) = u$.

A continuous operator between two Banach spaces is called a *compact operator* (or a *completely continuous operator*) if images of bounded sets are precompact (that is, their closures are compact).

Theorem A.2 (Schauder fixed point theorem [98]) *Let V be a compact convex set in a Banach space X, and let $\mathcal{F}$ be a continuous operator of V into itself. Then $\mathcal{F}$ has at least one fixed point in V, that is, $\mathcal{F}(u) = u$ for some $u \in V$.*

Corollary A.3 (Schauder fixed point theorem) *Let V be a closed convex set in a Banach space X, and let $\mathcal{F}$ be a continuous operator of V into itself such that the image $\mathcal{F}(V)$ is precompact. Then $\mathcal{F}$ has at least one fixed point in V.*

Now we review the Sobolev *embedding theorems* and interpolation inequalities. We use the symbol $X \overset{C}{\hookrightarrow} Y$ to indicate that the Banach space X is embedded compactly in the Banach space Y. We always assume that $\Omega \subset \mathbb{R}^n$ is a bounded open set. Theorems of this part can be found in [1, 16, 111].

Theorem A.4 *Let Ω be of class C^1, k be a positive integer, and $1 \leqslant p \leqslant \infty$.*

(1) *If $kp < n$, then $W_p^k(\Omega) \overset{C}{\hookrightarrow} L^q(\Omega)$ for all $1 \leqslant q < np/(n-kp)$, and the embedding constant depends only on n, k, p, q and Ω.*

(2) *If $kp = n$, then $W_p^k(\Omega) \overset{C}{\hookrightarrow} L^q(\Omega)$ for all $1 \leqslant q < \infty$, and the embedding constant depends only on n, k, q, and Ω.*

(3) *If $kp > n$, then $W_p^k(\Omega) \overset{C}{\hookrightarrow} C^{k-[\frac{n}{p}]-1+\alpha}(\overline{\Omega})$ for all $0 < \alpha < \alpha_0$, where*

$$\alpha_0 = \begin{cases} \left[\frac{n}{p}\right] + 1 - \frac{n}{p} & \text{when } n/p \text{ is not an integer}, \\ \text{any } \alpha^* \in (0,1) & \text{when } n/p \text{ is an integer}, \end{cases}$$

and the embedding constant depends only on n, k, p, α and Ω.

Remark A.5 *When we use $\overset{\circ}{W}_p^k(\Omega)$ instead of $W_p^k(\Omega)$, conclusions of Theorem A.4 hold without assuming that Ω is a C^1 domain.*

In the following theorems, the embedding constants depend only on n, p, d^{-1}, T^{-1} and $\partial\Omega$, where $d = \text{diam}(\Omega)$ is the diameter of Ω.

Theorem A.6 *Let $p \geqslant 1, n \geqslant 2$ and Ω be of class C^2.*

(1) *Assume $p < n + 2$. Then for $q = p(n + 2)/(n + 2 - p)$, we have $\|Du\|_{q, Q_T} \leqslant C\|u\|_{W_p^{2,1}(Q_T)}$ for all $u \in W_p^{2,1}(Q_T)$.*

(2) *If $p < 1 + n/2$, then $W_p^{2,1}(Q_T) \hookrightarrow L^q(Q_T)$ with $q = p(n + 2)/(n + 2 - 2p)$.*

(3) *If $p = 1 + n/2$, then $W_p^{2,1}(Q_T) \hookrightarrow L^q(Q_T)$ for all $1 \leqslant q < \infty$.*

Theorem A.7 *Let Ω be of class C^2.*

(1) *If $1 + n/2 < p < n + 2$, then $W_p^{2,1}(Q_T) \hookrightarrow C^{\alpha, \alpha/2}(\overline{Q}_T)$ with $\alpha = 2 - (n + 2)/p$.*

(2) *If $p = n + 2$, then $W_p^{2,1}(Q_T) \hookrightarrow C^{\alpha, \alpha/2}(\overline{Q}_T)$ for any $0 < \alpha < 1$.*

(3) *If $p > n + 2$, then $W_p^{2,1}(Q_T) \hookrightarrow C^{1+\alpha, (1+\alpha)/2}(\overline{Q}_T)$ with $\alpha = 1 - (n + 2)/p$.*

Theorem A.8 *Let k, l be constants and $0 \leqslant l < k$. Then $C^{k,k/2}(\overline{Q}_T) \overset{C}{\hookrightarrow} C^{l,l/2}(\overline{Q}_T)$.*

Theorem A.9 (Interpolation inequality [111, Theorems 2.11.4 and 2.11.5]) *Let k, m be integers, Ω be of class C^m with $m \geqslant 1$, $1 \leqslant p, r \leqslant \infty$ and $u \in W_p^m(\Omega) \cap L^r(\Omega)$.*

(1) *Let $0 \leqslant k < m$ and $q \geqslant nr/(n + rk)$ such that $W_p^m(\Omega) \hookrightarrow W_q^k(\Omega)$. If $k/m \leqslant \theta \leqslant 1$ and*

$$k - n/q = \theta(m - n/p) - n(1 - \theta)/r,$$

then

$$\|u\|_{W_q^k(\Omega)} \leqslant C\|u\|_{W_p^m(\Omega)}^{\theta}\|u\|_r^{1-\theta}.$$

(2) *Assume $0 < \theta \leqslant 1$. If μ satisfies*

$$0 < \mu \leqslant \theta(m - n/p) - n(1 - \theta)/r,$$

and this inequality is strict when μ is an integer, then

$$\|u\|_{C^\mu(\overline{\Omega})} \leqslant C\|u\|_{W_p^m(\Omega)}^{\theta}\|u\|_r^{1-\theta}.$$

Bibliography

[1] Adams, R.A., and J.J.F. Fournier. 2003. *Sobolev Spaces, 2nd Edition.* New York: Academic Press.

[2] Agmon, S. 1962. On the eigenfunctions and on the eigenvalues of general elliptic boundary value problems. *Comm. Pure Appl. Math.* 15: 119-147.

[3] Ahn, I., S. Baek, and Z.G. Lin. 2016. The spreading fronts of an infective environment in a man-environment-man epidemic model. *Appl. Math. Model.* 40:7082-7101.

[4] Alikakos, N.D. 1979. L^p bounds of solutions of reaction-diffusion equations. *Comm. Partial Differ. Equat.* 4:827-868.

[5] Alikakos, N.D. 1979. An application of the invariance principle to reaction diffusion equations. *J. Differ. Equat.* 33:201-225.

[6] Amann, H. 1971. On the existence of positive solutions of nonlinear elliptic boundary value problems. *Ind. Univ. Math. J.* 21:125-146.

[7] Amann, H. 1986. Quasilinear parabolic systems under nonlinear boundary conditions. *Arch. Ration. Mech. An.* 92:153-192.

[8] Amann, H. 1988. Parabolic evolution equations and nonlinear boundary conditions. *J. Differ. Equat.* 72:201-269.

[9] Aronson, D., and H.F. Weinberger. 1978. Multidimensional nonlinear diffusion arising in population genetics. *Adv. Math.* 30:33-76.

[10] Ball, J. 1977. Remarks on blow-up and nonexistence theorems for nonlinear evolution equations. *Quart. J. Math. Oxford.* 28:473-486.

[11] Bandle, C., and H. Brunner. 1998. Blow-up in diffusion equations, a survey. *J. Comp. Appl. Math.* 97:3-22.

[12] Bunting, G., Y.H. Du, and K. Krakowski. 2012. Spreading speed revisited: Analysis of a free boundary model. *Netw. Heterog. Media* 7:583-603.

[13] Cao, J.F., Y.H. Du, F. Li, and W.T. Li. 2019. The dynamics of a Fisher-KPP nonlocal diffusion model with free boundaries. *J. Funct. Anal.* 277:2772-2814.

[14] Castro, A., and A.C. Lazer. 1982. Remarks on periodic solutions of parabolic equations suggested by elliptic theory. *Boll. Univ. Mat. Ital. B* 6:1089-1104.

[15] Cazenave, T., and A. Haraux. 1998. *An Introduction to Semilinear Evolution Equations.* Translated by Y. Martel. Oxford: Clarendon Press.

[16] Chen, Y.Z. 2003. *Second Order Parabolic Differential Equations* (in Chinese). Beijing: Peking University Press.

[17] Chlebík, M., and M. Fila. 1999. From critical exponents to blow-up rates for parabolic problems. *Rend. Mat. Appl.* 19:449-470.

[18] Deng, K., and H.A. Levine. 2000. The role of critical exponents in blow-up theorems: the sequel. *J. Math. Anal. Appl.* 243:85-126.

[19] Deng, K., and C.L. Zhao. 2001. Blow-up for a parabolic system coupled in an equation and a boundary condition. *Proc. Roy. Soc. Edinburgh Sec. A* 131:1345-1355.

[20] Ding, W., Y.H. Du, and X. Liang. 2017. Spreading in space-time periodic media governed by a monostable equation with free boundaries, Part 1: Continuous initial functions. *J. Differ. Equat.* 262:4988-5021.

[21] Dockery, J., V. Hutson, K. Mischaikow, and M. Pernarowski. 1998. The evolution of slow dispersal rates: a reaction-diffusion model. *J. Math. Biol.* 37:61-83.

[22] Du, Y.H. 2006. *Order Structure and Topological Methods in Nonlinear PDEs Vol. 1: Maximum Principle and Applications.* Singapore: World Scientific.

[23] Du, Y.H., Z.M. Guo, and R. Peng. 2013. A diffusive logistic model with a free boundary in time-periodic environment. *J. Funct. Anal.* 265:2089-2142.

[24] Du, Y.H., and S. B. Hsu. 2004. A diffusive predator-prey model in heterogeneous environment. *J. Differ. Equat.* 203:331-364.

[25] Du, Y.H., and Z.G. Lin. 2010. Spreading-vanishing dichotomy in the diffusive logistic model with a free boundary. *SIAM J. Math. Anal.* 42:377-405.

[26] Du, Y.H., and Z.G. Lin. 2014. The diffusive competition model with a free boundary: Invasion of a superior or inferior competitor. *Discrete Contin. Dyn. Syst.-Ser. B* 19:3105-3132.

[27] Du, Y.H., and B.D. Lou. 2015. Spreading and vanishing in nonlinear diffusion problems with free boundaries. *J. Eur. Math. Soc.* 17:2673-2724.

[28] Du, Y.H., H. Matsuzawa, and M.L. Zhou. 2014. Sharp estimate of the spreading speed determined by nonlinear free boundary problems. *SIAM J. Math. Anal.* 46:375-396.

[29] Du, Y.H., M.X. Wang, and M.L. Zhou. 2017. Semi-wave and spreading speed for the diffusive competition model with a free boundary, *J. Math. Pure Appl.* 107:253-287.

[30] Du, Y.H., L. Wei, and L. Zhou. 2018. Spreading in a shifting environment modelled by the diffusive logistic equation with a free boundary. *J. Dyn. Diff. Equat.* 30:1389-1426.

[31] Evans, L.C. 2002. *Partial Differential Equations.* Graduate Studies in Mathematics, 19. Providence, RI: American Mathematical Society.

[32] Fujita, H. 1966. On the blowing up of solutions of the Cauchy problem for $u_t = \Delta u + u^{1+\alpha}$. *J. Fac. Sci. Univ. Tokyo Sect. A. Math.* 16:105-113.

[33] Fila, M., and J. Filo. 1996. Blow-up on the boundary: a survey. *Banach Center Publ.* 33:67-78.

[34] Fila, M., and H.A. Levine. 1996. On critical exponents for a semilinear parabolic system coupled in an equation and a boundary condition. *J. Math. Anal. Appl.* 204:494-521.

[35] Fila, M., and P. Quittner. 1999. The blow-up rate for a semilinear parabolic system. *J. Math. Anal. Appl.* 238:468-476.

[36] Friedman, A. 1964. *Partial Differential Equations of Parabolic Type.* Englewood Cliffs: Prentice-Hall.

[37] Galaktionov, V.A. 2004. *Geometric Sturmian Theory of Nonlinear Parabolic Equations and Applications.* New York: Chapman & Hall/CRC.

[38] Garroni, M. G., and Menaldi, J.L. 1992. *Green functions for second order parabolic integrodifferential problems.* Longman Scientific & Technical.

[39] Ge, J., K. Kim, Z.G. Lin, and H.P. Zhu. 2015. A SIS reaction-diffusion-advection model in a low-risk and high-risk domain. *J. Differ. Equat.* 259:5486-5509.

[40] Gilbarg, G., and N.S. Trudinger. 2001. *Elliptic Partial Differential Equations of the Second Order.* New York: Springer-Verlag.

[41] Gu, H., B.D. Lou, and M.L. Zhou. 2015. Long time behavior for solutions of Fisher-KPP equation with advection and free boundaries. *J. Funct. Anal.* 269:1714-1768.

[42] Guo, D.J. 2004. *Nonlinear Functional Analysis, Second Edition* (in Chinese). Jinan: Shandong Science and Technology Press.

[43] Guo, J.S., and C.-H. Wu. 2012. On a free boundary problem for a two-species weak competition system. *J. Dyn. Diff. Equat.* 24:873-895.

[44] Guo, J.S., and C.-H. Wu. 2015. Dynamics for a two-species competition-diffusion model with two free boundaries. *Nonlinearity* 28:1-27.

[45] Hayakawa, K. 1973. On the nonexistence of global solutions of some semilinear parabolic equations. *Proc. Jpn. Acad.* 49:503-525.

[46] Hess, P. 1991. *Periodic-Parabolic Boundary Value Problems and Positivity.* Harlow: Longman Scientific & Technical.

[47] Hu, B. 1996. Remarks on the blowup estimate for solutions of the heat equation with a nonlinear boundary condition. *Differ. Integral Equ.* 9: 891-901.

[48] Hu, B. 2011. *Blow-up Theories for Semilinear Parabolic Equations.* Lecture Notes in Mathematics, vol. 2018. Berlin: Springer.

[49] Hu, B., and H.M. Yin. 1994. The profile near blowup time for solutions of the heat equation with a nonlinear boundary condition. *Trans. Am. Math. Soc.* 346:117-135.

[50] Hu, B., and H.M. Yin. 1996. Critical exponents for a system of heat equations coupled in a nonlinear boundary condition. *Math. Meth. Appl. Sci.* 19:1099-1120.

[51] Huang, H.M., and M.X. Wang. 2015. The reaction-diffusion system for an SIR epidemic model with a free boundary. *Discrete Contin. Dyn. Syst.-Ser. B* 20:2039-2050.

[52] Hutson, V., K. Mischaikow, and P. Poláčik. 2001. The evolution of dispersal rates in a heterogeneous time-periodic environment. *J. Math. Biol.* 43:501-533.

[53] Kalantarov, V.K., and O.A. Ladyzhenskaya. 1977. Formation of collapses in quasilinear equations of parabolic and hyperbolic types. *Zap. Nauchn. Sem. Leningrad. Otdel. Mat. Inst. Steklov. (LOMI).* 69:77-102.

[54] Kaneko, Y., H. Matsuzawa, and Y. Yamada. 2020. Asymptotic profiles of solutions and propagating terrace for a free boundary problem of nonlinear diffusion equation with positive bistable nonlinearity. *SIAM J. Math. Anal.* 52:65-103.

[55] Kaplan, S. 1963. On the growth of solutions of quasilinear parabolic equations. *Comm. Pure Appl. Math.* 16:305-333.

[56] Kawai, Y., and Y. Yamada. 2016. Multiple spreading phenomena for a free boundary problem of a reaction-diffusion equation with a certain class of bistable nonlinearity. *J. Differ. Equat.* 261:538-572.

[57] Keller, H., and D. Cohen. 1967. Some positone problems suggested by nonlinear heat generation. *J. Math. Mech.* 16:1361-1376.

[58] Kobayashi, K., T. Siaro, and H. Tanaka. 1977. On the blow-up problem for semilinear heat equations. *J. Math. Soc. Jpn.* 29:407-424.

[59] Ladde, G.S., V. Lakshmikantham, and A.S. Vatsala. 1985. *Monotone Iterative Techniques for Nonlinear Differential Equations.* Boston: Pitman.

[60] Ladyzenskaya, O.A., V.A. Solonnikov, and N.N. Ural'ceva. 1968. *Linear and Quasilinear Equations of Parabolic Type.* Providence, R.I: Amer. Math. Society.

[61] Lazer, A.C. 1982. Some remarks on periodic solutions of parabolic differential equations. In *Dynamical Systems II*, ed. A. Bednarek, and L. Cesari, 227-246. New York: Academic Press.

[62] Levine, H.A. 1973. Some nonexistence and instability theorems for solutions of formally parabolic equations of the form $Pu_t = -Au + F(u)$. *Arch. Ration. Mech. An.* 51:371-386.

[63] Levine, H.A. 1990. The role of critical exponents in blow up theorems. *SIAM Rev.* 32:262-288.

[64] Levine, H.A., and L.E. Payne. 1974. Nonexistence theorems for the heat equation with nonlinear boundary conditions and for the porous medium equation backward in time. *J. Differ. Equat.* 16:319-334.

[65] Levine, H.A., and L.E. Payne. 1974. Some nonexistence theorems for initial-boundary value problems with nonlinear boundary constraints. *Proc. Am. Math. Soc.* 46:277-284.

[66] Li, F., X. Liang, and W.X. Shen. 2016. Diffusive KPP equations with free boundaries in time almost periodic environments: II. Spreading speeds and semi-wave. *J. Differ. Equat.* 261:2403-2445.

[67] Li, F.J., B.C. Liu, and S.N. Zheng. 2007. Simultaneous and non-simultaneous blow-up for heat equations with coupled nonlinear boundary fluxes. *Z. Angew. Math. Phys.* 58:717-735.

[68] Li, H.L., P.Y.H. Pang, and M.X. Wang. 2012. Qualitative analysis of a diffusive prey-predator model with trophic interactions of three levels. *Discrete Contin. Dyn. Syst.-Ser. B* 17:127-152.

[69] Li, L., S.Y. Liu, and M.X. Wang. 2020. A viral propagation model with a nonlinear infection rate and free boundaries. *Sci. China Math.* https://doi.org/10.1007/s11425-020-1680-0.

[70] Li, L., W.J. Sheng, and M.X. Wang. 2020. Systems with nonlocal vs. local diffusions and free boundaries. *J. Math. Anal. Appl.* 483:123646.

[71] Lieberman, G.M. 1986. Mixed boundary value problems for elliptic and parabolic differential equations of second order. *J. Math. Anal. Appl.* 113: 422-440.

[72] Lieberman, G.M. 2005. *Second Order Parabolic Differential Equations, Revised Edition.* Singapore: World Scientific.

[73] Lieberman, G.M. 2013. *Oblique Derivative Problems for Elliptic Equations.* Singapore: World Scientific.

[74] Lin, Z.G., and C.H. Xie. 1998. The blow-up rate for a system of heat equations with nonlinear boundary conditions. *Nonlinear Anal.* 34:767-778.

[75] Lin, Z.G., and H.P. Zhu. 2017. Spatial spreading model and dynamics of West Nile virus in birds and mosquitoes with free boundary. *J. Math. Biol.* 75:1381-1409.

[76] Liu, B.C., and F.J. Li. 2012. Non-simultaneous blowup in heat equations with nonstandard growth conditions. *J. Differ. Equat.* 252:4481-4502.

[77] Liu, S.Y., H.M. Huang, and M.X. Wang. 2019. Asymptotic spreading of a diffusive competition model with different free boundaries. *J. Differ. Equat.* 266:4769-4799.

[78] Liu, S.Y., H.M. Huang, and M.X. Wang. 2020. A free boundary problem for a prey-predator model with degenerate diffusion and predator-stage structure. *Discrete Contin. Dyn. Syst.-Ser. B* 25:1649-1670.

[79] Liu, S.Y., and M.X. Wang. 2020. Existence and uniqueness of solution of free boundary problem with partially degenerate diffusion. *Nonlinear Anal.: RWA* 54:103097.

[80] Lou, Y. 2006. On the effects of migration and spatial heterogeneity on single and multiple species. *J. Differ. Equat.* 223:400-426.

[81] Marcus, M., and L. Véron. 1998. The boundary trace of positive solutions of semilinear elliptic equations: the subcritical case. *Arch. Ration. Mech. An.* 144:201-231.

[82] Miklavčič, M. 1998. *Applied Functional Analysis and Partial Differential Equations.* Singapore: World Scientific.

[83] Monobe, H., and C.H. Wu. 2016. On a free boundary problem for a reaction-diffusion-advection logistic model in heterogeneous environment. *J. Differ. Equat.* 261:6144-6177.

[84] Nazarov, A.I. 2012. A centennial of the Zaremba-Hopf-Oleinik lemma. *SIAM J. Math. Anal.* 44(1):437-453.

[85] Ni, W.J., and M.X. Wang. 2016. Long time behavior of a diffusive competition model. *Appl. Math. Lett.* 58:145-151.

[86] Ni, W.J., J.P. Shi, and M.X. Wang. 2020. Global stability of nonhomogeneous equilibrium solution for the diffusive Lotka-Volterra competition model. *Calc. Var. Partial Differ. Equ.* 59:132.

[87] Pang, P.Y.H., and M.X. Wang. 2004. Strategy and stationary pattern in a three-species predator-prey model. *J. Differ. Equat.* 200:245-274.

[88] Pao, C.V. 1992. *Nonlinear Parabolic and Elliptic Equations.* New York: Plenum Press.

[89] Pazy, A. 1983. *Semigroup of Linear Operators and Applications to Partial Differential Equations.* New York: Springer-Verlag.

[90] Peng, R., and X.Q. Zhao. 2013. The diffusive logistic model with a free boundary and seasonal succession. *Discrete Contin. Dyn. Syst.* 33:2007-2031.

[91] Protter, M.H., and H.F. Weinberger. 1984. *Maximum Principles in Differential Equations.* New York: Springer-Verlag.

[92] Quittner, P. 1996. Global existence of solutions of parabolic problems with nonlinear boundary conditions. *Banach Center Publ.* 33:309-314.

[93] Quittner, P., and P. Souplet. 2007. *Superlinear Parabolic Problems: Blow-up, Global Existence and Steady States.* Basel: Birkhäuser.

[94] Rossi, J.D., and P. Souplet. 2005. Coexistence of simultaneous and nonsimultaneous blow-up in a semilinear parabolic system. *Differ. Integral Equ.* 18:405-418.

[95] Rothe, F. 1982. Uniform bounds from bounded L^p-functionals in reaction-diffusion equations. *J. Differ. Equat.* 45:207-233.

[96] Samarskii, A.A., V.A. Galaktionov, S.P. Kurdyumov, and A.P. Mikhailov. 1995. *Blow-up in Quasilinear Parabolic Equations.* Berlin: De Gruyter.

[97] Sattinger, D.G. 1972. Monotone methods in nonlinear elliptic and parabolic equations. *Indiana Univ. Math. J.* 21:979-1000.

[98] Schauder, J. 1930. Der Fixpunktsatz in Funktionalräumen. *Studia Math.* 2:171-180.

[99] Smoller, J. 1994. *Shock Waves and Reaction-Diffusion Equations, 2nd Edition.* New York: Springer-Verlag.

[100] Souplet, P., and S. Tayachi. 2004. Optimal condition for non-simultaneous blow-up in a reaction-diffusion system. *J. Math. Soc.* Japan 56(2):571-584.

[101] Tsutsumi, M. 1972. Existence and nonexistence of global solutions for nonlinear parabolic equations. *Publ. Res. Inst. Math. Sci. Kyoto Univ.* 8:221-229.

[102] Walter, W. 1975. On existence and nonexistence in the large of solutions of parabolic differential equations with a nonlinear boundary condition. *SIAM J. Math. Anal.* 6:85-90.

[103] Wang, J.P., and M.X. Wang. 2020. Free boundary problems with nonlocal and local diffusions I: Global solution. *J. Math. Anal. Appl.* 490:123974.

[104] Wang, J.P., and M.X. Wang. 2020. Free boundary problems with nonlocal and local diffusions II: Spreading-vanishing and longtime behavior. *Discrete Contin. Dyn. Syst.-Ser. B* 25(12):4721-4736.

[105] Wang, M.X. 1993. *Nonlinear Partial Differential Equations of Parabolic Type* (in Chinese). Beijing: Science Press.

[106] Wang, M.X. 2000. Global existence and finite time blow-up for a reaction-diffusion system. *Z. Angew. Math. Phys.* 51:160-167.

[107] Wang, M.X. 2001. Blow-up estimates for a semilinear reaction-diffusion system. *J. Math. Anal. Appl.* 257:46-51.

[108] Wang, M.X. 2001. Blow-up estimates for semilinear parabolic systems coupled in an equation and a boundary condition. *Sci. China Ser. A* 44:1465-1468.

[109] Wang, M.X. 2006. *Semigroups of Operators and Evolution Equations* (in Chinese). Beijing: Science Press.

[110] Wang, M.X. 2009. *Basic Theory of Partial Differential Equations* (in Chinese). Beijing: Science Press.

[111] Wang, M.X. 2013. *Sobolev Spaces* (in Chinese). Beijing: Higher Education Press.

[112] Wang, M.X. 2014. On some free boundary problems of the prey-predator model. *J. Differ. Equat.* 256:3365-3394.

[113] Wang, M.X. 2016. A diffusive logistic equation with a free boundary and sign-changing coefficient in time-periodic environment. *J. Funct. Anal.* 270:483-508.

[114] Wang, M.X. 2018. Note on the Lyapunov functional method. *Appl. Math. Lett.* 75:102-107.

[115] Wang, M.X. 2019. Existence and uniqueness of solutions of free boundary problems in heterogeneous environments. *Discrete Contin. Dyn. Syst.-Ser. B* 24:415-421.

[116] Wang, M.X., and P.Y.H. Pang. 2008. Global asymptotic stability of positive steady states of a diffusive ratio-dependent prey-predator model. *Appl. Math. Lett.* 21:1215-1220.

[117] Wang, M.X., and P.Y.H. Pang. 2009. Qualitative analysis of a diffusive variable-territory prey-predator model. *Discrete Contin. Dyn. Syst.-Ser. B* 23:1061-1072.

[118] Wang, M.X., and Y. Zhang. 2016. The time-periodic diffusive competition models with a free boundary and sign-changing growth rates. *Z. Angew. Math. Phys.* 67(5), Art.132(1-24).

[119] Wang, M.X., and Y. Zhang. 2017. Note on a two-species competition-diffusion model with two free boundaries. *Nonlinear Anal.* 159:458-467.

[120] Wang, M.X., and Y. Zhang. 2018. Dynamics for a diffusive prey-predator model with different free boundaries. *J. Differ. Equat.* 264:3527-3558.

[121] Wang, M.X., and J.F. Zhao. 2014. Free boundary problems for a Lotka-Volterra competition system. *J. Dyn. Diff. Equat.* 26:655-672.

[122] Wang, M.X., and J.F. Zhao. 2017. A free boundary problem for the predator-prey model with double free boundaries. *J. Dyn. Diff. Equat.* 29:957-979.

[123] Wu, C.-H. 2015. The minimal habitat size for spreading in a weak competition system with two free boundaries. *J. Differ. Equat.* 259:873-897.

[124] Ye, Q.X., Z.Y. Li, M.X. Wang, and Y.P. Wu. 2011. *Introduction to Reaction-Diffusion Equations* (in Chinese). Beijing: Science Press.

[125] Zhao, M., Y. Zhang, W.T. Li, and Y.H. Du. 2020. The dynamics of a degenerate epidemic model with nonlocal diffusion and free boundaries. *J. Differ. Equat.* 269:3347-3386.

Index

Printed in the United States
by Baker & Taylor Publisher Services